AF517118

# Plant Protection in Horticulture

## Volume 3

Bacterial and Viral Diseases and their Management

OTHER VOLUMES

**Plant Protection in Horticulture**

Vol. 1 : Insect, Mite and Vertebrate Pests and their Management
ISBN: 978-81-7233-615-8

Vol. 2 : Fungal Diseases and their Management
ISBN: 978-81-7233-616-5

*BY THE SAME AUTHOR*

- Diseases of Horticulture Crops: Nematode Problems and their Management ISBN: 978-81-7233-543-4
- Integrated Nematode Management in Horticultural Crops ISBN: 978-81-7233-523-6
- Organic Farming for Sustainable Horticulture: Principles and Practices ISBN: 978-81-7233-501-4

# Plant Protection in Horticulture

## Volume 3

## Bacterial and Viral Diseases and their Management

**P. Parvatha Reddy**

Former Director and Former Head
Division of Entomology & Nematology
Indian Institute of Horticultural Research
Bangalore-560 089

Published by:

SCIENTIFIC PUBLISHERS (INDIA)
5A, New Pali Road, P.O. Box 91
Jodhpur 342 001, India
E-mail: info@scientificpub.com
Website: www.scientificpub.com

Print:2025



ISBN: 978-81-72336-18-9
eISBN: 978-93-87913-00-4

Printed in India

*Affectionately dedicated to*
*my sisters*

Smt. Yerribasamma
Smt. Rajamma

# PREFACE

Horticulture in India is fast emerging as a major commercial venture, because of higher remuneration per unit area and the realization that consumption of fruits and vegetables is essential for health and nutrition. In the last one decade, export potential of horticultural crops has significantly increased attracting even multinationals into floriculture, processing and value added products.

During 2005-06 coverage of area under various horticultural crops in the country was 19.237 million hectares (which is about 13.08% of the total cultivated area) with production of 185.207 million MT. The horticulture sector contributed around 28% of GDP in agriculture. India is the largest producer in the world of cashew nuts, coconuts, tea, ginger, turmeric and black pepper. India is the second largest producer of fruits and vegetables (after China) with a production of 58.74 and 109.05 million MT and contributes 10.9 and 11.9% share in global fruit and vegetable production, respectively. India is the second largest producer of flowers after China. About 126,000 hectare of area is under floriculture, producing 694,000 million tonnes of loose flowers annually (2005-2006). The production of cut flowers has increased to 2,762 million numbers. India is still a marginal player in the world floriculture trade indicating strong potential that can be exploited in this sector. India is the third largest producer of coconut, fourth largest producer and consumer of rubber and sixth largest producer of coffee in the world.

Productivity of horticultural crops in India is relatively low compared to other countries. Of the several factors responsible for lower productivity of horticultural crops, bacterial and viral/mycoplasmal/phytoplasmal diseases are considered as important limiting factors. Some disease problems are of national importance which causes significant losses to horticultural crops. Diseases of horticultural crops continue to cause losses of about 10% of the crop yields worth more than Rs. 15,000 crores annually. More than 9,600 MT of technical grade fungicides are used annually to manage the diseases in India.

The information on bacterial and viral/mycoplasmal/phytoplasmal diseases of horticultural crops (fruits, vegetables, plantation, spice, tuber, ornamental, medicinal and aromatic crops) is very much scattered. There is no book at present which comprehensively and exclusively deals with the above aspects on horticultural crops. The present book deals with integrated disease management in horticultural crops in detail using regulatory, physical, cultural, chemical, biological, host plant resistance and integrated methods. The book is extensively illustrated with excellent quality photographs enhancing the quality of publication. The book is written in lucid style, easy to understand language along with adoptable recommendations involving eco-friendly components of IDM.

In view of greater emphasis being given by the Government of India to horticulture by establishing horticulture mission and exclusive State Horticultural Universities, there is an urgent need for good text books to teach courses on different disciplines of horticulture. In this context the present book is of immense help to all those concerned with diseases of horticultural crops.

This book is a practical guide to practicing farmers of horticultural crops. Further, it is a useful reference to policy makers, research and extension workers and students. The material can also be used for teaching under-graduate and post-graduate courses. Suggestions to improve the contents of the book are most welcome (E-mail: reddy_parvatha@yahoo.com).

I profusely thank Dr. K. Srinivasan, Chief Agricultural Economist, TAFE, Chennai; Dr. N. Ramachandran, Indian Institute of Horticultural Research, Bangalore; and Dr. V.K. Das and Dr. D.K. Ghosh, National Research for Citrus, Nagpur; who have kindly made available many of the photographs included in this book. The publisher, Scientific Publishers (India), Jodhpur, deserves commendation for their professional contribution.

Bangalore **P. Parvatha Reddy**

August, 15th, 2009

# CONTENTS

## Chapter 1

# HORTICULTURAL CROPS

### 1.1. Importance of Horticultural Crops

Agriculture is the mainstay of the Indian economy, and horticulture is a crucial component thereof. Horticulture development has been accorded high priority in the XI$^{th}$ Five Year Plan. The impact of enhanced investment in horticulture has been highly encouraging in terms of vastly improved quality production and export potential - an increase from 96.1 million tonnes in 1990-91 to 185.2 million tonnes during 2005-06 (Bijay Kumar, 2007) (Table 1.1).

**Table 1.1. Area, production and productivity of horticultural crops in India (2005-06) (Bijay Kumar, 2007)**

| Horticultural crops | Area (million ha) | Production (million MT) | Productivity (MT/ha) | % Share of hortil. crops |
|---|---|---|---|---|
| Fruits | 5.510 | 58.740 | 10.7 | 31.74 |
| Vegetables | 7.164 | 109.050 | 15.2 | 58.93 |
| Flowers-loose | 0.126 | 0.694 | --- | 0.38 |
| Flowers-cut | --- | 2762.000* | --- | --- |
| Mushrooms | --- | 0.040 | --- | 0.03 |
| Spices | 2.858 | 5.108 | 1.8 | 2.76 |
| Plantation crops | 3.283 | 11.263 | 3.4 | 6.09 |
| Nuts | 0.121 | 0.134 | 1.1 | 0.07 |
| Aromatic & Medicinal | 0.266 | 0.220 | --- | --- |
| Total | 19.237 | 185.207 | 9.6 | --- |

*Million numbers

India is the second largest producer of fruits next only to China contributing 10.9% of the total world production. Total production of fruits during 2005-06 has been estimated at 58.74 million MT from 5.51million ha. (Bijay Kumar, 2007). India is the largest producer of mango, banana, and papaya. India produces 41.7% of the world mangoes, and 25.7% of the bananas. In grapes, India has recorded the highest productivity per unit area in the world. The per capita consumption of fruits has increased from 40 to 85g/day within the last one decade.

India is the second largest producer of vegetables next only to China contributing 11.9% of the total world production. Vegetables occupy an area of 7.164 million ha with a production of 109.05 million tonnes (Bijay Kumar, 2007). India produces 13.6% of the world onions. In vegetables, India occupies the first position in the production of peas and okra, second in onion, brinjal, cabbage, cauliflower and third in potato. The per capita consumption of vegetables has increased from 96 to 175 g/day within the last one decade.

Floriculture is estimated to cover an area of 126,000 ha with production of 694,000 MT of loose flowers. The area under cut flowers has increased significantly in recent years and the production has reached 2762 million numbers.

India has emerged as the largest producer of spices, tea, cashew and areca nut. Coconuts sustain an area of 2.029 million ha with a production of 101.90 million tonnes. Areca nut is cultivated in 0.381 million ha with a production of 0.483 million tonnes. Cashew nuts are grown in an area of 0.843 million ha with a production of 0.579 million tonnes. The annual production of spices is 5.108 million tonnes from 2.858 million ha.

Horticulture has emerged as an important sector for diversification of agriculture. Presently, horticulture has established its credibility in improving income through increased productivity, generating employment and in enhancing exports besides providing protective food, household nutritional security, precious raw material for various processing and ancillary industries and finally in enhancing the aesthetic aspects of life. The focused attention on investment in horticulture during the last decade has been rewarding in terms of increased production and productivity of horticultural crops with manifold export potential. Horticulture has played a significant role in women empowerment, providing employment opportunities through mushroom cultivation, floriculture, processing, nursery raising, vegetable seed production etc.

## 1.2. Export of Horticultural Produce

Horticulture in India is fast emerging as a major commercial venture, because of higher remuneration per unit area and the realization that consumption of fruits and vegetables is essential for health and nutrition. In the last one decade, export potential of horticultural crops has significantly increased attracting even multinationals into floriculture, processing and value added products.

The present annual growth rate of horticulture is more than 6.5% which contributes 24.5% from the mere 8% of area to the GDP and 54.55% to the export earnings in the agriculture sector (Table 1.2).

**Table 1.2. Export of horticultural produce from India (2005-06) (Bijay Kumar, 2007)**

| Horticultural product | Export quantity (MT) | Value (Rs. Crores) | % Share of value |
|---|---|---|---|
| Floriculture & seeds | | | |
| Floriculture | 35186.00 | 299.41 | 7 |
| Fruit & vegetable seeds | 7472.9 | 92.83 | 2 |
| Total | 42658.90 | 392.24 | |
| Fruits & vegetables | | | |
| Fresh onions | 960507.32 | 708.16 | 18 |
| Other fresh vegetables | 217280.59 | 267.69 | 6 |
| Walnuts | 5256.56 | 114.47 | 3 |
| Fresh mangoes | 69606.60 | 128.11 | 3 |
| Fresh grapes | 54049.87 | 214.61 | 5 |
| Other fresh fruits | 158339.26 | 225.67 | 5 |
| Total | 1465040.20 | 1658.71 | |
| Processed fruits & vegetables | | | |
| Dried & preserved vegetables | 124495.71 | 364.11 | 9 |
| Mango pulp | 134613.20 | 364.24 | 9 |
| Pickle & Chutneys | 135381.85 | 260.98 | 6 |
| Other processed fruits and vegetables | 107335.24 | 370.21 | 27 |
| Grand Total | 943568.59 | 2454.60 | |

The major destinations for export of different horticultural products and potential international markets for fresh and processed products are presented in Table 1.3.

**Table 1.3. Major destinations for export of different horticultural products (Chamber, 2006)**

| Horticultural crops | Major destinations for export |
|---|---|
| Fruits | UAE, Saudi Arabia, Sri Lanka, Bangla Desh, The Netherlands, USA, Malaysia |
| Fresh grapes | UK, UAE, The Netherlands, Bangla Desh, Germany, Sri Lanka |
| Pomegranate | UAE, Saudi Arabia, Sri Lanka, Kuwait |
| Fresh guava | Saudi Arabia, Netherlands, USA, UAE, Kuwait |
| Gooseberries | UAE, Qatar |
| Mango pulp | UK, Canada, Bangle Desh, Bahrain, Netherlands, Japan, Saudi Arabia |
| Mango slices | UK, Saudi Arabia, Japan, USA |
| Dried grapes | UK, Malaysia, Netherlands, UAE |

| | |
|---|---|
| Mango pickles | Saudi Arabia, UAE, UK, Canada |
| Mango chutney | Oman, USA |
| Guava jam & jelly | UK, Germany, Hong Kong, USA, Netherlands, Saudi Arabia |
| Dried tamarind | Egypt, UAE, Pakistan, Saudi Arabia, UK |
| Dried figs | Nepal, Bangla Desh |
| Vegetables | UAE, UK, Saudi Arabia, Sri Lanka, USA, Singapore |
| Fresh vegetables | UAE, Saudi Arabia, Sri Lanka, USA |
| Fresh onion | UAE, Singapore, Sri Lanka, Bangla Desh, Malaysia |
| Fresh tomato | Bangle Desh, Nepal, Sri Lanka |
| Gherkin | France, USA, Spain, Belgium |
| Dried vegetables | UAE, Sri Lanka, USA, UK, Singapore |
| Dehydrated onion | Germany, Romania, Sri Lanka, USA, Malaysia, Netherlands, UK |
| Vegetable seeds | USA, The Netherlands, France, Italy, Japan, Pakistan |
| Floriculture | USA, Japan, The Netherlands, UK, Italy |

Horticulture today is not merely a means of diversification but forms an integral part of food and nutritional security, as also an essential ingredient of economic security. Adoption of horticulture both by small and marginal farmers brought prosperity in many regions of the country of which Maharashtra, Karnataka and Andhra Pradesh are the prime examples. A Golden Revolution through significant production increases in horticultural crops is already on the horizon.

Since India is now a signatory member of WTO, it will require its produce and products to be competitive both in the domestic and export markets. This would call for the use of improved HYV/F1 hybrids, recommended agro-techniques and plant protection measures, hi-tech intervention and scientific post-harvest management.

After the success of 'Green Revolution' in agriculture, India is now moving towards the **'Golden Revolution'** comprising fruits, vegetables and flowers. Commercialization of horticulture along with post harvest management, processing and marketing assume great significance and add value not only in the context of farmers' income, but also its contribution to the agriculture GDP. Commercialization of horticulture is largely dependent upon use of appropriate technology and market oriented production.

Chapter 2

# FRUIT CROPS

A large variety of fruits are grown in India. Of these, banana, citrus, pineapple, papaya, sapota, jack fruit and avocado (tropical fruits); mango, grape, guava, litchi and loquat (subtropical fruits); apple, pear, peach, plum, apricot, almond and walnut (temperate fruits) and amla, ber, pomegranate, annona, fig and phalsa (arid zone fruits) are important. The leading fruit growing states are Maharashtra, Andhra Pradesh, Tamil Nadu, Kerala, Gujarat, Karnataka, Bihar, Uttar Pradesh, West Bengal and Orissa.

India is the second largest producer of fruits next only to China contributing 10.9% of the total world production. The present area under fruit crops is estimated to be about 5.510 million hectares with an estimated production of 58.740 million tonnes (Tables 2.1 and 2.2). Fruit crops and nuts contribute 31.81% share to the total production of horticultural crops. The nut crops are grown in 0.1209 million hectares with an annual production of 0.1335 million tonnes (Table 2.3). India is the largest producer of mango, banana and papaya. India produces 40.1% of the world mangoes, 24.5% of the bananas and 28.8% of world papaya. In grapes, India has recorded the highest productivity per unit area (25.4 tonnes/ha) in the world. The per capita consumption of fruits has increased from 40 to 85 g/day within the last one decade.

**Table 2.1. Major fruit producing countries of the world (Bijay Kumar, 2007)**

| Country | Area (in '000 ha) | Production (in '000 MT) | Productivity (in MT/ha) | % share of world production |
|---|---|---|---|---|
| China | 10456 | 89355 | 8.5 | 16.7 |
| India | 5510 | 58740 | 10.7 | 10.9 |
| Brazil | 2386 | 36952 | 15.5 | 6.9 |
| USA | 1215 | 28144 | 23.2 | 5.2 |
| Italy | 1219 | 18134 | 14.9 | 3.4 |
| Spain | 1806 | 15537 | 8.6 | 2.9 |
| Indonesia | 1167 | 15406 | 13.2 | 2.9 |
| Mexico | 1128 | 14903 | 13.2 | 2.8 |

| | | | | |
|---|---|---|---|---|
| Iran | 1354 | 13848 | 10.2 | 2.6 |
| Philippines | 1073 | 13110 | 12.2 | 2.4 |
| Turkey | 1074 | 12998 | 12.1 | 2.4 |
| Others | 24781 | 219465 | 8.9 | 40.9 |
| Total | 53170 | 536591 | | |

**Table 2.2. Area, production and productivity of fruit crops in India (2005-06) (Bijay Kumar, 2007)**

| Fruit crop | Area ('000 ha) | Production ('000 MT) | Productivity (MT/ha) | % share of India production |
|---|---|---|---|---|
| Apple | 235 | 1756 | 7.5 | 3.0 |
| Banana | 565 | 18702 | 33.1 | 31.8 |
| Citrus | 749 | 6326 | 8.4 | 10.8 |
| Grapes | 64 | 1631 | 25.4 | 2.8 |
| Guava | 182 | 1823 | 10.0 | 3.1 |
| Litchi | 63 | 381 | 6.0 | 0.6 |
| Mango | 2021 | 12538 | 6.2 | 21.3 |
| Papaya | 73 | 2317 | 31.7 | 3.9 |
| Pineapple | 85 | 1353 | 15.8 | 2.3 |
| Pomegranate | 116 | 849 | 7.3 | 1.4 |
| Sapota | 140 | 1117 | 8.0 | 1.9 |
| Others | 1217 | 9947 | 8.2 | 16.9 |
| Total | 5510 | 58740 | 10.7 | |

**Table 2.3. Area, production and productivity of nut crops in India (2005-06) (Bijay Kumar, 2007)**

| Nut crop | Area ('000 ha) | Production ('000 MT) | Productivity (MT /ha) |
|---|---|---|---|
| Almond | 21.1 | 14.6 | 0.7 |
| Walnut | 99.8 | 118.9 | 1.2 |
| Total | 120.9 | 133.5 | |

India has been exporting fruits and fruit products since several decades. However, in the recent past export promotional activities received greater attention (Table 2.4). Mango and grapes are the main items in fresh fruit export and some other fruits like pomegranate, sapota, banana, citrus, litchi, apple and strawberry also entered fresh fruit export market in limited quantities.

**Table 2.4. Export of fruit produce from India (2005-06) (Bijay Kumar, 2007)**

| Fruit products | Quantity (in MT) | Value (in Rs. Million) |
|---|---|---|
| Fruit & vegetable seeds | 7472.90 | 928.3 |

| | | |
|---|---|---|
| Walnuts | 5256.56 | 1144.7 |
| Fresh mangoes | 69606.60 | 1281.1 |
| Fresh grapes | 54049.87 | 2146.1 |
| Other fresh fruits | 158339.26 | 2256.7 |
| Mango pulp | 134613.20 | 3642.4 |
| Pickle & Chutneys | 135381.85 | 2609.8 |
| Other processed fruits & vegetables | 107335.24 | 3702.1 |

The major potential international markets for fresh and processed fruit products are presented in Table 2.5.

**Table 2.5. Potential international markets for fresh and processed fruits**

| Fresh & processed fruits | Potential international markets |
|---|---|
| Fresh grapes | UK, Netherlands, Sri Lanka, UAE |
| Pomegranate | UAE, Saudi Arabia, Sri Lanka, Kuwait |
| Fresh guava | Saudi Arabia, Netherlands, USA, UAE, Kuwait |
| Gooseberries | UAE, Qatar |
| Mango pulp | UK, Canada, Bangla Desh, Bahrain, Netherlands, Japan, Saudi Arabia |
| Mango slices | UK, Saudi Arabia, Japan, USA |
| Dried grapes | UK, Malaysia, Netherlands, UAE |
| Mango pickles | Saudi Arabia, UAE, UK, Canada |
| Mango chutney | Oman, USA |
| Guava jam & jelly | UK, Germany, Hong Kong, USA, Netherlands, Saudi Arabia |
| Dried tamarind | Egypt, UAE, Pakistan, Saudi Arabia, UK |
| Dried figs | Nepal, Bangla Desh |

## 2.1. BANANA

Globally, banana (*Musa* spp.) is one of the maximum distributed fruit crop grown in more than 130 countries in 4.306 million hectares with an annual production of 76.436 million tonnes (Table 2.6). It is the fourth most important food crop after rice, wheat and maize. It is a dessert fruit for millions and is used in different regions of the world as staple food. As a diet, banana is filling, easy to digest, nearly fat free, rich source of carbohydrate with calorific value of 67 calories/100g and is free from sodium, making it salt-free diet and also has zero cholesterol. Hence, regular use of it is also considered to be good for heart. Banana also contains several minerals such as Calcium, Magnesium, Potassium and Phosphorus. Banana fruit is particularly rich in Vitamin-C and contains significant amount of other vitamins such as Vitamin-A.

**Table 2.6. Major banana producing countries of the world (Bijay Kumar, 2007)**

| Country | Area (in '000 ha) | Production (in '000 MT) | Productivity (in MT/ha) | % share of world production |
|---|---|---|---|---|
| India | 565 | 18702 | 33.1 | 24.5 |
| Brazil | 491 | 6703 | 13.6 | 8.8 |
| China | 274 | 6670 | 24.4 | 8.7 |
| Philippines | 418 | 6298 | 15.1 | 8.2 |
| Ecuador | 221 | 6118 | 27.7 | 8.0 |
| Indonesia | 315 | 5178 | 16.4 | 6.8 |
| Coasta Rica | 41 | 2353 | 57.2 | 3.1 |
| Mexico | 77 | 2250 | 29.2 | 2.9 |
| Thailand | 141 | 1865 | 13.2 | 2.4 |
| Colombia | 65 | 1765 | 27.2 | 2.3 |
| Others | 1698 | 18534 | 10.9 | 24.2 |
| Total | 4306 | 76436 | 33.1 | |

Banana is an important fruit crop having great socio-economic significance in India as it supports livelihood of millions of people. India is the largest producer of banana in the world with an annual production of 18.702 million tonnes from an area of 0.565 million ha (Table 2.7). Although, banana occupies 10.3 per cent of the area under fruits it contributes 31.8 per cent of the total fruit production in the country. The national productivity of banana is very impressive (33.1 tonnes per ha). Its productivity in commercial zones especially in Maharashtra, Gujarat, Tamil Nadu, Karnataka, Andhra Pradesh and Bihar is highly impressive and comparable to the productivity in other countries.

**Table 2.7. Area, production and productivity of leading banana growing states of India (2005-06) (Bijay Kumar, 2007)**

| State | Area (in '000 ha) | Production (in '000 MT) | Productivity (in MT/ha) |
|---|---|---|---|
| Tamil Nadu | 94.6 | 4647.6 | 49.1 |
| Maharashtra | 73.2 | 4608.5 | 63.0 |
| Gujarat | 49.2 | 2498.8 | 50.8 |
| Andhra Pradesh | 61.1 | 1528.7 | 25.0 |
| Karnataka | 56.4 | 1399.1 | 24.8 |
| Bihar | 28.0 | 920.0 | 34.3 |
| Kerala | 56.2 | 443.1 | 7.9 |
| Assam | 47.0 | 699.4 | 14.9 |
| Madhya Pradesh | 15.4 | 615.6 | 40.0 |
| West Bengal | 27.8 | 544.9 | 19.6 |
| Others | 56.2 | 756.9 | 13.5 |
| Total | 565.1 | 18701.9 | 33.1 |

Research and development efforts made in the last two decades have resulted in commendable increase in productivity (144 per cent in the last decade) and overall production of banana in the country. The improved production technologies include higher plant densities (4500 plants per ha), improved crop geometries coupled with adoption of drip irrigation, more judicious fertilization management through balanced use of organic and inorganic fertilizers, appropriately timed split application of inorganics, use of proper sized rhizomes and well timed plantings, commercial production of *in vitro* plants and development of effective integrated pest management strategies. By adoption of these improved technologies, productivity of several individual farmers in different parts of the country has resulted in substantial increase (60 to 65 tonnes/ha).

### 2.1.1. Bacterial Diseases

#### 2.1.1.1. Bacterial Wilt or Moko Disease, *Ralstonia solanacearum*

Bacterial wilt of banana was first recorded in Guyana in 1840. Decline in yield of 74% in plantain and 64% in banana were attributed to Moko disease in Guyana from 1963-73. The export of banana in Trinidad was virtually eliminated in the early 1960's due to this disease. It is observed in many countries where banana is grown extensively. Bacterial wilt in banana crops is a serious problem in South and Central America, the Carribean islands and the Philippines. In India, the disease is reported on banana cvs. Robusta and Poovan from Tamil Nadu causing 75-100% yield losses.

**i) Symptoms:** The disease is characterized by rapid wilting and collapse of leaves (Fig. 2.1), premature ripening of fruits, discolouration of the vascular strands and wilting and blackening of suckers. In Cavendish cv., the lower leaves develop a yellowish tinge which quickly spread to all other leaves. Sometimes second or third leaf or both the leaves collapse near the junction of petiole and lamina. In severe cases, plants collapse before the emergence of bunches. The presence of yellow fingers in otherwise green stem often indicates the occurrence of moko disease. Internally, a firm dry rot is found in the fruit. Fruit stalks also show discoloured vascular strands.

There is often confusion in the differentiation of Moko disease and Panama wilt of banana, caused by the Fusarium fungus. With Moko disease, bananas at all growth stages can be infected, vascular tissue has a black to brown discolouration, fingers ripen prematurely and split and the pulp is discoloured.

**ii) Pathogen:** *Ralstonia solanacearum* is the causal pathogen of bacterial wilt. It has been classified into three races based on their host range. Race 2 of *R. solanacearum* is pathogenic to triploid banana and *Heliconia* (a relative of *Musa* spp.).

This organism is a Gram negative, non-spore forming, rod-shaped, aerobic bacterium. Optimum growth is at 30-35°C and it does not grow above 41°C. In culture, colonies are cream coloured and slimy. When tissue from the base of infected stems is placed in water, ooze exudes.

**Fig. 2.1.** Moko wilt of banana (Courtesy, K. Srinivasan, TAFE, Chennai).

**iii) Host Range:** Most crops belonging to the Family Solanaceae are host to *R. solanacearum*. Common crops affected by bacterial wilt include eggplant, capsicum, tomato, potato, tobacco, sweet potato, banana, ginger, onion, peanut, mung bean, cashew, papaya, cassava and sesame.

**iv) Epidemiology**

- The pathogen infects roots through wounds and colonizes the vascular tissue causing plugging of the xylem and leaf wilting. There is often an association between nematode infection and bacterial wilt, where the nematodes create wounds in the root tissue to allow an entry point for the bacterium to infect the plant.
- The bacterium survives in the soil and can maintain infectious populations over several years. Alternate weed hosts may also play a role in survival and over-seasoning.
- Temperature is a major determinant in the distribution of this pathogen, which is widespread in tropical, sub-tropical and warm temperate regions where the mean soil temperature is greater than 15°C.
- Wet soil increases the incidence of disease and water movement contributes to the dissemination of inoculum with water movement.
- The incidence and rate of wilting therefore increases with high temperatures and soil moisture.
- Continuous cropping of susceptible plants will also favour infection.
- The pathogen is disseminated by:
    - Contaminated farming equipment, in soil on tyres and footwear.
    - Drainage water carrying inoculum through the soil.
    - Infected seed, especially in the case of banana.
    - Seedlings raised in infected soil, spreading the pathogen to new areas.

- Cultural operations such as interculturing.
- Moko disease of banana is spread mechanically, by insects and on infected plants.

**v) Management**

- Strict phytosanitary measures are necessary to prevent the spread of the disease. Better drainage can reduce the disease incidence.
- Fallowing for 6 to 12 months is needed to manage the disease and the area should be kept weed-free during fallow period.
- Growing cover crops like kudzu (*Pueraria phaseoloides*) and sorghum is helpful in reducing the soil inoculum.
- Flower visiting insect pollinators are major transmission agents. Hence removal of male buds when the hands begin to turn upwards can reduce the disease incidence.
- *Heliconia* spp. and several other weeds serve as alternate hosts and these plants should be removed from the plantation.
- Bacterization of suckers with *Pseudomonas fluorescens* and use of antagonistic crops such as garlic and sorghum help in reducing the disease.
- Rotation with maize or rice for more than 2 years.
- Using healthy seed/certified seed and seedlings, stored under dry cool conditions and planting in cooler conditions.
- Maintaining clean machinery may aid in minimizing spread of the disease.
- Deep burying of inoculum with tillage, or repeated disk ploughing of dry soil to expose inoculum to the sun can be performed as a control practice.
- Fallows of 18-months have been shown to control the bacterial wilt of banana, with a shorter period for dry soils.
- All rhizome tissue and weeds must be removed for an effective fallow. Since major cause of disease spread in bananas is through the replanting of infected suckers, new plantings should be started with tissue cultured plants to restrict the spread of the pathogen.
- Selecting resistant cultivars of crop.

### 2.1.1.2. Soft Rot, *Erwinia carotovora*

The *Erwinia* genus is generally responsible for soft rots in banana plants.

**i) Symptoms:** Bananas being propagated from freshly cut rhizome tissue or suckers may show poor sprout emergence, stunting and yellowing (Fig. 2.2). Incidence is highest in wet humid areas where the material has been treated with hot water prior to planting.

**Fig. 2.2.** Soft rot of banana.

**ii) Pathogen:** *E. carotovora* is cultured on Stewarts MacConkey-pectate media, where it forms craters in the media. The organism is rod-shaped, Gram negative, approximately 0.7-1.5 um long and has peritrichous flagella. They are non-spore-forming and facultative anaerobic. The soft rot pathogens can be differentiated with biochemical tests.

**iii) Host Range:** Hosts of this pathogen include potato, lettuce, Brassicas, bananas, onions, maize and succulent plants. Head and rhizome rots of banana are caused by *Erwinia carotovora* sub sp. *carotovora* and *E. chrysanthemi*. Soft rot of cucumber and melon also occurs but as a post-harvest disease.

**iv) Epidemiology**

- The soft rot pathogen, *Erwinia carotovora* sub sp. *carotovora* can be disseminated into new infection sites by irrigation water or an insect invasion of the crop.
- The pathogen survives in the soil and enters the host through hydathodes, lenticels or wounds. Infections in banana usually arise from wounds to the rhizome. Enzyme secretion causes a breakdown of cell walls.
- Colonized tissues become water-soaked and secondary organisms cause decay.
- Soft rot increases when the humidity is high, soils are moist and temperatures are above 22°C. *Diptera* spp. are known insect vectors.
- The disease is more prevalent in the wet seasons, and where excess nitrogen has been applied.

**v) Management**

- Avoid wounding plant tissue during cultivation, harvest, packing and storage.

- The primary inoculum in soft rot of banana is the sucker, and there are a number of control measures effective in reducing disease severity. Suckers, especially those that have been cut should only be planted in well-drained soil. Avoid washing suckers and take care not to damage them prior to planting.
- By avoiding excess irrigation, the soil remains aerobic thus non-conducive to the growth of the pathogen.
- Use resistant varieties.
- Frequently clean and disinfect farm equipment. Use sterilized knives and trimming equipment especially in the case of bananas.
- Store produce in conditions favourable for the rapid healing of injured tissue.
- Seedling treatment (tissue cultured) with *Bacillus subtilis/ Pseudomonas fluorescens* followed by 3 soil drenches at 15 days interval starting from 20 DAP gave 100% and 87.5% disease control and increased fruit yield by 85.71% and 70% over control, respectively.

### 2.1.2. Viral Diseases

#### 2.1.2.1. Bunchy Top Virus

It is the most severe and widely prevalent disease of banana in many countries of the world. The disease was first recorded in Fiji in 1889 on Cavendish banana and many plantations were abandoned because of the severity of it during 1894. It is most severe and widely prevalent in Australia, Borneo, Burundi, Egypt, Gabon, India, Pacific Islands, Sri Lanka, Taiwan and the Philippines. In India, this disease is prevalent in the states of Kerala, Tamil Nadu, Andhra Pradesh, Karnataka and west coast of Maharashtra. Many elite cvs. like Virupaksha (hill banana), a choice clone in lower Palney hills in Tamil Nadu; Nendran and Palanthodan, popular and most widely grown in Kerala; Hari-Sal and Lal Velchi, having high export value in Vasai area of the west coast of Maharashtra have been almost wiped out by this disease. The disease intensity ranges from 20.0 to 32.7% in Kerala and Karnataka resulting in considerable economic losses. The drastic non-bearing nature of bunchy top infected plants aroused great concern making it a problem of national importance.

**i) Symptoms:** The first characteristic symptoms appear as irregular, nodular, green streaks along the secondary veins on the underside of lower portion of leaf blade, midrib and petiole. The streaks are about 0.75 mm wide and vary in length up to 2.5 cm. The subsequent leaves show the same symptoms and are progressively dwarfed. They also show marginal chlorosis and curling. The leaves are brittle in texture and petioles are incompletely elongated. Pale, white streaks may be seen along the length of lamina. In subsequent leaves, unfurling process is premature but slow so that several leaves are unrolling at the same time. The leaves become smaller and eventually the crown of the plant becomes composed of stunted leaves, the rosette or bunchy top (Fig. 2.3). The emergence of the bunches is choked by the pseudo stem and may split instead of emerging in the normal passion. The bunches are reduced in size and the fruits lose their market value.

**Fig. 2.3.** Bunchy top on banana (Courtesy, K. Srinivasan, TAFE, Chennai).

**ii) Transmission:** The pathogen is transmitted by mechanical means. In nature, the banana aphid, *Pentalonia nigronervosa* transmit the pathogen in a persistent manner. A single aphid is enough to transmit the disease. Once the aphid acquires the pathogen it remains viruliferous throughout its life span.

**iii) Epidemiology:** The disease spreads very quickly in the hilly areas because the hill banana is cultivated as a perennial crop and the aphid population in the hills is high due to conducive weather conditions prevailing in these areas. The disease spreads faster in rainy season than in summer season. The vector survives on large (*Amomum subulatum*) and small cardamom (*Elettaria cardamomum*), *Colocasia antiquorum* and *Alocasia macror-hiza*. The optimum number of aphids required for maximum transmission is 20. Acquisition feeding period is 3 hr. The transmission threshold is 30 min. The aphids may remain infective for up to 13 days after feeding on infected material.

**iv) Management**

- The suckers should be selected from disease-free gardens for planting and strict quarantine measures are necessary to prevent introduction of bunchy top to new areas.
- Remove the affected plants periodically along with rhizomes and destroy. Orchard should be kept free of weeds.
- Apply 40 g of carbofuran 3G granules per pit before planting disease-free plants. Spray the crop with systemic insecticides viz., 0.1%

phosphamidon or 0.2% oxydemeton methyl or 0.1% monocrotophos at 3 weeks interval for at least 3 times. Injection of monocrotophos at 1 ml/plant (1 ml diluted in 4 ml of water) at 45 days interval from the third month till flowering is very effective.

- Banana cvs. Kanchikala and Venattukunnan are reported to be resistant to bunchy top.

### 2.1.2.2. Mosaic or Infectious Chlorosis

It is also called as heart rot or virus sheath rot and was first reported from New South Wales, Australia in 1929. It has now been recorded in many other countries of the Old and New worlds. The disease was observed from India during 1949 from some districts of Maharashtra and Gujarat. The incidence of the disease has been reported from all banana growing areas of the country causing 30-40% losses in Jalgaon district (Maharashtra) alone.

**i) Symptoms:** The virus is characterized by the presence of typical mosaic-like or discontinuous linear streaking in bands extending from margin to midrib (Fig. 2.4). Rolling of leaf margins, twisting and bunching of leaves at the crown and a rigid erectness in newly emerged leaves are also observed. The presence of dead or dying suckers is noticed in advanced cases referred as heart rot resulting from rotting of heart leaf and central portion of pseudo stem due to association of some bacteria.

**Fig. 2.4.** Infectious chlorosis on banana (Courtesy, K. Srinivasan, TAFE, Chennai).

**ii) Epidemiology:** The primary spread is through primarily infected daughter suckers from diseased plants and the secondary transmission of the disease is through *Aphis gossypii*, but not through *Pentalonia nigronervosa*.

Primary infection develops in plants grown from suckers of previously infected suckers and secondary infection is contracted at any stage of growth. Primarily infected plants develop severe mosaic symptoms in young growth showing broadly streaked chlorotic or yellowish green bands and patches or chlorotic mottling distributed in patches over the leaf lamina. The infected leaves are narrower and smaller than normal ones. The plants are stunted in growth to such an extent that they do not produce bunches. Such plants act as reservoir of virus.

The mosaic virus is stylet-borne in its vector, the melon aphid, *Aphis gossypii*, and is sap transmissible to cucumber. It appears to be strain of cucumber mosaic virus (Patel and Mali, 1983).

**iii) Management**

- The banana orchards should be kept free from weeds as the virus survives in them during off-season. Suckers from infected clumps should not be used for planting. Growing pumpkin, cucumber and other cucurbits between the rows of banana crop should be avoided.
- Dry heat treatment of suckers at 40°C for one day inactivated the virus.
- Spraying 0.03% methyl demeton at 3 to 4 weeks interval controls the vector and reduces the spread of the disease.
- Banana cvs. Ela Vazhi, Athiya Kol and Kadu Bale are not susceptible to the disease.

### 2.1.2.3. Bract Mosaic Virus (BBrMV)

Banana bract mosaic virus (BBrMV) causes significant yield losses in banana and abacá in the Philippines, India and Sri Lanka. BBrMV could potentially be confused with Sugarcane mosaic virus (syn. Abaca mosaic virus), another potyvirus infecting both banana and abacá in the Philippines and inciting similar foliar symptoms.

**i) Symptoms:** Mosaic symptoms on the male flower bracts are diagnostic and distinct from those caused by all other known viruses of banana. Mosaic patterns, stripes and spindle-shaped streaks may also be visible on the base of the pseudo stem when the leaf sheaths are removed (Fig. 2.5). Infection is often associated with an increase in pigmentation on the pseudo stem, and sometimes symptoms are chlorotic on a red background and sometimes reddish, yellow or chlorotic on a green background. Leaf symptoms, consisting of spindle-shaped lesions and streaks running parallel to the veins, are not always evident but can occur on recently infected young plants.

Suckers exhibit usual reddish brown streaks at emergence and separation of leaf sheath from central axis. Clustering of leaves at crown with a "traveler's palm" appearance can be seen. Elongated peduncle and half filled hands are its characteristic symptoms. The symptom expression of this virus disease is clearly seen in banana cvs. Monthan and Plantain groups.

**ii) Pathogen:** BBrMV has flexuous virus particles, each 750 nm × 11 nm and containing a single coat protein (CP) of 39 kDa. Several partial genomic sequences are available on GenBank, covering the CP and the C-terminus of the nuclear inclusion body (NIb) protein.

**iii) Transmission:** BBrMV is transmitted by at least three species of aphids (*Aphis gossypii*, *Rhopaloshiphum maidis* and *Pentalonia nigrone-*

*rvosa*) in the non-persistent manner. Infections are also perpetuated by vegetative propagation of plants.

**Fig. 2.5.** Stripes and spindle-shaped streaks at the base of the banana pseudo stem

**iv) Management**

- Disease-free planting material should always be used for new planting.
- The diseased plants should be removed as and when noticed to avoid the spread of the disease.

## 2.2. CITRUS

Citrus fruits (*Citrus* spp.) rank third in area and production after banana and mango with an estimated production of 6.323 million tonnes from an area of 0.749 million hectares (Table 2.10). There are many commercially grown *Citrus* spp. such as mandarins (*Citrus reticulata*), sweet orange (*C. sinensis*), acid lime (*C. aurantifolia*), lemon (*C. limon*), grape fruit (*C. paradisi*), pumello (*C. grandis*), etc.

**Table 2.8. Major lime/lemon producing countries of the world (2005-06) (Bijay Kumar, 2007)**

| Country | Area (in '000 ha) | Production (in '000 MT) | Productivity (in MT/ha) | % share of world production |
|---|---|---|---|---|
| Mexico | 137 | 1807 | 13.2 | 14.0 |
| India | 210 | 1786 | 8.5 | 13.8 |
| Argentina | 47 | 1393 | 29.5 | 10.8 |
| Brazil | 50 | 1031 | 20.5 | 8.0 |
| Spain | 45 | 945 | 20.9 | 7.3 |

| | | | | |
|---|---|---|---|---|
| USA | 24 | 870 | 36.7 | 6.7 |
| China | 51 | 701 | 13.7 | 5.4 |
| Iran | 41 | 615 | 15.0 | 4.8 |
| Italy | 29 | 603 | 20.8 | 4.7 |
| Turkey | 20 | 600 | 30.0 | 4.6 |
| Others | 226 | 2557 | 11.3 | 19.8 |
| Total | 880 | 12909 | | |

**Table 2.9. Major sweet orange producing countries of the world (2005-06) (Bijay Kumar, 2007)**

| Country | Area (in '000 ha) | Production (in '000 MT) | Productivity (in MT/ha) | % share of world production |
|---|---|---|---|---|
| Brazil | 806 | 17853 | 22.2 | 27.8 |
| USA | 298 | 9252 | 31.0 | 14.4 |
| India | 539 | 4540 | 8.4 | 7.1 |
| Mexico | 317 | 4113 | 13.0 | 6.4 |
| China | 359 | 2447 | 6.8 | 3.8 |
| Spain | 139 | 2376 | 17.1 | 3.7 |
| Italy | 104 | 2261 | 21.8 | 3.5 |
| Iran | 148 | 2253 | 15.3 | 3.5 |
| Indonesia | 68 | 2214 | 32.6 | 3.4 |
| Egypt | 88 | 1789 | 20.3 | 2.8 |
| Others | 1125 | 15283 | 13.5 | 23.6 |
| Total | 3991 | 64282 | | |

**Table 2.10. Area, production and productivity of leading citrus growing states of India (2005-06) (Bijay Kumar, 2007)**

| States | Area (in '000 ha) | Production (in '000 MT) | Productivity (in MT/ha) |
|---|---|---|---|
| Andhra Pradesh | 166.2 | 2234.9 | 13.4 |
| Maharashtra | 248.8 | 1469.0 | 5.9 |
| Punjab | 26.0 | 365.4 | 14.0 |
| Gujarat | 30.5 | 303.3 | 9.9 |
| Madhya Pradesh | 16.7 | 266.9 | 16.0 |
| Rajasthan | 12.7 | 237.7 | 18.7 |
| Karnataka | 11.0 | 231.5 | 21.0 |
| Orissa | 26.5 | 205.3 | 7.7 |
| Manipur | 27.0 | 167.7 | 6.2 |
| Assam | 18.7 | 159.5 | 8.5 |
| Uttarakhand | 26.1 | 122.9 | 4.7 |

| | | | |
|---|---|---|---|
| Bihar | 16.8 | 112.3 | 6.7 |
| Others | 121.6 | 449.7 | 3.7 |
| Total | 748.8 | 6323.2 | |

**Table 2.11. Area, production and productivity of leading lime/lemon growing states of India (2005-06) (Bijay Kumar, 2007)**

| States | Area (in '000 ha) | Production (in '000 MT) | Productivity (in MT/ha) |
|---|---|---|---|
| Andhra Pradesh | 53.6 | 750.5 | 14.0 |
| Gujarat | 30.5 | 303.3 | 9.9 |
| Maharashtra | 35.7 | 136.2 | 3.8 |
| Karnataka | 8.1 | 198.5 | 24.4 |
| Uttarakhand | 26.1 | 122.9 | 4.7 |
| Bihar | 16.8 | 112.4 | 6.7 |
| Assam | 11.0 | 68.1 | 6.2 |
| Rajasthan | 2.7 | 16.8 | 6.3 |
| Madhya Pradesh | 1.1 | 17.3 | 16.0 |
| Others | 24.0 | 60.0 | 2.5 |
| Total | 209.5 | 1786.0 | |

**Table 2.12. Area, production and productivity of leading sweet orange growing states of India (2005-06) (Bijay Kumar, 2007)**

| States | Area (in '000 ha) | Production (in '000 MT) | Productivity (in MT/ha) |
|---|---|---|---|
| Andhra Pradesh | 134.2 | 1477.9 | 11.0 |
| Maharashtra | 91.6 | 611.6 | 6.7 |
| Karnataka | 1.1 | 18.5 | 17.3 |
| Madhya Pradesh | 0.2 | 3.9 | 16.2 |
| Rajasthan | 0.5 | 9.5 | 17.9 |
| Nagaland | 0.1 | 0.5 | 5.0 |
| Others | 0.1 | 0.2 | 2.1 |
| Total | 227.8 | 2122.1 | |

**Table 2.13. Area, production and productivity of leading mandarin orange growing states of India (2005-06) (Bijay Kumar, 2007)**

| States | Area (in '000 ha) | Production (in '000 MT) | Productivity (in MT/ha) |
|---|---|---|---|
| Maharashtra | 121.5 | 721.2 | 5.9 |
| Madhya Pradesh | 15.4 | 245.7 | 16.0 |
| Assam | 7.7 | 91.3 | 11.9 |

| | | | |
|---|---|---|---|
| Rajasthan | 6.8 | 88.8 | 13.0 |
| Meghalaya | 8.2 | 34.2 | 4.2 |
| Karnataka | 1.6 | 11.3 | 7.1 |
| West Bengal | 3.6 | 32.5 | 9.2 |
| Mizoram | 5.7 | 10.5 | 1.8 |
| Tripura | 2.8 | 18.2 | 6.5 |
| Others | 20.5 | 59.4 | 2.9 |
| Total | 193.7 | 1313.1 | |

### 2.2.1. Bacterial Diseases

#### 2.2.1.1. Canker, *Xanthomonas campestris* pv. *citri*

Citrus canker was first reported from U.K. and the U.S.A. in 1933 on herbarium specimens of *Citrus medica* collected at Dehra Dun during 1827-1831. The disease is known to occur in almost all citrus growing areas. The disease is very severe on acid lime, lemon and grapefruit.

**i) Symptoms:** The disease affects all the above ground parts-leaves, twigs, older branches, thorns and fruits (Fig. 2.6). Lesions first appear as small, round, watery, translucent spots on lower surface of the leaves and then on the upper surface. As the disease progresses, spots become white or grayish and give a rough corky crater appearance. The lesions are surrounded by yellowish halo. Elongated lesions are formed on twigs. On larger branches, the cankers are irregular, rough and more prominent. Cankers on fruits are similar to those on leaves except that the yellow halo is absent and crater-like depressions in the centre are more pronounced. The lesions on fruits remain confined to the rind, but sometimes, it causes cracks and fissures on the skin.

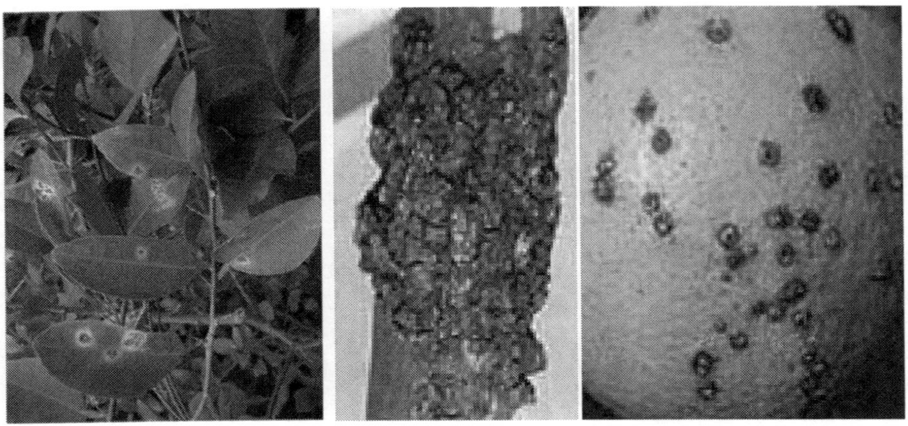

**Fig. 2.6.** Bacterial canker on acid lime leaves, fruit and bark (Courtesy: K. Srinivasan, TAFE, Chennai;V.K. Das, NRCC, Nagpur).

**ii) Epidemiology:** The disease is carried from one season to another in the cankerous leaves, twigs and branches which form the main source of inoculum. The pathogen survives up to 5 months in the infected leaves (Rao and Hingorani, 1963). Chakravarti *et al.* (1966) reported that the bacterium can survive up to 76 months in infected twigs. The pathogen enters the host through stomata, lenticels and wounds. Temperatures between 20° and 30°C with evenly distributed rains are most suitable for the disease development. Presence of free moisture on the host surface for 20 min is essential for successful infection (Ramakrishnan, 1954). The pathogen from cankers is disseminated mainly by wind splashed rains. Dissemination through leaf miner (*Phyllocnistis citrella*) is reported by Nirvan (1961). The long distance dissemination takes place through infected planting material.

**iii) Management**

- Pruning of infected twigs before monsoon and application of 1% Bordeaux mixture are found effective for protecting the leaves and fruits.
- Two prunings along with 4 sprays of 0.3% copper oxychloride or 1% Bordeaux mixture was found effective (Kishun and Chand, 1987).
- Pruning of infected twigs before monsoon and spraying 3 times with 500 ppm streptomycin sulphate + 0.2% copper oxychloride at 20 days interval was found effective to manage the disease (Ravi Kumar *et al.*, 2001).
- Spraying of Bacterimycin (Dinitro-dibromo propane mono tri diol) 0.5 g/litre of water at appearance of disease symptoms on leaves is effective.
- Application of neem cake solution on the foliage can reduce the canker (Reddy and Rao, 1980).
- Acid lime cv. 'Rasraj' is resistant to canker. Seedless lime is also resistant to canker (Kishun and Chand, 1987).
- Application of 2 phylloplane antagonists i.e. *Bacillus subtilis* and *Aspergillus terreus* isolated from citrus leaves were effective in reducing the disease incidence.

### 2.2.1.2. Greening, *Liberobacter asiaticum*

The presence of greening disease in India was reported from different citrus growing regions of India. The disease is widespread all over the country including Sikkim State. In Punjab, the disease incidence was as high as 64.7% (Kapur *et al.*, 1988). The disease is caused by Gram negative bacterium found in phloem sieve elements of infected tissue.

**i) Symptoms:** In new flushes, the leaf tissue immediately adjacent to mid-vein turns consistently yellow after the leaf reaches maturity. Yellowing of leaves often spreads along the main lateral veins and, in extreme cases, interveinal tissue also turns yellow. Secondary growth is upright, short with small leaves and multiple buds and show die-back symptoms. Leaf symptom pattern indicates zinc or iron deficiency. Later, whole lamina including veins becomes chlorotic with a few dark green islands (Fig. 2.7). The pathogen causing greening disease was considered to be Gram – ve bacterium by Naidu and Govindu (1981).

**Fig. 2.7.** Greening on citrus
(Courtesy: K. Srinivasan, TAFE, Chennai; V.K. Das, NRCC, Nagpur).

**ii) Epidemiology:** The greening disease is transmitted by infected buds and through citrus psylla, *Diaphorina citri* in the orchards. Even a single psylla was capable of spreading the pathogen (Raychaudhuri *et al.*, 1969). It can be successfully transmitted through leaf patch or bark patch or wedge grafting or by *D. citri*. The initial symptoms of greening appear in 20 to 45 days after feeding by the psyllid.

### iii) Management

- The disease can be managed by removal of greening affected trees and replanting with disease-free budded plants raised on improved rootstock.
- Locate citrus nurseries where psyllid vector is virtually absent.
- Certified pathogen-free bud wood should be used for propagating nursery stock. The bud wood can be exposed to 47°C for 2 to 4 hours (Thind *et al.*, 1987).
- Cheema *et al.* (1986b) reported that the greening pathogen could be eliminated by dipping the bud sticks in 500 ppm Ledermycin or

Pencillin extract in combination with 500 ppm of carbendazim before budding.

- The psyllids can be effectively managed by spraying with 0.05% monocrotophos or 0.15% metasystox at periodical intervals. Spraying with 500 ppm of streptocycline at fortnightly interval reduces the disease incidence by inhibiting the multiplication of the pathogen.

### 2.2.2. Viral Diseases

#### 2.2.2.1. Tristeza Virus

Tristeza, also called as quick decline, was suggested to describe the appearance of the diseased citrus trees. The disease is widespread in Argentina, Australia, Brazil, India, Indonesia, Java, Malaysia, Pakistan, Panama, Sri Lanka, Thailand, Uruguay, USA and Venezuela. It is endemic and causes damage to mandarins, sweet oranges, acid lime, grapefruit, etc., throughout India.

**i) Symptoms:** The infected trees usually show general decline. The leaves lose dark green luster, become dull and chlorotic, curl length-wise and upwards and drop exposing twigs which start drying up from tip downwards. Honey-combing, a fine pitting of the inner face of the bark in susceptible rootstock portion below the bud union, is also common (Fig. 2.8). In grapefruit and acid lime, which are highly susceptible to the disease, vermiform or linear pits appear in the woody cylinder.

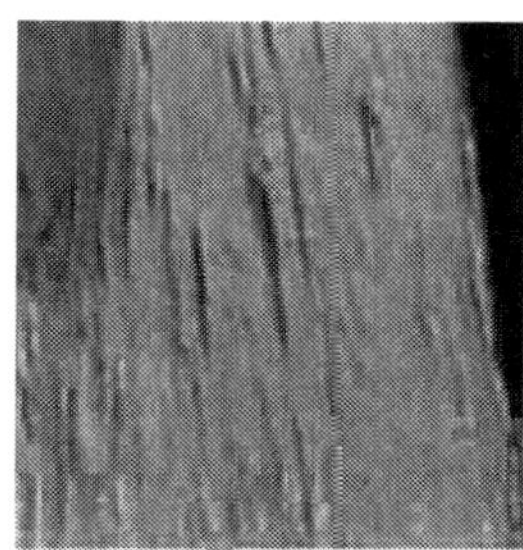
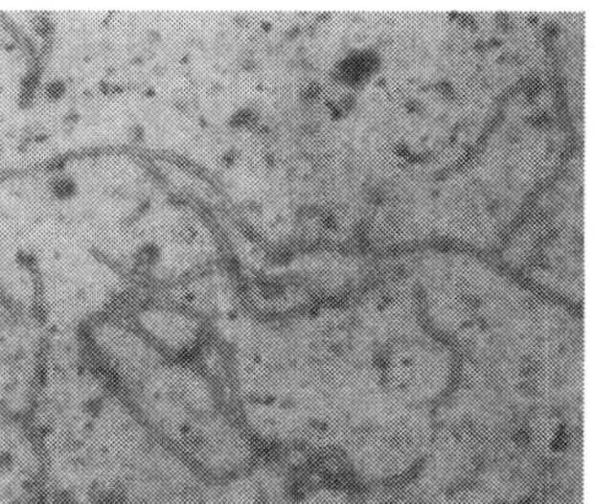

**Fig. 2.8.** Stem pitting due to tristeza virus infection in citrus (Courtesy: K. Srinivasan, TAFE, Chennai; D.K. Ghosh, NRCC, Nagpur).

In acid lime, another characteristic symptom is vein clearing in young leaves which is seen intermittently when viewed against light (Cheema *et al.*, 1986a). Where rootstocks are susceptible and the scion tissues allow virus to multiply, the bud union is injured, phloem sieve undergo necrosis and carbohydrate transport from top to roots is interrupted resulting in the starvation of roots. Most of the feeder roots are killed, water can not reach the top and trees start wilting. Infected trees bear heavily and get exhausted.

**ii) Etiology:** The disease is caused by virus particles which are long flexuous rods of 10 x 2000 μm in size (Fig. 2.8). The existence of 3 strains of tristeza, namely mild, severe and seedling yellows were reported (Balaraman and Ramakrishnan, 1977).

**iii) Epidemiology:** Spread of the virus generally occurs through the use of infected bud wood by the nurserymen (Reddy and Murti, 1985). Under field conditions, it is transmitted by aphid vector. The most efficient vector is the black citrus aphid, *Toxoptera citricidus* which is abundantly present in India (Balaraman and Ramakrishnan, 1979). Other species of aphids which can transmit the virus include *T. aurantii, Aphis gossypii, A. craccivora, Dactynotus jaceae* and *Myzus persicae.*

**Fig. 2.9.** A - Protected Nagpur mandarin mother plant being used as source of virus-free scion for implementing citrus bud wood certification programme. B – Virus-free planting material of citrus ready for distribution to the farmers (Courtesy: D.K. Ghosh, NRC for Citrus, Nagpur).

**iv) Management**

- Use of disease-free planting material on tolerant rootstocks such as Rough lemon and Cleopatra mandarin offer a great scope to live with tristeza virus (Fig. 2.9).
- Use of cross protected plants with mild strain and use of nucellar seedlings which are virus-free and true to type are some of the recommended practices to control the disease.
- Periodic spraying with insecticides like 0.05% monocrotophos or 0.15% metasystox reduces the secondary spread of tristeza in the orchard.

### 2.2.2.2. Mosaic Badna Virus

Yellow mosaic disease of citrus is common in India especially on sweet orange. The causal virus having size of 130 x 30 nm and ds DNA as its genome has been named as citrus yellow mosaic bacilliform virus as a member of the badna virus group. The incidence of disease varies from 10-70% in Sathgudi and Chinni sweet orange of South India. In Central India, the disease incidence ranged from a maximum of 13.7% in Mosambi sweet orange, as compared to 5.8% in acid lime and 3.2% in Nagpur mandarin as detected by PCR. The virus is transmissible by graft and mechanical inoculation from among citrus species and is also dodder transmissible. Under experimental conditions, it was transmitted by mealy bug, *Planococcus citri*.

**i) Symptoms:** Bright yellow mottling on the leaves of sweet orange and pummelo are characteristic symptoms of infection (Fig. 2.10). The field symptoms on citrus species include loss of pigmentation in the form of rings (diameter ranging from 2.1 to 24.2 mm) with green tissue in the middle of mature leaves which may be one to several per leaf. These symptoms are prominent on Kinnow mandarin and Malta orange. Blisters on twigs and leaves are also associated with the disease. In Kinnow mandarin, the quality of fruits is adversely affected due to reduction in fruit size, juice, total soluble solids and vitamin C content. Severely affected plants show die-back and decline symptoms.

**Fig. 2.10.** Yellow mosaic symptoms on Sathgudi sweet orange due to citrus mosaic badna virus (Courtesy: D.K. Ghosh, NRCC, Nagpur).

### 2.2.2.3. Exocortis Viroid

Exocortis viroid is present in almost all the citrus growing regions of the world. It is highly transmissible by tools from tree to tree.

**i) Symptoms:** Field symptoms range from mild bark cracking to very severe bark scaling on Trifoliate and Rangpur lime rootstocks (Fig. 2.11), accompanied by various degrees of stunting of the tree. Twigs and branches of infected Trifoliate or Rangpur lime may show yellow streak like blotches especially on younger greenish twigs. The temperature ranging from 32-40°C during the day and 27-30°C during the night is ideal for symptom expression on indicator plants. Severe symptoms associated with exocortis appear 8-12 weeks after inoculation in the form of classical leaf epinasty.

**Fig. 2.11.** Severe bark splitting on Rangpur lime rootstock due to exocortis viroid infection (Courtesy: D.K. Ghosh, NRCC, Nagpur).

### 2.2.2.4. Witches' Broom

One of the most recent diseases of citrus, which is seen in isolated pockets of the world, is witches' broom disease. The disease was first reported on lime from Oman and UAE in 1986. In India, it was first observed in the Nagpur district in eastern Maharashtra in 1995. Subsequently it was seen in other major acid lime growing states of Andhra Pradesh, Tamil Nadu and Karnataka.

**i) Symptoms:** Symptoms include small chlorotic leaves, highly proliferated shoots, and shortened internodes (Fig. 2.12). Leaves drop prematurely and infected twigs are distorted. In advanced stages, infected branches develop die-back.

**Fig. 2.12.** Phytoplasma induced witches' broom symptom in acid lime (Courtesy: D.K. Ghosh, NRCC, Nagpur).

## 2.3. PAPAYA

Papaya (*Carica papaya*) is one of the most important fruit crops valued for its rich source of vitamins and nutrient content. It is native to tropical America and its place of origin is said to be in Southern Mexico and Costa Rica. It is believed to have been introduced into India during the 16$^{th}$ century. It is grown both in tropical and sub-tropical countries of the world like Brazil, Australia, USA, Taiwan, Puerto Rico, Peru, Pakistan, Bangla Desh, Malaysia, India and various parts of Central and South Africa. The ripe papaya fruits are used for table purpose, raw fruits are cooked and used as vegetable. Immature fruits are used for the extraction of proteolytic enzyme 'papain', which is mainly used as a drug in digestive disorders and in the treatment of ulcers and diphtheria. It has also varied industrial uses viz., in tanning industry, degumming of silk, etc.

India is the largest producer of papaya in the world. Papaya is grown in 73,000 ha with annual production of 2.317 million tonnes (Table 2.14). Productivity of papaya is the highest (31.7 t/ha) (Table 2.15) among the fruit crops, which has attracted the growers for its commercial cultivation resulting in the spectacular increase in area and production in last few decades. In the last decade, phenomenal increase in area and production is recorded from Southern states owing to demand of fruits for industrial use.

**Table 2.14. Major papaya producing countries of the world (2005-05) (Bijay Kumar, 2007)**

| Country | Area (in '000 ha) | Production (in '000 MT) | Productivity (in MT/ha) | % share of world production |
|---|---|---|---|---|
| India | 73 | 2317 | 31.7 | 28.8 |
| Brazil | 33 | 1574 | 48.3 | 19.5 |
| Nigeria | 99 | 834 | 8.4 | 10.4 |
| Mexico | 19 | 709 | 37.5 | 8.8 |
| Indonesia | 24 | 549 | 22.8 | 6.8 |
| Ethiopia | 13 | 259 | 20.5 | 3.2 |
| Congo | 13 | 216 | 17.3 | 2.7 |
| Peru | 14 | 171 | 12.5 | 2.1 |
| Philippines | 9 | 147 | 15.5 | 1.8 |
| Colombia | 6 | 138 | 24.7 | 1.7 |
| Others | 81 | 1142 | 14.0 | 14.2 |
| Total | 383 | 8055 | | |

**Table 2.15. Area, production and productivity of leading papaya growing states (2005-06) (Bijay Kumar, 2007)**

| States | Area (in '000 ha) | Production (in '000 MT) | Productivity (in MT/ha) |
|---|---|---|---|
| Andhra Pradesh | 12.9 | 1009.2 | 78.2 |
| Gujarat | 7.7 | 323.0 | 41.8 |

| | | | |
|---|---|---|---|
| West Bengal | 9.5 | 263.7 | 27.7 |
| Karnataka | 3.5 | 230.1 | 66.1 |
| Assam | 8.1 | 130.5 | 16.1 |
| Chattisgarh | 4.2 | 119.1 | 28.2 |
| Kerala | 18.9 | 85.5 | 4.5 |
| Tamil Nadu | 0.3 | 53.7 | 182.0 |
| Madhya Pradesh | 0.7 | 19.0 | 27.0 |
| Others | 7.3 | 83.4 | 11.4 |
| Total | 73.1 | 2317.2 | 31.7 |

On an average, 1000 g of ripe papaya fruit contains 2500 I.U. of Vitamin A, 8 I.U. of Vitamin B, 33 Sherman units of Vitamin B2 and 70 mg of Vitamin C. Besides, it contains papain, a proteolytic enzyme which acts as papsin and is useful in digestive disorders.

### 2.3.1. Bacterial Diseases

#### 2.3.1.1. Internal Yellowing, *Enterobacter cloacae*

**i) Symptoms:** Internal yellowing affects the internal flesh of ripening fruit without displaying any external symptoms. The disease is characterized by soft, yellow-discoloured flesh with diffuse, spreading margins and an offensive, rotting odor in ripening papaya fruit. Infected tissue, which are observed only when fruit are cut open, occur in areas around and including portions of the seed cavity, usually near the calyx and middle sections of the fruit. Occasionally, vascular tissue near the stem-end become infected and appear yellow to yellowish-tan in colour.

Fruit ripening is a factor in the incidence of internal yellowing with ripe and overripe fruit having the highest incidence of infection. Symptom expression also varies with fruit ripening. In mature but not fully-ripe fruit, infected tissue appear as "pockets" of yellow-discolored flesh, while in ripe fruit, infected areas do not have distinct margins and appear "diffuse".

During the period 1984-1986, disease incidence in Hawaii of the hot-water treated fruit was as high as 9.7%. Since then, recent surveys indicate that this level is often surpassed, depending on the time of year and the packing house.

**ii) Biology:** *Enterobacter cloacae* is a gram-negative, rod-shaped bacterium that has peritrichous flagella, measures 0.3-0.6 x 0.8-2.0 µm, is oxidase-negative, catalase-positive, and is facultative anaerobic. The bacterium is positive for ß-galactosidase, arginine dihydrolase, ornithine decarboxylase, citrate utilization, nitrate reduction, and Voges-Proskauer reaction. Production of lysine decarboxylase, hydrogen sulfide, urease, tryptophan deaminase and indole are all negative. Acid is produced from many carbon sources. Reactions to many of these biochemical tests may be obtained using API 20E strips (Analytab Products). In addition, *E. cloacae* is negative for phenylalanine deaminase and pectate degradation.

*Enterobacter cloacae* grows well on standard bacteriological media on which yellow pigment and purple stain are not produced. Colonies are creamy tan on yeast extract-dextrose-calcium carbonate agar, and dark pink to burgundy with translucent margins on tetrazolium chloride agar. Miller-Schroth medium is useful for initial screenings because of the positive reaction (orange colonies) of *E. cloacae* on this medium.

**iii) Behaviour:** *Enterobacter cloacae* has been isolated from papaya flowers, homogenates of papaya seeds, and the crop and mid-gut of the oriental fruit fly (*Dacus dorsalis*). Recent studies claiming an apparent attraction of *D. dorsalis* to *E. cloacae,* suggest that fruit flies may possibly be involved in the transmission of the bacterium to papaya. It is suspected that after *E. cloacae* is transmitted to papaya flowers by fruit flies or other insects, the pathogen remains quiescent during fruit development until symptoms are expressed when the fruit are fully ripe.

Although *E. cloacae* was recovered from hot-water treatment tanks, it is believed that these tanks are not a major source of infection in processed papaya fruit because of the relatively rare occurrence of internal yellowing symptoms in the vascular tissue near the stem-end of the fruit.

A report of *E. cloacae* isolated from homogenates of papaya seeds in 1972 suggests that this organism may have been present in a non-pathogenic form for many years. Monthly samplings from five papaya packing houses, that processed fruit from different areas on the island of Hawaii, indicate that the incidence of internal yellowing is sporadic and may be affected by environmental factors.

**iv) Management:** Control of *E. cloacae* is currently limited to post harvest hot water quarantine treatments which effectively reduce the incidence of internal yellowing in papaya fruit. Without the hot water treatment, incidence of internal yellowing may be as high as 43%, depending on the packing house and the time of year.

The hot water treatments may be inducing resistance in the fruit to the bacterial pathogen, but the biochemical and physiological mechanisms involved are not known.

### 2.3.2. Viral Diseases

#### 2.3.2.1. Mosaic Virus

The disease was first reported by Capoor and Varma (1948) from erstwhile Bombay state in 1947. The mosaic is prevalent in Andhra Pradesh, West Bengal, Bihar, Karnataka, Madhya Pradesh, Maharashtra, Orissa, Punjab, Rajasthan and Uttar Pradesh. It may cause crop damage as high as 90% in severe cases.

**i) Symptoms:** Even though all stages of plant growth are susceptible to mosaic, one year old plants are more prone to the disease. Initial symptoms are light chlorotic spots, vein clearing, puckering and profuse mottling, especially in younger leaves followed by marked reduction in growth. The younger leaves show intense yellowing, chlorosis with dark green patches and large blister-like islands of dark green areas on the leaf blade (Fig. 2.13). The leaves are much reduced in size, distorted and modify into shoe string-like structures. The number of lobes of a leaf is increased. In young plants, growth is arrested and they fail to bear the fruits resulting in death of plants. The older leaves show marginal curling, turn brown, gradually bend and are shed prematurely, making the plant with only a small cluster of distorted leaves at their tops. The dark green spots and elongated streaks become prominent on petiole and stem of the diseased plants. The symptoms are masked during summer months but are discernible in winter months. The fruits produced on diseased plants develop innumerable circular water-soaked lesions with solid spots in the centre. In severe cases, the fruit size is reduced with deformed shape.

**Fig. 2.13.** Mosaic virus on leaves of papaya (Courtesy: K. Srinivasan, TAFE, Chennai).

**ii) Epidemiology:** The virus is mainly transmitted by mechanical inoculation, grafting and by a number of aphids. *Myzus persicae* is reported to be the most efficient vector (Capoor and Varma, 1958). *Aphis gossypii, A. craccivora, A. malvae, Rhopalosiphum maidis* and *M. persicae* transmit the virus in a non-persistent manner (Cheema and Reddy, 1985).

**iii) Management**

- Planting disease-free seedlings, rouging of diseased plants and spraying groundnut oil to inhibit transmission helps in reducing the incidence of mosaic.
- *Carica cauliflora* was found immune to mosaic disease. The leaf extract was found to contain an inhibitory factor.
- The plants remained free from mosaic until flowering stage by dipping of healthy seedlings in the leaf extract of *Carica cauliflora* diluted in water in the ratio of 3: 1 or 4: 1 for 4 hr.

### 2.3.2.2. Ring Spot Virus

It is a devastating virus disease. It occurs in India, Africa, Kenya, Nigeria, Tanzania and Uganda. Ring spot virus, a member of poty virus group, is commonly prevalent in India. Khurana (1968) first reported this disease from Gorakhpur. It was later recorded from Rajasthan, Maharashtra, Bihar and Karnataka.

In Karnataka, the disease is severe in northern parts where commercial cultivation of papaya is undertaken. The area under papaya has come down drastically by about 60% in Gulbarga, Bidar and Bijapur districts.

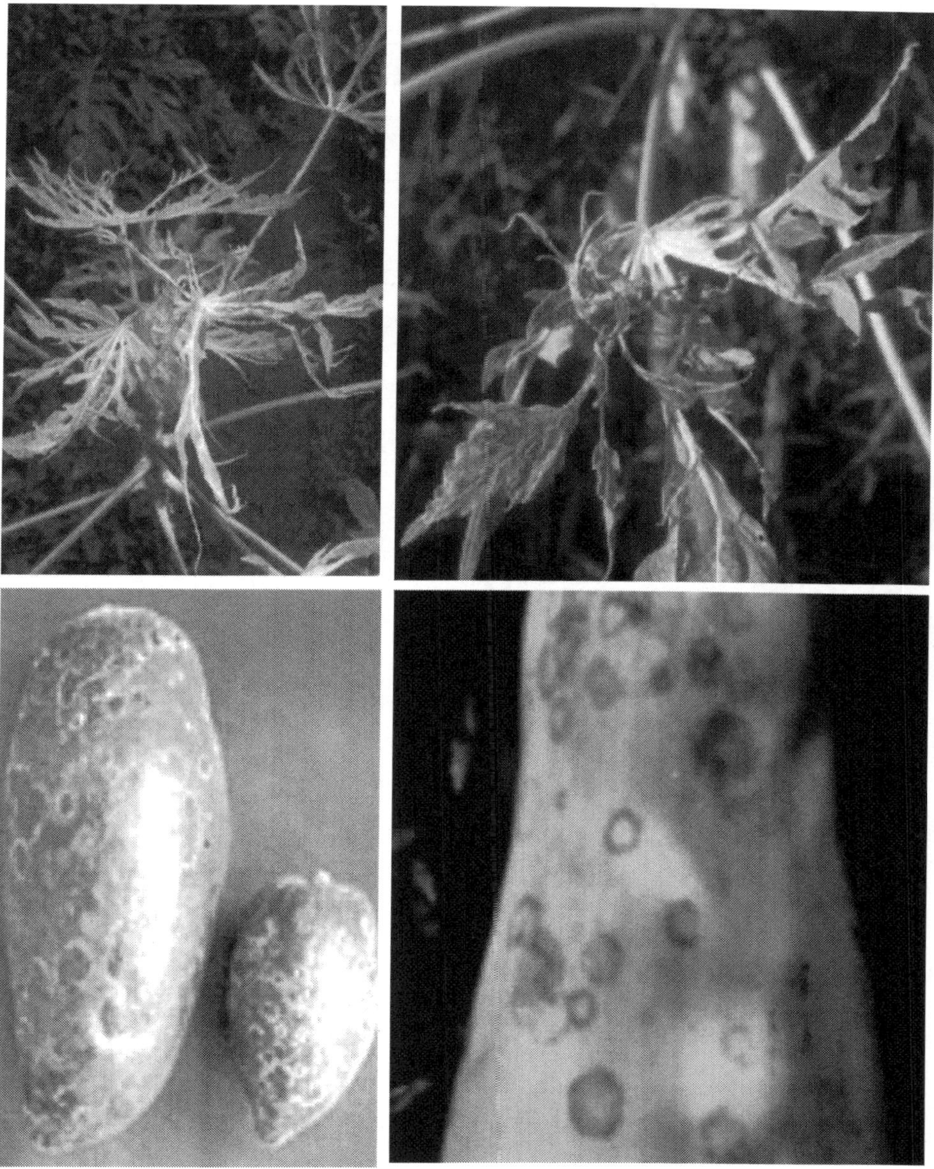

**Fig. 2.14.** Ring spot virus on leaves and fruits of papaya (Courtesy: K. Srinivasan, TAFE, Chennai).

**i) Symptoms:** The characteristic symptoms include vein clearing, puckering or bulging of the leaf tissue between the secondary veins and veinlets on the upper surface of the terminal leaves. The margins and distal parts of young leaves may roll downwards and inwards. The virus induces mosaic, mottling, dark green blisters, necrosis of chlorotic areas, leaf distortion resulting in shoe string symptoms followed by stunting of the plants (Fig. 2.14). Mosaic or mottle symptoms, dark green spots and oily or water-soaked streaks are seen on stems of young plants. The fruits are smaller, deeply lobed, lop-sided and have circular concentric rings. Diseased fruits contain 40% less sugar content and deterioration in latex quality.

**ii) Epidemiology:** The virus is transmitted mechanically and by aphids. *Aphis gossypii* and *Myzus persicae* are the most efficient vectors. Additional vectors of ring spot include *Acrythosiphon pisum, A. craccivora, A. citricola* and *Rhopalosiphum maidis*.

**iii) Management**

- Papaya nursery should be raised under insect proof net to avoid early infection of seedlings.
- Rouging must be followed right from the seedling stage. Infected plants should be uprooted and burnt.
- Yellow sticky traps should be placed along the borders to trap the aphid vectors.
- Rows of barrier crops like maize and sorghum should be raised along the borders to prevent horizontal movement of aphids.
- Spray 0.4% Nimbicidine, groundnut oil at 20 ml/l to reduce the disease incidence.
- Avoid cultivation of cucurbits in and around papaya orchards, which are also a host of PRSV.
- Papaya cv. Rainbow is resistant to ring spot virus.

### 2.3.2.3. Leaf Curl Virus

**i) Symptoms:** The characteristic disease symptoms include severe curling, crinkling and distortion of leaves accompanied by vein clearing and reduction in leaf size. The leaf margins are rolled downward and inward to form inverted cup followed by thickening of veins (Fig. 2.15). The infected leaves become leathery and brittle and petioles get twisted in a zig zag manner. The interveinal areas are raised on the upper surface due to hypertrophy which gives rugosity to the leaves. In advanced stages, defoliation takes place and the plant growth is arrested. The affected plants fail to flower or bear fruits.

**ii) Epidemiology:** The disease is transmitted by grafting and whitefly, *Bemisia tabaci* (Nariani, 1956).

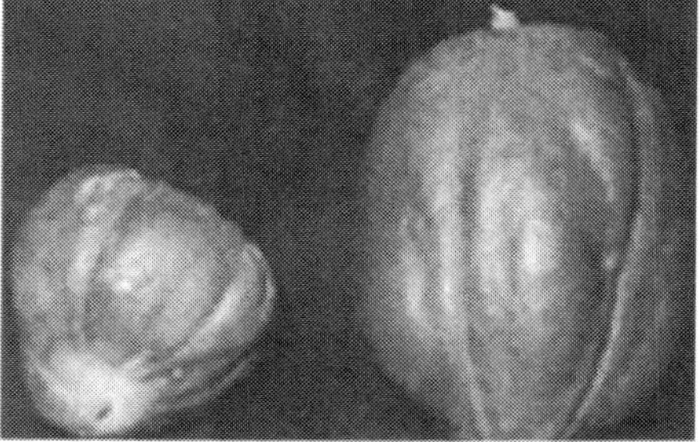

**Fig. 2.15.** Leaf curl virus on leaves and fruits of papaya (Courtesy: K. Srinivasan, TAFE, Chennai).

### iii) Management

- Destruction of infected plants in the nursery helps in preventing its spread to the main field.
- Rouging of affected plants in the field reduces the incidence of the disease.
- Avoiding growing of tomato, tobacco and Cape gooseberry (alternate hosts) in the vicinity of papaya plantation helps to keep the disease under check (Capoor, 1967).
- Check the infestation of insect vector (white flies) by spraying with 0.1% malathion or metasystox at 10-12 days interval.

## 2.4. PINEAPPLE

Pineapple (*Ananas comosus*) is grown in 85,400 ha with annual production of 1.353 million tonnes (Tables 2.16 and 2.17). There has been phenomenal increase in production and productivity (15.8 MT/ha) due to adoption of improved production technologies especially in Karnataka, West Bengal and Kerala. However, productivity is still low in North Eastern States (8-10 MT/ha) owing to traditional methods of cultivation compared to West Bengal (26.38 MT/ha).

**Table 2.16. Major pineapple producing countries of the world (2005-05) (Bijay Kumar, 2007)**

| Country | Area (in '000 ha) | Production (in '000 MT) | Productivity (in MT/ha) | % share of world production |
|---|---|---|---|---|
| Brazil | 62 | 2292 | 37.1 | 12.7 |
| Thailand | 98 | 2183 | 22.2 | 12.1 |
| Philippines | 49 | 1788 | 36.3 | 9.9 |
| Coasta Rica | 27 | 1605 | 59.9 | 8.9 |
| China | 76 | 1442 | 19.0 | 8.0 |
| India | 85 | 1353 | 15.8 | 7.5 |

| | | | | |
|---|---|---|---|---|
| Indonesia | 80 | 925 | 11.6 | 5.1 |
| Nigeria | 121 | 917 | 7.6 | 5.1 |
| Kenya | 14 | 600 | 42.0 | 3.3 |
| Mexico | 15 | 552 | 36.2 | 3.1 |
| Others | 268 | 4386 | 16.4 | 24.3 |
| Total | 896 | 18043 | | |

**Table 2.17. Area, production and productivity of leading pineapple growing states (2005-06) (Bijay Kumar, 2007)**

| States | Area (in '000 ha) | Production (in '000 MT) | Productivity (in MT/ha) |
|---|---|---|---|
| West Bengal | 13.4 | 379.2 | 26.3 |
| Assam | 14.6 | 241.0 | 16.5 |
| Manipur | 2.6 | 134.5 | 52.1 |
| Kerala | 12.8 | 109.3 | 8.6 |
| Bihar | 4.2 | 108.0 | 25.5 |
| Tripura | 5.2 | 106.4 | 20.5 |
| Karnataka | 10.6 | 88.6 | 8.4 |
| Meghalaya | 9.5 | 83.7 | 8.8 |
| Arunachal Pradesh | 7.9 | 37.0 | 4.7 |
| Nagaland | 2.0 | 24.8 | 12.4 |
| Tamil Nadu | 0.7 | 21.7 | 33.2 |
| Others | 2.0 | 18.9 | 9.7 |
| Total | 85.4 | 1353.1 | 15.8 |

### 2.4.1. Viral Diseases

#### 2.4.1.1. Yellow Spot Virus

The yellow spot virus has been shown to be identical to the tomato spotted wilt virus. It infects over a 100 species of plants including peppers, tomato, tobacco, eggplant, potato, broad bean, spinach, chicory and peas. A number of wild plants, including some common weeds such as the black jack (*Bidens pilosa*), *Emilia sonchifolia* and *Datura stramonium*, are also hosts of this virus.

**i) Symptoms:** One of the symptoms in the fruit is a blackened, dry cavity in the side of the fruit due to one or more "eyes" having died. This is known as "dead eye". Infection of very young fruit results in an irregular arrangement of fruit lets as some fail to develop. Attacked fruits may also fail to develop a crown. Thrips feeding on the crown of fruits results in concentric ring patterns developing on crown leaves. These spots enlarge and the infection spreads into the fruit itself, by which time the crown will often have dried out. The flesh of fruits thus infected will be discoloured and necrotic below the butt of the crown. The whole fruit may eventually become affected.

**ii) Transmission:** Thrips are vectors of this virus. When host plants of the virus are grown near to pineapple plants, the incidence of the yellow spot virus is considerably enhanced due to thrips migrating into the pineapple field. Wind carries thrips to long distances and thus also plays an important role in the transmission of the virus.

**iii) Management**

- Control weeds in and around pineapple fields. This is very important for disease control because the presence of certain weeds leads to increased number of thrips.
- Field management offers potential effective control. Cut off the fruits showing early symptoms of infection. This will prevent the spread of infections into the fruits.
- Where this disease is a problem, avoid planting near host plants of the thrips and the virus. Even if thrips are not strongly inclined to migrate from their favoured host plants, they will do so if disturbed, for instance by human passing, animals or machinery.
- Control thrips.

## 2.5. KARONDA

Karonda, *Carissa carandus* belonging to the Family Apocynaceae is an evergreen spiny shrub or small tree. It is widely grown in India, South Africa and Malaya. Fruits are sub acidic and sweet when ripe and rich in vitamin C and minerals. Fruits are eaten fresh but more useful for pickling, chutney and for culinary purposes. It also makes excellent jelly.

### 2.5.1. Bacterial Diseases

#### 2.5.1.1. Canker, *Xanthomonas compestris* pv. *carissae*

The disease has been reported from Maharashtra and Rajasthan.

**i) Symptoms:** The disease first appear on lower surface of leaves as small, round, water-soaked, translucent dots, gradually extending, turning dark brown and surrounded by halo. Spots on under surface of the leaves enlarge and become raised. The upper surface may be slightly raised. The central tissue of the spots finally collapse and form a depression. Twigs show similar symptoms. Flowers and fruits do not show symptoms. The corky growth appear on leaves and fruits due to tissue disintegration.

**ii) Disease Cycle:** The pathogen survives in diseased plant debris. The bacterium enters the host mainly through stomata but can also enter through wounds. The disease is favoured by a mean temperature of 24°C and relative humidity of 56%.

**iii) Management**

- Removal of all the diseased leaves and pruning of the shrub at a height of 45 cm from ground level are recommended for the management of the disease.
- Spraying with 200 ppm phytomycin or tetracycline or streptomycin also helps to reduce the disease incidence.

## 2.6. STRAWBERRY

Strawberry (*Fragaria vesca*) is an attractive, luscious, tasty and nutritious fruit with a distinct and pleasant aroma and delicate flavour. It is rich in vitamin C and iron and is mainly consumed as fresh. Jam and syrup are also prepared from strawberry. It is cultivated in tropical and sub-tropical areas round the year. It is commercially cultivated in Himachal Pradesh, Uttar Pradesh, Maharashtra, West Bengal, Nilgiri Hills, Delhi, Haryana, Punjab and Rajasthan.

### 2.6.1. Bacterial Diseases

#### 2.6.1.1. Angular Leaf Spot, *Xanthomonas fragariae*

Angular leaf spot (ALS) is a bacterial disease caused by *Xanthomonas fragariae*, a pathogen highly specific to wild and cultivated strawberry, *Fragaria ananassa* (Legard *et al.*, 2003). ALS is an important disease on winter strawberry production worldwide. This is a potentially devastating disease of strawberry. The rapid spread of ALS is influenced by the increasing rate of interchange of plant material.

**i) Symptoms**

**a) Leaves:** Angular leaf spot appears first as tiny water-soaked spots (lesions) on the lower leaf surface. These enlarge to form angular lesions, restricted by small leaf veins (Fig. 2.16). The young spots are usually best viewed on the underside of the leaf and appear translucent when looked at with a light source behind them (Fig. 2.16), and dark green when viewed normally. This difference is an important distinguishing characteristic in identifying the disease. Lesions eventually become visible on the upper leaf surface, and appear as irregular, reddish brown spots, which may become necrotic (dead). Spots on the upper surface may grow together to cover large leaf areas, causing infected leaves to appear scorched or blighted (Fig. 2.16). At this point, symptoms closely resemble those of leaf spot, caused by *Mycosphaerella fragariae* and leaf scorch, caused by *Diplocarpon earliana*. The necrotic tissue becomes dry and brittle, breaking off, giving leaves a frayed or ragged look. Heavily infected leaves may die if the bacterial infection moves into major veins.

**b) Petioles and Crown:** Systemic infections may occur, with all types of vascular tissue or the crown being infected. In severe cases, a vascular decline similar to that caused by *Phytophthora cactorum* or anthracnose crown rot may develop. Water soaking at the base of newly emerging leaves may be the only visible symptoms to be expressed before the plant suddenly dies. The bacterium may be isolated from roots, crowns, petioles, stolons, and daughter plants.

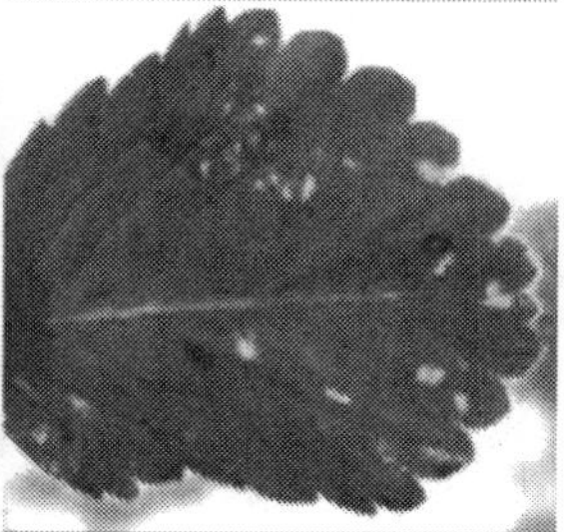

**Fig. 2.16.** Angular leaf spots and scorching on strawberry leaves.

**c) Fruit:** When infections of *X. fragariae* become systemic, the berry cap (calyx) may also be infected. The modified leaves of the berry cap (sepals) darken and dry. This reduces the marketability of the fruit significantly in some instances.

**d) Signs**: When moist conditions are present, the undersides of infected leaves will have a viscous bacterial exudate. This dries to form a whitish scaly film. This scaly exudate is also important in identifying the disease.

**ii) Pathogen:** This bacterium is not free living in soil. It can, however, over winter in soil on previously infected plant material. Transmission is by splashing water. It is host-specific and highly resistant to degradation—it can persist in the soil for long periods of time. It is killed by methyl bromide/ chloropicrin mixture used as a pre-plant fumigant, so it is very likely that most initial infections in fields that have been fumigated originate from contaminated plants. Lesions on the leaf surface serve as a source for secondary inoculum and cells are dispersed by splashing rain or overhead irrigation. The disease is favoured by cool, moist days with cold nights near freezing.

*Xanthomonas fragariae* can cause vascular collapse, although this is uncommon in California. This symptom initially appears as a water-soaked area at the base of newly emerged leaves. Shortly after, the whole plant suddenly dies, much like plants infected with crown rot. *Xanthomonas fragariae* is also associated with strawberry blossom blight in California.

**iii) Disease Cycle:** Inoculum for primary spring infections in new growth comes primarily from infected transplants or systemically infected over wintered plants and dead leaves. This bacterium is resistant to adverse conditions such as desiccation, and can survive for long periods in dry leaf debris or buried leaves in soil. Bacteria exuded from the undersides of leaves under high moisture conditions serve as the secondary source of inoculum in plantings. Angular leaf spot bacteria are carried from plant to plant by splashing water from rain or overhead irrigation, as well as harvesting operations. The motile bacterial cells may enter the plant through drops of dew, guttation droplets, rain or irrigation water.

**iv) Spread:** The primary source of inoculum in a new field is contaminated transplants (Mass 1998). Secondary inoculum comes from bacteria that exude from lesions under high moisture conditions. Bacteria can survive on dry infested leaves and tissue buried in the soil for up to 1 year (Roberts *et al.,* 1997; Strand 1993). The pathogen can be spread easily by harvesting operations when wet and cool conditions favour the production of bacterial exudates. The pathogen also can be dispersed by rain and overhead sprinkler irrigation. If the disease invades the vascular system of the plant, the disease will be difficult to control. Affected plants may wilt and die.

**v) Conditions Favouring Infection:** Moderate to low daytime temperatures and night time temperature below freezing are needed. High humidity is also a key factor.

The development of the disease is favoured by warm days (20°C) and cold nights (2-4°C) (Howard *et al*., 1985).

Bacteria produced in leaf spots are spread under wet conditions to the bloom and this is where the greatest economic damage occurs. Flowers may be killed outright or sections of the hull may become infected and turn black. For many producers the first indication that a field is infested with angular leaf spot is when otherwise normal fruit are harvested which have dark or black hulls.

Ideal conditions for this disease, in the period up to and during bloom, are wet cool days with night time temperatures near zero. When these conditions occur, losses can be significant. Under the hot, dry conditions of midsummer the disease becomes harder to detect, but leaf spots are easily found again in September and October.

**vi) Management :** The best way to control ALS is to use pathogen-free transplants. Copper-based products can provide effective control of the disease in some instances, but low rates of copper should be used since phytotoxicity has been documented with repeated sprays. Thoroughly cover all above ground plant parts with spray, especially undersides of leaves, as this is where bacterial inoculum forms and is dispersed. Growers should avoid harvesting and moving equipment through infected fields when the plants are wet. Minimizing the use of overhead sprinklers during plant establishment and for freeze protection will also reduce the spread of the disease.

Apply nitrogen fertilizers only at renovation to reduce succulent new leaf tissue which is more susceptible. Carefully space runner plants in matted-row culture and control weeds in all plantings to improve air circulation and reduce drying time for leaves. Remove older or infected leaves before setting runners in new plantings. Removing and burning all debris at renovation (after harvest) helps to reduce over wintering inoculum of the bacterium, as it does not survive alone in soil. Rotate crops to avoid infesting fields, and avoid overhead irrigation when possible.

## 2.7. MANGO

India is the largest producer of mango in the world with an annual production of 12.538 million tonnes from an area of 2.021 million ha (Tables 2.18 and 2.19). Mango occupies 36.7 per cent of the area under fruits and it contributes 21.3 per cent of the total fruit production in the country.

**Table 2.18. Major mango producing countries of the world (2005-06), Bijay Kumar, 2007)**

| Country | Area (in '000 ha) | Production (in '000 MT) | Productivity (in MT/ha) | % share of world production |
|---|---|---|---|---|
| India | 2021 | 12538 | 6.2 | 40.1 |
| China | 434 | 3676 | 8.5 | 11.8 |
| Thailand | 285 | 1800 | 6.3 | 5.8 |
| Mexico | 183 | 1679 | 9.2 | 5.4 |
| Pakistan | 165 | 1606 | 9.7 | 5.1 |
| Indonesia | 273 | 1413 | 5.2 | 4.5 |
| Brazil | 84 | 1348 | 16.0 | 4.3 |
| Philippines | 172 | 1003 | 5.8 | 3.2 |
| Nigeria | 138 | 812 | 5.9 | 2.6 |
| Egypt | 35 | 380 | 10.9 | 1.2 |
| Others | 578 | 4996 | 8.6 | 16.0 |
| Total | 4369 | 31251 | | |

**Table 2.19. Area, production and productivity of leading mango growing states (2005-06) (Bijay Kumar, 2007)**

| States | Area (in '000 ha) | Production (in '000 MT) | Productivity (in MT/ha) |
|---|---|---|---|
| Andhra Pradesh | 399.3 | 3194.3 | 8.0 |
| Uttar Pradesh | 251.5 | 2673.3 | 10.6 |
| Karnataka | 124.5 | 1236.8 | 9.9 |
| Bihar | 140.2 | 1222.7 | 8.7 |
| Gujarat | 96.0 | 772.1 | 8.0 |
| Maharashtra | 444.5 | 638.6 | 1.4 |
| Tamil Nadu | 125.1 | 537.8 | 4.3 |
| West Bengal | 70.1 | 513.3 | 7.3 |
| Kerala | 88.0 | 511.1 | 5.8 |
| Orissa | 125.3 | 428.8 | 3.4 |
| Others | 156.2 | 809.1 | 5.2 |
| Total | 2020.6 | 12537.9 | 6.2 |

### 2.7.1. Bacterial Diseases

#### 2.7.1.1. Canker, *Xanthomonas campestris* pv. *mangiferaeindicae*

Bacterial canker was reported for the first time on mango from Pune, India (Patel *et al.*, 1948a, b). It is known to be present in Australia, Brazil, Japan, Kenya, Malaysia, Pakistan, South Africa, Sudan, Venezuela, Taiwan, Reunion Island, Somali Land and many other countries. In India, it is widely prevalent throughout the country, especially severe in North India. The loss ranges from 10 to 80% depending upon severity and stage of infection. Its incidence and intensity ranged from 0.52 to 60.0% and 15.0 to 90.0%, respectively (Kishun, 1997).

**i) Symptoms:** The disease attacks leaves, petioles, twigs, stems and fruits. Small water-soaked lesions appear in groups on any part of the leaf blade. They increase in size and turn brown to black surrounded by yellow halo (Fig. 2.17). Several lesions may coalesce to form large necrotic patches which are often rough and raised. As the disease further advances, the infected leaves drop off. On young fruits, water-soaked lesions develop which also turn dark brown to black. Infected fruits may develop cracks in the skin and those badly affected drop prematurely. The canker stage also invites the secondary infection from saprophytic fungi.

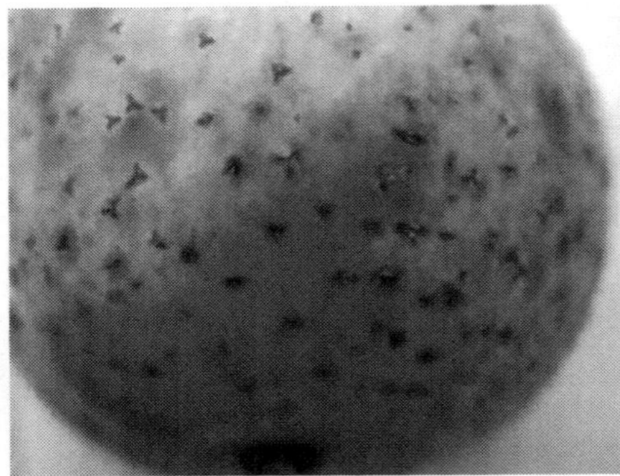

Fig. 2.17. Bacterial black spot (*Xanthomonas campestris* pv. *mangiferaeindicae*) on mango leaves and fruit.

**ii) Disease Cycle:** Bacterium survives in infected parts on the tree. Although the cankers are reduced by fall of infected leaves, the pathogen was found to survive up to 8 months in these leaves (Kishun, 1981a). Twig cankers initiate the infection on fruits. Bacterium is also found pathogenic on *Spondias mangifera, Anacardium occidentale, Acanthospermum hispidum, Lantana camara, Ficus glomerata, Solanum turbum, Psophocarpus fellagonolobus* and *Caesalpina minosoides* (Kishun, 1981). Role of mango stones in survival of pathogen has been established by Kishun and Chand (1988). Pathogen has also been found to survive on weed hosts (Kishun and Chand, 1988). Disease spread is rapid during rains. In new area, the disease spreads through infected planting material.

**iii) Epidemiology:** The bacterium survives in resting phase on different parts of host plant such as twigs, smaller branches, seeds of fruit and as phylloplane microflora on weeds. The long distance spread is through the use of infected planting material. The disease is favoured by high relative humidity (90%) and temperature of 25-30°C. Rainfall during flowering/berry development causes fruit infection. The disease is also known to be transmitted by ashy weevil, leaf webber and bugs as non-persistent bacterium.

**iv) Management**

**a) Cultural Methods:** Orchard sanitation and seedling certification are reported as preventive measures against the disease (Kishun and Sohi, 1983). Planting of stones extracted from healthy fruits for rootstocks is suggested (Kishun, 1995). Wind breaks have been found useful to certain extent in protection of mango orchards from the disease (Pruvost *et al.*, 1990).

**b) Chemical Methods:** Spraying of Agrimycin-100 at 2000 ppm proved effective under field conditions (Sheikh *et al.*, 1995). Two sprays of Streptocycline at 200-300 ppm at 20 days interval (Bose and Singh, 1980) and three sprays at 10 days interval (Misra, 1995) reduced fruit infection. Kishun and Sohi (1984) reported that monthly spraying of 0.1% carbendazim was found effective against the disease. Carbendazim at 0.1% was also found effective when injected in 3- to 5-year-old mango trees (Kishun, 1988b). Copper fungicides such as Bordeaux mixture, copper oxychloride (Lim *et al.*, 1989) and copper hydroxide (Pruvost *et al.*, 1989) are reported effective against bacterial canker disease.

**c) Biological Methods:** Antagonistic bacteria like *Bacillus coagulans* (Kishun, 1998), *B. subtilis* and *B. amyloliquifaciens* (Pruvost and Luisetti, 1991a) and fluorescent pseudomonads (Tzeng *et al.*, 1994) have been found effective biocontrol agents against the disease. Kishun (1991) has isolated specific bacteriophage of the pathogen which can be useful in the management of bacterial canker.

Garg and Kasera (1984) have reported that essential oil from *Anacardium occidentale* was effective against the pathogen.

**d) Integrated Methods**

- Collection and burning of infected leaves and twigs and orchard sanitation and seedling certification have been recommended as preventive measures against the disease (Kishun and Sohi, 1983).
- Copper sprays are the only method of combating the disease and are not always successful when disease pressure is high. One or two post-harvest copper sprays to cover the post-harvest flush and tail end of the rain season are effective in reducing inoculum (disease) pressure for the following season.
- Dipping the fruits in 250 ppm agrimycin-100 proved effective. Two sprays of 1% Bordeaux mixture or 200-300 ppm of streptocycline or 0.2% copper oxychloride at 20 days interval reduced fruit infection.

- Kishun and Sohi (1984) reported that the disease can be reduced by monthly spraying of 0.1% carbendazim. Its stem injection in 3 to 5 years old seedlings was also found effective (Kishun, 1985). Heidi, Kensington, Sensation and Tommy Atkins are tolerant to Bacterial black spot.
- Mango cvs. Bombay Green, Jehangir, Fazli and Suvarnarekha are reported to be resistant (Kishun and Chand, 1988).
- *Bacillus coagulans* is a potential bioagent which reduced the disease incidence to the extent of 75%.
- Prune off diseased twigs and establish windbreaks around the orchard.
- Monitor for the disease weekly.

## 2.8. GRAPEVINE

Grapevine (*Vitis vinifera*) is one of the important commercial crops grown in different climatic zones of the world for table delicacies and wine preparation. It is fairly good source of minerals like calcium, phosphorus and iron and vitamins like $B_1$ and $B_2$. Grapevine is grown in 64,300 ha with annual production of 1.630 million tonnes (Tables 2.20 and 2.21). In grapes, India has recorded the highest productivity (25.4 t/ha) in the world.

**Table 2.20. Major grape producing countries of the world (2005-06), Bijay Kumar, 2007)**

| Country | Area (in '000 ha) | Production (in '000 MT) | Productivity (in MT/ha) | % share of world production |
|---|---|---|---|---|
| Italy | 755 | 8554 | 11.3 | 12.6 |
| USA | 378 | 7088 | 18.7 | 10.5 |
| France | 855 | 6790 | 7.9 | 10.0 |
| China | 453 | 6592 | 14.5 | 9.7 |
| Spain | 1161 | 6054 | 5.2 | 8.9 |
| Turkey | 530 | 3850 | 7.3 | 5.7 |
| Iran | 315 | 2964 | 9.4 | 4.4 |
| Argentina | 212 | 2830 | 13.4 | 4.2 |
| Chile | 178 | 2250 | 12.6 | 3.3 |
| Australia | 153 | 2027 | 13.2 | 3.0 |
| South Africa | 113 | 1683 | 14.9 | 2.5 |
| India | 64 | 1631 | 25.4 | 2.4 |
| Others | 2222 | 15468 | 7.0 | 23.0 |
| Total | 7385 | 67696 | | |

**Table 2.21. Area, production and productivity of leading grape growing states (2005-06) (Bijay Kumar, 2007)**

| States | Area (in '000 ha) | Production (in '000 MT) | Productivity (in MT/ha) |
|---|---|---|---|
| Maharashtra | 45.1 | 1275.0 | 28.3 |
| Karnataka | 10.4 | 193.2 | 18.5 |
| Tamil Nadu | 2.6 | 84.8 | 32.6 |
| Andhra Pradesh | 1.9 | 40.3 | 21.2 |
| Punjab | 1.1 | 30.2 | 27.0 |
| Haryana | 0.1 | 3.3 | 33.0 |
| Madhya Pradesh | 0.1 | 2.6 | 25.0 |
| Mizoram | 0.5 | 0.6 | 1.2 |
| Jammu & Kashmir | 0.2 | 0.3 | 1.5 |
| Others | 2.2 | 0.3 | 0.1 |
| Total | 64.3 | 1630.7 | 25.4 |

Currently, about 20,000 tonnes of fresh grapes are exported, accounting for just 2% of the total production. Out of this, about 12,000 tonnes are exported to the Middle East markets and the rest to U.K, Europe and few South-East Asian markets. Among these markets, U.K. is the most paying besides being reliable. Middle East markets are not dependable and are also less paying.

### 2.8.1. Bacterial Diseases

#### 2.8.1.1. Canker, *Xanthomonas campestris* pv. *viticola*

The disease was first noticed on grape cv. Anab-e-Shahi at Tirupati in Andhra Pradesh. It appeared in an epiphytotic form on grape cv. Thompson Seedless during 1984 in Sangli and Sholapur districts of Maharashtra State. The disease is prevalent in Karnataka, Andhra Pradesh, Maharashtra and Tamil Nadu. The disease appeared in severe form during 1999-2002 in North Karnataka.

**i) Symptoms:** The disease starts as small water-soaked spots surrounded by yellowish halo on the lower surface of leaves. These spots enlarge in size in due course and become dark brown and angular. Sometimes these spots coalesce and form larger patches. Leaves also show vein infection. Lesions are brown to black, elongated and cankerous on petioles and canes. In advance stage of infection, stunting, cracking and abnormal growth of canes are noticed. Lesions on berries are brown to black, cankerous and severely affected ones become small and shriveled. Generally, cankerous symptoms are more prominent on all parts of the plant; hence the disease was renamed as bacterial canker.

**ii) Epidemiology:** The bacterium survives up to 65 and 35 days in infected dry leaves and in moist soil, respectively. Mango, neem and

*Phyllanthus maderaspatensis* may serve as alternate hosts. The temperature range of 25° to 30°C appears to be more favourable for the disease development. Free water on leaf from dew, irrigation or rain is more important for pathogenesis. The disease frequency increased under maximum rainy conditions.

**iii) Management**

- Use of disease-free cuttings for planting helps to reduce the disease incidence.
- Destruction of infected plant parts reduces the inoculum density.
- Excess use of irrigation water should be curbed.
- Late October pruning helps to reduce the incidence of canker.
- Spraying of antibiotics at 3000 ppm starting from second leaf stage at an interval of 15 days controls the disease (Chand and Kishun, 1988).

### 2.8.1.2. Crown Gall, *Agrobacterium vitis*

Crown gall is an important disease of wine grapes when they are grown in cold climates. Incidence may range from a few vines in a vineyard to 100% of the vines. Gall development may result in girdling of vines and reduced vigour and yield.

**i) Symptoms:** Gall formation on the aerial parts of the vine is the most common symptom associated with this disease. The bacterium which causes crown gall may be present in plants that do not show any symptoms. Galls are usually noticed as swellings near the base of the vine and up the trunk (Fig. 2.18). Galls on roots of grape are not typical, however, the bacteria can induce a localized necrosis of roots. Young galls are soft, creamy to greenish in colour, with no bark or covering. As they age, the tissue darkens to brown. The surface becomes open and the texture becomes moderately hard and very rough. The surface tissue of the galls turns black as it dies, but the bacterium remains alive in the vine.

**ii) Disease Cycle:** The crown gall bacterium survives systemically inside the grapevine. The primary source of diseased vines appears to be through propagation of diseased wood. The bacterium remains inside the vine until there is an injury to the trunk and only then invades the outer part of the trunk where it causes rapid cell multiplication and distortion of tissue producing galls.

Most grape cultivars are susceptible to crown gall. Major outbreaks are generally restricted to the varieties that are most susceptible to winter injury. The amount of crown gall present from year to year appears to be related to the severity of the preceding winter and the maturity of the vines.

The crown gall bacterium may also survive in vineyard soils in vine debris, but the importance of soil-borne inoculum in replanted vineyards is not well understood. It is believed that the majority of infections are a result of contaminated planting material.

**Fig. 2.18.** Aerial crown gall on grapevine

### iii) Management

### a) Before Planting

- Select sites with good air and water drainage.
- Avoid planting in frost-prone areas.
- Select rootstocks that are resistant to crown gall. Certain rootstocks such as Courderc 3309, 101-14 Mgt, and Riparia Gloire are resistant, whereas Teleki 5C and 110 Richter are susceptible. Resistant rootstocks can reduce the amount of crown gall that appears on susceptible scions.
- Select hardy varieties where possible. In general, *V. vinifera* cultivars are all highly susceptible.
- Prepare sites with nutrients and lime before planting if necessary to avoid vine stress due to poor nutrition or low pH.
- Consider pest control programs for nematodes and phylloxera before planting through the use of soil fumigation or rootstocks. Feeding damage by these pests provides sites for the entry of crown gall bacteria.
- Plant old vineyard areas where crown gall was present only after grapevines have been removed for at least 2 years. This is important because crown gall bacteria can survive in the remnants of the old grape plants until the debris decomposes. Success in reducing crown gall from the soil by leaving soil fallow or rotating to other crops may

vary depending on the amount of vine debris that is left in the soil, and how fast it breaks down.

- Do not propagate wood taken from galled vines.
- Specify that plant material you purchase be propagated from crown gall free plants. Producers must promote the production of certified indexed crown gall free vines.
- Hot water treatment of vines is effective in reducing crown gall infection levels in planting materials.

**b) After Planting**

- Avoid injury to vines (winter, mechanical and human). Remove suckers when shoots are small (3 to 6 cm) to reduce trunk damage and promote rapid healing of wounds. Removing larger shoots before they harden will result in clean and small scars. Hill young vines with 30 cm or more of soil or other material to protect them from cold temperatures. Remove this material carefully to avoid mechanical damage.
- Prune to delay bud break (late pruning) on varieties prone to early bud break.
- In France, application of $K_2O$ instead of nitrogen fertilizers is used to improve resistance of the vines to cold.
- Biological control of grape crown gall has been effective in trials, and is a promising management tool for the future.

### 2.8.2. Viral Diseases

#### 2.8.2.1. Fan Leaf Virus (GFLV)

Fan leaf is caused by a virus that also causes yellow mosaic and vein banding. All three are transmitted by the nematode *Xiphinema index*.

**i) Symptoms:** Leaf symptoms that resemble a fan are very conspicuous on Mission and French Colombard (Fig. 2.19). Infected vines have shortened and more irregular internodes. Lateral sprout development, double nodes, and stem fasciations cause a bushy appearance. Many berries shatter, and others do not develop beyond shot size. The disease affects berry set in grapes and can result in yield reductions of up to 80%, making it a highly important disease from an economic perspective (Walker and Yin, 1998).

**ii) Transmission:** *X. index* transmits the fan leaf yellow mosaic vein banding virus (GVFV-GVYMV) disease of grape (Fig. 2.17). Leftover roots of dead and old plants in vineyards act as source of viral inoculum of dagger nematode year after year. *X. americanum* has been found transmitting tobacco mosaic virus, grape yellow vine virus, peach rosette mosaic virus and tomato ring spot virus.

**iii) Spread:** This virus is transmitted from grape to grape by the longidorid nematode *Xiphinema index*. The virus is not transmitted through grape seeds and has no natural weed hosts

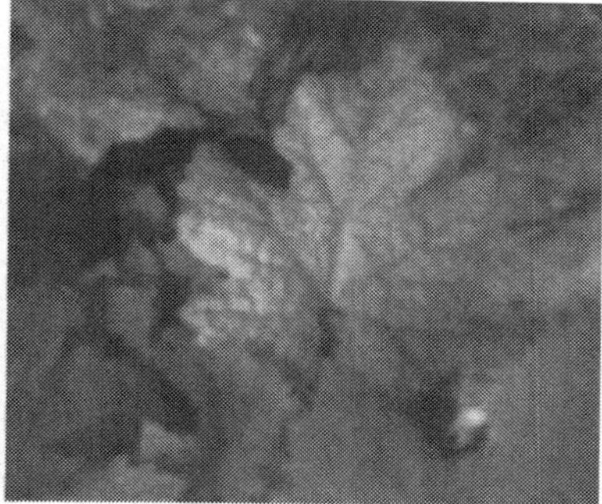

**Fig. 2.19.** Grapevine infected with fan leaf virus disease.

### iv) Management

Plant virus-free nursery material in nematode free soils. Plant disease free material obtained from a certification scheme within the industry. Diseased vines should be removed and destroyed.

In France, to save grapevines from attack by *Xiphinema index* and subsequently infection by grapevine fan leaf virus, a 7-year rotation with cereal and alfalfa crop has been recommended (Lamberti, 1981).

Control nematode vectors through chemical treatments or soil fumigants. DD at 250 gallons/acre was suitable recommendation for heavy soils against *X. index*, while only 200 gallons/acre was found to be effective in lighter soils (Rankin *et al.*, 1970).

### 2.8.2.2. Leaf Roll

Grapevine leaf roll is widely regarded as one of the most devastating viral diseases of grapevine in South Africa. Damage directly attributed to the disease is nearly impossible to determine, but probably runs into millions of rand annually. The etiology of leaf roll disease is not fully understood, but it is believed to be caused by a complex of viruses. At least nine distinct viruses, all members of the closterovirus group, have been associated with the disease. The most prominent of these is grapevine leaf roll associated virus III, a long, flexuous virus that is restricted to the phloem tissues of the grapevine.

Economical losses, however, are incurred by substantial reduction in yield, while poor fruit quality, e.g. reduced sugar content, has also been linked to these symptoms. A yield loss of approximately 20% occurs each year for as long as the diseased vines are held in the vineyard. Since leaf roll does not kill vines, producers may be reluctant to remove affected vines because of the perennial nature of the grape and the cost of re-establishing vineyards

**i) Symptoms:** Disease symptoms are fairly characteristic, especially in red cultivars, where a typical interveinal reddening and downward rolling of leaf lamina occur (Fig. 2.20). The leaves roll characteristically; yield and quality of the grapes are reduced. White cultivars develop only slight yellowish discolourations of the leaves

**ii) Spread:** This virus is usually transmitted by an insect vector known as the mealy bug, *Planococcus ficus*. The mealy bug is a serious pest on grapevines. It has spread rapidly in South Africa and has been found in most of the grape growing regions. Once established in a region, it is difficult to eradicate. The mealy bug can feed on roots, trunks, canes, leaves and fruit. It is more tenacious than other, more common mealy bugs because it can have up to six generations per year and can over winter under the bark. The mealy bug does damage to grapevines by exuding honey dew, which makes the leaves and fruits clusters sticky and also attracts ants. Sooty mould grows on the honey dew, which can make the fruit unmarketable.

**Fig. 2.20.** Leaf roll of grapevine: Left-Infected leaf; Right-Healthy leaf.

**iii) Management :** The natural spread in commercial vineyards is slow. To reduce the spread of the infection, removal of infected plants and control of the vector are required. Plant disease free material obtained from a certification scheme within the industry.

## 2.9. APPLE

Apple (*Malus pumila*) is the most important temperate fruit of the North-Western Himalayan region. It is predominantly grown in Jammu and Kashmir, Himachal Pradesh and hills of Uttarakhand accounting for about 90% of the total production. Its cultivation has also been extended to Arunachal Pradesh, Sikkim, Nagaland and Meghalaya in North-Eastern region and Nilgiri hills in Tamil Nadu. Apple is grown in an area of 2,34,700 ha with annual production of 1.755 million tonnes (Tables 2.22 and 2.23).

**Table 2.22. Major apple producing countries of the world (2005-06) (Bijay Kumar, 2007)**

| Country | Area (in '000 ha) | Production (in '000 MT) | Productivity (in MT/ha) | % share of world production |
|---|---|---|---|---|
| China | 1851 | 24020 | 13.0 | 38.7 |
| USA | 153 | 4409 | 18.8 | 7.1 |

| | | | | |
|---|---|---|---|---|
| Iran | 201 | 2662 | 13.2 | 4.3 |
| Turkey | 121 | 2570 | 21.2 | 4.1 |
| Italy | 57 | 2192 | 38.4 | 3.5 |
| Poland | 170 | 2075 | 12.2 | 3.3 |
| France | 47 | 1857 | 39.1 | 3.0 |
| Russian Federation | 384 | 1773 | 4.6 | 2.9 |
| India | 235 | 1756 | 7.5 | 2.8 |
| Chile | 37 | 1350 | 37.0 | 2.2 |
| Argentina | 45 | 1272 | 28.3 | 2.0 |
| Others | 1531 | 16109 | 10.5 | 26.0 |
| Total | 4832 | 62044 | | |

**Table. 2.23. Area, production and productivity of leading apple growing states (2005- 06) (Bijay Kumar, 2007)**

| States | Area (in '000 ha) | Production (in '000 MT) | Productivity (in MT/ha) |
|---|---|---|---|
| Jammu & Kashmir | 107.9 | 1093.3 | 10.1 |
| Himachal Pradesh | 88.6 | 540.4 | 6.1 |
| Uttarakhand | 28.8 | 112.3 | 3.9 |
| Arunachal Pradesh | 9.1 | 9.6 | 1.1 |
| Others | 0.3 | 0.0 | 0.0 |
| Total | 234.7 | 1755.6 | 7.5 |

### 2.9.1. Bacterial Diseases

#### 2.9.1.1. Crown Gall, *Agrobacterium radiobacter* pv. *tumefaciens* and Hairy Root, *A. rhizogenes*

The crown gall has worldwide distribution and is reported to occur in South Africa, New Zealand, Australia, Asia, South America, USA, Canada, etc.

#### i) Symptoms

**a) Crown Gall:** Globular, elongated or irregular galls are produced at or near the graft union. Galls vary in size from 0.6 to 10.0 cm in diameter and in extreme cases may go up to 30 cm or more. The young galls are soft, composed of light coloured frosty tissue resembling fresh callus growth. With the enlargement of galls, the outer tissue becomes darker and often turns black.

**b) Hairy Root:** The disease is mostly encountered in the nursery. First evidence of the disease is a convoluted tumor production composed of light coloured frosty masses of tissue. Root primordials appear from this tissue which gives rise to fleshy roots. These roots are thick which gives the woolly appearance, a characteristic of the disease.

**ii) Epidemiology:** The bacteria enter the plants through the wounds and multiply in the watery intercellular medium produced by the rupture of the cells. Following the infection, the plant cells transform into autonomously proliferating tumor cells due to unregulated cell division.

**iii) Management**

- Destruction of infested material by uprooting and burning, adoption of rigorous sanitation in the nursery and rotation of nursery sites are recommended.
- The disease incidence can be reduced by practicing budding instead of grafting (most infections appear at grafting point).
- Maize and other grain crops should be planted for several years prior to replanting of the nursery block.
- Entire root system of healthy grafted plants should be dipped in 1% copper sulphate solution for 90 min prior to transplanting. Antibiotics such as pencillin and streptomycin have also been employed.

### 2.9.2. Viral Diseases

#### 2.9.2.1. Apple Mosaic Ilarvirus (ApMV)

Apple mosaic virus is one of the oldest known and most widespread apple viruses. The same virus can cause line pattern symptoms in plum and rose mosaic disease. Apple mosaic virus is related to Prunus necrotic ringspot virus.

**i) Symptoms:** Apple trees infected with apple mosaic virus develop pale to bright cream spots on spring leaves as they expand. Apple leaves tend to be mottled with white to creamy yellow flecks, spots, and patches (Fig. 2.21). These spots may become necrotic after exposure to summer sun and heat. Most commercial cultivars are affected, but vary in severity of symptoms. Except in severe cases, infected trees can still produce a crop; yield reductions vary from 0 to 50 percent. In some cultivars, bud set is severely affected. Large numbers of leaves may drop from the tree early. Extreme variation occurs in symptom severity as a result of cultivar sensitivity and strain virulence.

**ii) Management:** Roguing of infected trees and the use of certified disease-free planting stock are effective control measures. The bud wood can be rendered virus-free by keeping the bud wood at 36°C for 4 weeks or by hot water treatment at 50°C for 8 min. Seedling plants maintained at 37°C and 40°C for 27 and 20 days, respectively inactivates the virus. Virus-free mother plants should be indexed for graft and bud wood.

**Fig. 2.21.** Mosaic virus on apple leaves.

### 2.9.2.2. Apple Chlorotic Leaf Spot Ilarvirus (ACLSV)

This disease causes symptoms on apples, quince, pears, plums, apricots, peaches and nectarines. It is present in a lot of fruit cultivars that do not exhibit symptoms of infection. It is present in a lot of different serological races in the different hosts. It is one of the most common fruit tree viruses worldwide. Although many cultivars are symptomless hosts, ACLSV can cause leaf symptoms in peach and fruit markings or graft incompatibility diseases in other stone fruits. These viruses are found erratically spread throughout the whole plant.

**i) Symptoms:** Symptoms caused by this virus group vary seasonally. Virulent isolates of ACLSV causes fruit necrosis symptoms like pseudo pox, virile on apricot or cherry fruit necrosis (Fig. 2.22). Symptoms of apple chlorotic leaf spot virus may include chlorotic leaf spots, leaf distortion, chlorotic rings and line patterns, reduced leaf size, and stunting (Fig. 2.22). Symptom expresssion is highly variable.

**Fig. 2.22.** Chlorotic leaf spot virus on leaves and fruits of apple.

**ii) Spread:** The virus is usually spread through contaminated plant material, mechanical inoculation as well as by grafting. This virus is not transmitted by seed.

**iii) Management:** ACLSV can be controlled by using virus-free rootstocks and scion material. Newly established plantings remain free of the virus where clean planting material has been used. Virus free material can be obtained through heat-treatment, selection of spontaneous virus free shoots, or meristem culture followed by indexing.

### 2.9.2.3. Apple Stem Grooving Virus (ASGV)

It is a common latent virus in apple.

**i) Symptoms:** As the designation "latent" implies, the virus is symptomless in most commercial cultivars, but may cause symptoms in certain cultivars, scion/rootstock combinations, and ornamental varieties. Apple stem grooving virus produces symptoms associated with 'Virginia Crab' such as chlorotic leaf spots, stem grooving and pitting, union necrosis, and swelling of the stem above the graft union. Although causing no marked symptoms in commercial apple cultivars, latent virus may have a detrimental effect on the growth and cropping of some cultivars.

**ii) Spread:** Very little, if any, spread is usually observed within apple orchards with infected trees. Most spread has been attributed to grafts. No natural vectors are yet known to spread this virus and no evidence of seed transmission has been reported.

**iii) Management :** Eliminating known infected trees and the use of certified disease free nursery material are effective control measures.

### 2.9.2.4. Apple Stem Pitting Virus (ASPV)

It is also a common latent virus on apple. Apple stem pitting virus is the causal agent of pear vein yellows.

**i) Symptoms:** As the designation "latent" implies, the virus is symptomless in most commercial cultivars, but may cause symptoms in certain cultivars, scion/rootstock combinations, and ornamental varieties. Apple stem pitting virus symptoms are associated with 'Virginia Crab', and include stem pitting of xylem that generally stops at the graft union. Apple stem pitting virus also causes fluted (grooved) fruit in 'Virginia Crab', and epinasty and decline in 'Spy 227'. Although causing no marked symptoms in commercial apple cultivars, latent virus may have a detrimental effect on the growth and cropping of some cultivars.

**ii) Spread:** Very little, if any, spread is usually observed within apple orchards with infected trees. Most spread has been attributed to grafts. No natural vectors are yet known to spread this virus and no evidence of seed transmission has been reported.

**iii) Management :** Eliminating known infected trees and the use of certified disease free nursery material are effective control measures.

## 2.10. PEACH AND PLUM

Peach (*Prunus persica*) is a temperate fruit rich in proteins, sugar, minerals and vitamins. In India, peach is grown in Himachal Pradesh,

Punjab, Haryana, Delhi, west Uttar Pradesh, Jammu and Kashmir and Assam.

### 2.10.1. Bacterial Diseases

#### 2.10.1.1. Crown Gall, *Agrobacterium radiobacter*

**i) Symptoms:** Small outgrowth appears on the stem and roots. At the young stage, the galls are soft, spherical, white or flesh coloured. The galls vary in size from 7 to 10 mm in diameter. On woody stems, the galls are hard and corky. They are generally knobby and knotty and become more clefty as they grow older. The affected plant may become stunted with chlorotic leaves.

**ii) Management**

Best possible approach for the control of crown gall is biocontrol. A non-pathogenic strain of *A. radiobacter* (strain 84) is employed to control crown gall and 99% success has been achieved in case of peach crown gall.

#### 2.10.1.2. Bacterial Leaf Spot, *Xanthomonas arboricola* pv. *pruni*

The disease is widespread throughout all fruit growing states. This is a plant disease, which occurs in both peach and nectarine trees. This disease, which in severe cases, may cause a peach or nectarine tree to lose all of its leaves. Bacterial leaf spot also causes discoloured, mottled and spotted fruit and can also be the catalyst for peach and nectarine trees to have a small crop or under grown fruits. It is one of the most devastating and destructive diseases of the stone fruit. This leaf spot disease is also a prime attacker of both plum and apricot trees.

**i) Symptoms:** Damage from bacterial leaf spot in peach and nectarine trees begins on the leaves. The peach and nectarine leaves develop brown spots, which are angular rather than circular. These spots can sometimes be black, but are most often a light to medium brown in colour (Fig. 2.23). The centres of the spots, which appear on the fruit tree leaves, may fall out, having disintegrated. However, often these spots simply cause the entire peach or nectarine leaf to turn yellowish, and then droop eventually dropping off and causing defoliation of the tree. The peach and nectarine fruits, themselves, may also suffer from a similar angular spotting pattern of disease and infection. Once the stone fruits become infected, they will show brown and black spots, which eventually become cracked and pitted in appearance (Fig. 2.23). Once the disease becomes severe, the spots will sink in as the fruits decays. If fruit trees are infected with bacterial leaf spot, appearance of lesions on the twigs of the tree can be noticed. Both peach and nectarine trees can develop these lesions, which will become oozy and sunken on their twigs. At the worst stage of infection, the entire peach or nectarine tree will defoliate leaving only infected branches and fruits unless the fruit has already dropped.

**a) Leaves:** Small spots 1 to 5 mm in diameter appear on the leaves. Spots are irregular to angular and have a deep purple to rusty-brown or black colour. In time, the centres dry and tear away, leaving ragged "shot-holes." When several spots merge, the leaf may appear scorched, blighted or ragged. Badly infected leaves may turn yellow and drop early. Early defoliation is most common on trees deficient in nitrogen or where the disease is further complicated by pesticide injury.

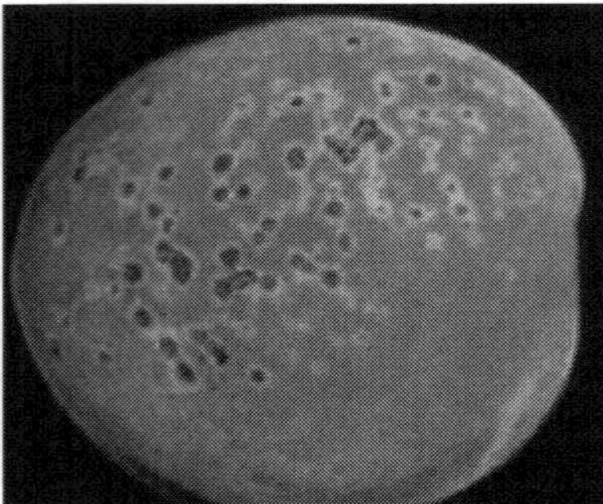

**Fig. 2.23.** Bacterial spot infected leaf and fruit of peach.

**b) Fruit:** Small, round olive-brown to black spots form on the fruit. They are usually sunken and frequently surrounded by a water-soaked margin. On peaches, spots usually form on the side exposed to the sun. Spots may slowly enlarge and merge to cover large irregular areas on the fruit. On some varieties the spots may exude a yellowish gum after rainy periods. Skin cracking and pitting may occur near the spots during fruit enlargement. Fruit infected at an early stage of development are usually the most malformed.

On plums, symptoms are different than on peaches. Large, sunken, black spots form on some varieties; on others, small pit-like lesions are common.

**c) Twigs:** On peaches, two distinct types of cankers damage twigs. "Spring cankers" develop on young twigs produced the previous summer. Spring cankers first appear as water-soaked, slightly darkened blisters about the time the first leaves appear. If these cankers encircle the twig, it will die. As the season progresses, the tissues over the blister-like lesions rupture and bacteria are released. These bacteria can be spread by wind-blown or splashing rain and can result in new infections. In time, spring cankers heal and become inactive.

"Summer cankers" develop on green twigs of the current season's growth. They usually occur later in the summer after leaf spots are evident. At first, they are water-soaked, dark purplish spots. In time, they enlarge, turn brown to purple-black, become slightly sunken and round to elliptical with water-soaked margins.

On certain plum and apricot varieties, twig cankers may continue to develop in two- and three-year old twigs. If the cankers are deep-seated, they can deform or kill the twigs.

**ii) Pathogen:** The bacterium *Xanthomonas arboricola* pv. *pruni* over winters in twigs that are infected late in the season about the time leaves are shed. The following spring, when environmental conditions are favourable, bacteria ooze out onto the surface of these twigs. The bacteria are then spread by wind-blown or splashing rain and can result in new infections throughout the growing season. The bacteria come in contact with healthy leaves, fruit and current-year twigs and enter the tissues through stomata or lenticels when surface moisture is present. Once inside healthy tissues, the bacteria multiply and disease develops.

Warm temperatures (21-29°C) with light rains, heavy dews or fogs and windy weather are most conducive for disease development and spread. The disease makes little progress when weather is hot and dry.

**iii) Host Range:** Bacterial spot affects peaches, nectarines, apricots, plums, prunes and cherries.

**iv) Disease Cycle:** The conditions which favour disease development include warm, moderate temperatures, frequent light rains, heavy dew, and considerable winds. If all, or most, of these conditions prevail, severe infection can be expected. When weather favours it, infection by these bacteria may occur anytime from shuck split until post-harvest. Hard, driving rains are more important in starting new infections, thus the disease can be proportionally more severe on one side of the tree than the other.

**v) Spread:** The bacterium over winters in the lesions, which develop on the trees. When spring begins, the bacterium begins seeping from the sunken lesions and is thus disseminated. Rainfall, splashing water and any form of watering can spread the bacterial ooze and infect nearby susceptible fruit trees. This organism is very fond of moist, wet regions and when heavy rainfall occurs the spread can be quick and deadly to peach and nectarine trees. If the disease becomes so severe that the peach and nectarine trees defoliate before early summer, it is likely that the entire peach or nectarine crop will have growth problems, as well as produce crops that are both poor in quality and small.

**vi) Management**

Bacterial leaf spot is, unfortunately, a tenacious disease in peach and nectarine trees and there is no adequate chemical control for this disease. However, in early spring when buds first open you can try a preventative method to suppress bacterial leaf spot by spraying with basic copper sulphate. Unfortunately, this only suppresses, and does not destroy the bacterial infection. Another method of temporary control is to prune out regions with infected lesions or to cut entire branches. This may stave off further infection of the tree, but it is likely once it rains, or water splashes any region where the bacterium still exists that it will continue to infect the tree.

- When planning an orchard, avoid low-lying or shaded sites with poor air circulation and soil drainage. Any practice that promotes faster drying of fruit and foliage will help reduce the risk of infection. Destroy nearby wild or neglected stone fruits (*Prunus* spp.). Buy and plant only vigorous, disease-free fruit trees from a reputable nursery.

- Prune trees annually to allow for better air circulation and to maintain tree vigour. If possible, prune during dry weather in the latter half of the dormant season.
- Select peach varieties with resistance to bacterial spot. The following varieties are somewhat resistant: Belle of Georgia, Biscoe, Candor, Comanche, Garnet Beauty, Harbrite, Harken, Late Sunhaven, Loring, Madison, Norman, Pekin, Raritan Rose, Redhaven, Redskin and Sunhaven. The following varieties are very susceptible: Babygold S, Blake, Elberta, Halehaven, Jersey Queen, Jerseyland, July Elberta, J.H. Hale, Kalhaven, Rio-Oso-Gem, Suncling, Suncrest and Sunhigh.
- Fertilize where needed to maintain vigour but not excessive shoot growth.
- Spray applications. At present, no spray program is completely effective for controlling bacterial spot. In the home orchard, spraying for bacterial spot is not considered practical.

### 2.10.1.3. Bacterial Canker, *Pseudomonas syringae* pv. *syringae*, *P. syringae* pv. *morsprunorum*

Two related bacteria, *Pseudomonas syringae* pv. *syringae* and *P. syringae.* pv. *morsprunorum* can cause bacterial canker. Both pathogens affect cherries, plums, peaches and prunes. The disease can occur on all stone fruit and is known as gummosis, blossom blast, spur and twig blight, sour sap and dieback

**i) Symptoms:** The disease attacks most parts of the tree. Cankers on trunks, limbs and branches exude gum during late spring and summer (Fig. 2.24). Leaves on the terminal portions of cankered limbs and branches may wilt and die in summer or early autumn if girdled by a canker. Occasionally, large scaffold limbs are killed. Leaf and fruit infections occur sporadically, but they can be of economic significance in years with prolonged wet, cold weather during or shortly after bloom. Leaf spots are dark brown, circular to angular, and sometimes surrounded with yellow halos (Fig. 2.24). The spots may coalesce to form large patches of dead tissue, especially at margins of leaves, or the centres of the necrotic spots may drop out, resulting in tattered leaves. Infected leaves may abscise during midseason. On fruit stems, lesions are elliptical and brown with water-soaked margins.

Infected leaf and flower buds may fail to open in spring, resulting in a condition referred to as “dead bud.” Small cankers often develop at the bases of these buds. Other infected buds open in spring but collapse in early summer, leaving wilted leaves and dried-up fruit. If blossom infection occurs, whole blossom clusters collapse as infection spreads into the fruit-bearing spurs. Blossom blight and spur blast are most likely in years when leaf and fruit infections are common.

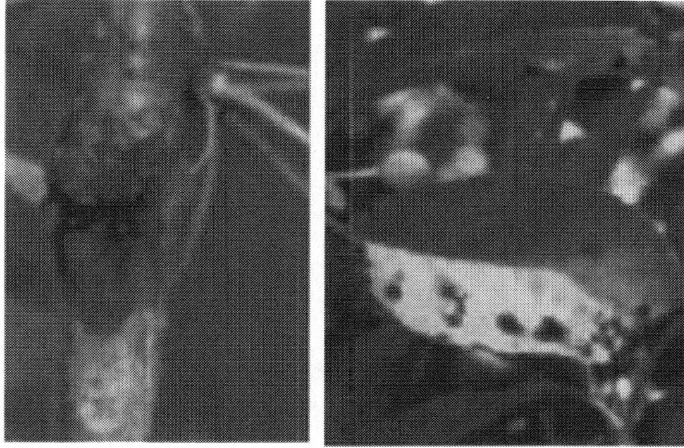

**Fig. 2.24.** Gumming is often associated with the canker disease.

**ii) Pathogen**: Two common genetic traits increase the bacteria's (*Pseudomonas syringae* pv. *Syringae)* ability to cause disease. Most produce a powerful plant toxin, syringomycin, which destroys plant tissues as bacteria multiply in a wound. Bacteria also produce a protein that acts as an ice nucleus, increasing frost wounds that bacteria easily colonize and expand.

**iii) Spread:** The bacteria can survive from one season to the next in bark tissue at canker margins, in apparently healthy buds and systemically in the vascular system. Bacteria multiply within these over wintering sites in the spring and are disseminated by rain to blossoms and to young leaves. Bacteria of both pathovars can live in an epiphytic phase on the surface of symptomless blossoms and leaves from bloom through leaf fall in autumn. After leaves abscise in autumn, the bacteria may enter the tree through fresh leaf scars.

Factors that weaken or injure the tree predispose it to developing cankers. Factors include wounds, frost damage, early pruning, soil pH, poor nutrition, and infection by other pathogens including *Verticillium, Nectria,* and nematodes. Sources of bacteria include old cankers, healthy buds, systemically within trees (with or without cankers), as epiphytes on leaf surfaces, weeds, grasses, and even soil. Wind, rain, insects, infected bud wood, and infected nursery stock can spread bacteria.

Outbreaks of bacterial canker are often associated with prolonged periods of cold, frosty, wet weather late in the spring or with severe storms that injure the emerging blossom and leaf tissues. Freezing can predispose the tissue to infection, but infection depends on the presence of wet weather during the thawing process. Free water on leaf surfaces and high relative humidity are required for at least 24 hours before significant leaf infection can occur following violent storms. Symptoms appear about 5 days later at temperatures between 21 and 27°C.

Bacterial canker is much more severe on cold-injured trees and on trees growing in sites with poor soil drainage. Therefore, practices that minimize wounding and freeze injury and maximize good cultural practices will reduce the incidence of bacterial canker. Pruning tools can transmit the bacteria so these should be disinfected between trees if bacterial canker is present.

**iv) Management:** Cankers should be pruned several inches below obvious infection during the winter when temperatures are freezing or below. The choice of cultivar is important, for some stone fruit are more susceptible than others to bacterial canker. Sprays during the growing season have not been effective in disease control. Chemical control of bacterial canker is based on protective copper sprays. These sprays, however, protect from initial infection but cannot prevent the canker phase once infection has begun. Also, there is reported resistance to copper in orchards that have consistently used copper. Stone fruit crops are sensitive to copper and injury is common

Copper-containing compounds may be of limited value for the control of bacterial canker because strains of *P. syringae syringae* resistant to copper are common in orchards with a history of copper usage. Also, copper injures most stone fruit crops. Even on the more tolerant crop species, it becomes more injurious as applications are repeated.

### 2.10.2.Viral Diseases

#### 2.10.2.1. Prunus Necrotic Ringspot Ilarvirus (PNRSV)

This virus infects and can cause disease in cherry, peach, and plum. It causes tatter leaf symptoms. Fruit yields may be decreased by 40 to 50 percent, depending on conditions. The virus is also a serious problem in the nursery, causing poor bud "take" (reduced 50-90 percent), poor scion "take", and reduced growth of trees.

**i) Symptoms:** The symptoms first appear on only one or a few branches in the first year. The following year, other branches may show symptoms while the original branch may not. Symptoms appear in acute or shock years, and then commonly disappear. In early spring, foliation and blooming of infected branches may be delayed, with leaves remaining small with depressed fine lines (etching) and partial-to-complete rings on the upper surface. Areas of symptomatic leaves may become necrotic (Fig. 2.25) and fall out; giving those leaves a tattered look. Green fruit may show arcs and rings. Occasionally, large areas of bark are killed and show gumming.

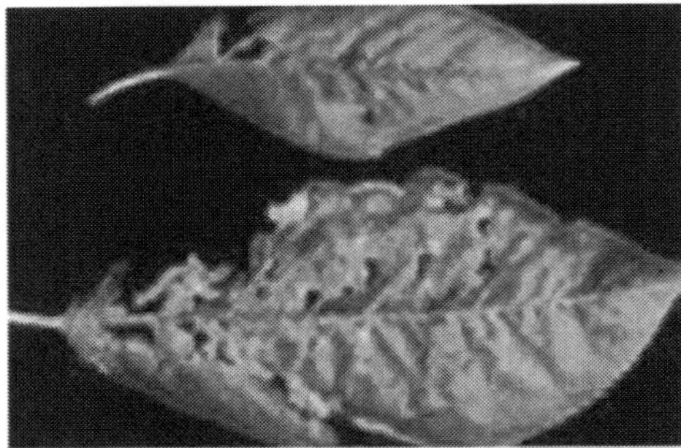

**Fig. 2.25.** Prunus necrotic ring spot virus infection on leaves.

**ii) Spread:** Movement of virus in the orchard occurs through transmission by pollen to seed and to pollinated plants. Once transmission occurs, symptoms appear in two weeks to two years. On older, well-established trees, the infection may remain latent, with no apparent symptoms. Symptom severity depends upon the age and general condition of the planting, and on temperature, with leaf symptoms developing best between 20-24°C, and shoot dieback being more severe at higher temperatures.

**iii) Management :** Making sure that existing orchards are tested on a regular basis and are best kept separated from other infected orchards. Keep in mind that bees can fly for long distances during pollination. Two methods of control are usually implemented i.e. Certification programs that supply disease free material and orchard management practices to prevent tree-to-tree spread. This entails prompt removal of diseased plants; avoid the use of commercial beehives that have been working the area and the use of virus-free wood for top working. Obtain virus free certified plant material from a reputable plant improvement organization.

Peach rootstocks grown from uncertified seed from peach canneries are important carriers of these viruses

### 2.10.2.2. Prune Dwarf Ilarvirus (PDV)

This disease is named for the symptoms it causes in 'Italian prune', *P. domestica.* The virus also causes yellows in sour cherry, alone or in combination with Prunus necrotic ringspot virus.

**i) Symptoms:** This virus causes stunting and leaf malformation in 'Italian Prune'. It is usually latent in peach and nectarine and may cause chlorotic spots and rings in sweet cherry (Fig. 2.26). If either rootstock or scion is infected, bud "take" drops 40 to 50 percent below normal for healthy stock.

**Fig. 2.26.** Prune dwarf virus.

**ii) Spread:** The virus is borne in pollen and seed. As high as 80 percent transmission in seed of *Prunus* species has been reported. When infected pollen pollinates a healthy tree, it may also infect that tree, although frequency of this type of transmission tends to be much lower than seed transmission.

**iii) Management :** Making sure that existing orchards are tested on a regular basis and are best kept separated from other infected orchards. Keep in mind that bees can fly for long distances during pollination. Two methods of control are usually implemented i.e. Certification programs that supply disease free material and orchard management practices to prevent tree-to-tree spread. This entails prompt removal of diseased plants; avoid the use of commercial beehives that have been working the area and the use of virus-free wood for top working. Obtaining virus free certified plant material from a reputable plant improvement organization.

Peach rootstocks grown from uncertified seed from peach canneries are important carriers of these viruses.

## 2.11. POMEGRANATE

Pomegranate (*Punica granatum*) is an ancient and favourable table fruit in tropical and sub-tropical regions of the world. It is commercially grown for its sweet-acidic fruits which provide a cool refreshing juice and is valued for its medicinal properties. Its popularity is also due to the ornamental nature of the plant especially when bearing bright red flowers throughout the year.

In India, it is considered as a crop of the arid and semi-arid regions because it can be grown even under conditions of severe drought and frost. Pomegranate is grown in an area of 1,16,400 ha with annual production of 0.849 million tonnes (Table 2.24). The fruit rind, juice, leaf and roots are used in the preparation of various Ayurvedic medicines. Fruit is good for stomach inflammations and heart pain, diarrhea and dysentery. Juice is useful for leprosy patients. The stem, root and bark are useful in destroying parasitic worms and cure of fever. The dry rind is useful in teeth and gum disorders. The fruit juice contains glucose and fructose. The juice and seed contain large quantities of tannin and agolic acid which are essential in curing several diseases.

**Table 2.24. Area, production and productivity of leading pomegranate growing states (2005-06) (Bijay Kumar, 2007)**

| States | Area (in '000 ha) | Production (in '000 MT) | Productivity (in MT/ha) |
|---|---|---|---|
| Maharashtra | 91.0 | 593.6 | 6.5 |
| Karnataka | 12.4 | 138.6 | 11.2 |
| Andhra Pradesh | 7.7 | 65.0 | 8.4 |
| Gujarat | 3.8 | 38.2 | 10.1 |
| Tamil Nadu | 0.4 | 11.1 | 25.0 |
| Rajasthan | 0.3 | 1.8 | 6.0 |
| Chhattisgarh | 0.1 | 0.6 | 6.7 |
| Himachal Pradesh | 0.6 | 0.2 | 0.4 |
| Total | 116.4 | 849.1 | 7.3 |

## 2. 11.1. Bacterial Diseases

### 2. 11.1.1. Bacterial Blight, *Xanthomonas axonopodis* pv. *punicae*

**i) Symptoms:** Small irregular, water-soaked spots 2-5 mm in diameter with necrotic centre of pinhead size appear on the leaves. Spots are translucent which later turn to light brown to dark brown and are surrounded by prominent water-soaked margins (Fig. 2.27). Spots coalesce to form larger patches. Severely infected leaves fall off. The bacterium attacks stems, branches and fruits also. On the stem, the disease starts as brown to black spots around the nodes resulting in girdling and cracking of nodes. Finally, the branches break down. Brown to black spots appears on the pericarp with L- or Y-shaped cracks. The spots on fruits are raised and oily in appearance. In severe cases, there will be extensive cracking of fruits (Fig. 2.27).

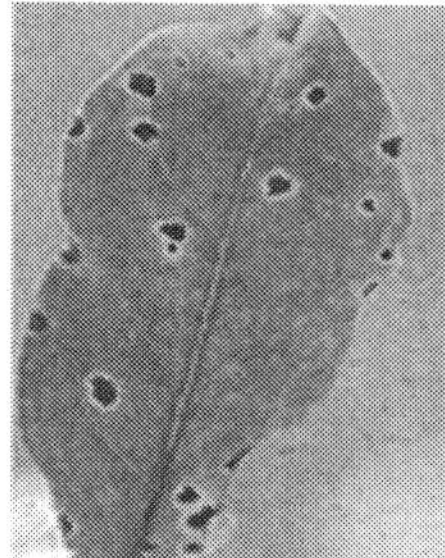

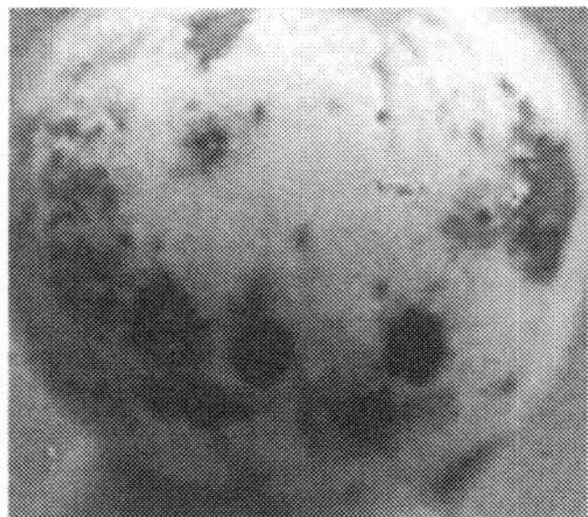

**Fig. 2.27.** Bacterial blight on leaf and fruits of pomegranate (Courtesy: K. Srinivasan, TAFE, Chennai).

**ii) Epidemiology:** High temperature and low humidity favour the disease. The bacterium survives on the tree. It can survive for 120 days on fallen leaves during off-season. The primary infection is through infected cuttings. The disease spreads through wind-splashed rains.

**iii) Management**

**a) Before Planting**

- Use disease-free seedlings for planting.
- Application of FYM/compost/vermicompost helps in building up resistance in plants to bacterial blight.

**b) After Planting**

- Practice field sanitation (collection and burning of diseased leaves, stem and fruits) to prevent the spread of the disease.
- Before pruning, spray 1% Bordeaux mixture on diseased leaves. Then spray with ethrel solution (2.0-2.5 ml/litre) to defoliate the diseased leaves. Collect and destroy the fallen leaves.
- Disinfect the pruning knife in Sodium hypochlorite solution (25 ml/litre) while pruning to prevent the spread of the disease.

- Dust the tree basins with bleaching powder at 20-25 kg/ha to kill the bacteria on left over leaves.
- After pruning, paste the diseased stems with Bromopal (Bacterinashak or Bitrenetol) at 0.5 g/litre or copper oxychloride (3 g/litre) mixed with red sandy loam soil.
- Take hasta bahar crop (pruning during September-October) if the disease severity is high.
- Give rest to trees during December to May and prune during June-July to reduce the disease severity.
- At the beginning of disease incidence stage, give 5 to 6 sprays with Bromopal (0.5 g/litre) or copper oxychloride (2 g/litre) at 10 days interval.
- After each spraying of bactericides, give mineral spray (1 g each of Cu $SO_4$, Mg $SO_4$, Ca $SO_4$ and Boron in 1 litre of water) to reduce the disease severity in plants.

## 2.12. BER

Ber, *Ziziphus mauritiana* (Syn. *Z. zuzuba*) is a native of Indian Subcontinent. It grows in tropical and sub-tropical regions, found all over India. Fruits are used as an ingredient in the preparation of 'Joshanda" (an Ayurvedic medicine used in chest trouble). The fruits are laxative and invigorating. The leaves are used to treat conjunctivitis. They are antihelminthic and good in stomatitis, gum bleeding, asthma and liver complaints. Bark is used for the treatment of diarrhea. The root is used as decoction in fever and its powder is applied to ulcers and old wounds. The kernels are sedative and as soporific which are used to stop vomiting. It is also used for abdominal pain in pregnancy.

### 2.12.1. Bacterial Diseases

#### 2.12.1.1. Canker, *Xanthomonas compestris* pv. *viticola*

**i) Management:** Use of disease-free cuttings, pruning after post-monsoon, destruction of infected plant parts, curb on excess use of irrigation water and late October pruning reduce the disease incidence. Spraying 3:3:50 Bordeaux mixture gave substantial reduction in disease.

## 2.13. PASSION FRUIT

### 2.13.1. Viral Diseases

#### 2.13.1.1. Passion Fruit Woodiness Potyvirus (PWV)

**i) Symptoms:** A number of virus diseases have been reported, notably the passion fruit woodiness potyvirus. Affected leaves show light and dark green mosaic pattern often with light yellow speckle. Sometimes small, yellow ring spots may develop on upper leaf surface. Infected fruits are small and misshapen with very hard rind and small pulp cavity. When affected fruit is cut, the inside rind tissue may have brown spots. Some strains of the virus cause cracking of affected fruits (Fig. 2.28).

Fig. 2.28. Passion fruit woodiness virus - fruit cracking.

**ii) Transmission:** They are spread by aphids (*Aphis gossypii* and *Myzus persicae*) and pruning knives. The virus has a wide host range including bananas, cucurbits and many weeds.

**iii) Management**

- Use virus-free planting material
- Clean pruning tools.
- Use resistant hybrids, or rootstocks of yellow passion fruit.
- Remove diseased vines from the field.
- Do proper weeding.
- Avoid planting bananas and cucurbits near passion fruit fields.

**Table 2.25. Management of bacterial diseases in fruit crops**

| Crop | Disease | Symptoms | Pesticide Dose/litre | Time of Application |
|---|---|---|---|---|
| Grapevine | Bacterial Canker *Xanthomonas campestris cv viticola* | Water soaked black spots | Copper oxychloride-3 g | Rainy season |
| Mango | Bacterial Canker *Xanthomonas campestris cv mangiferaeindicae* | Water soaked black spots on leaves, petiole and fruits | Copper oxychloride-3 g | Spraying with the onset of rain |
| | Bacterial Canker *Xanthomonas campestris cv mangiferaeindicae* | Water soaked black spots on leaves, petiole and fruits | Foliar spray with *Bacillus coagulans* | Spraying with the onset of rain |
| Citrus | Bacterial Canker *Xanthomonas campestris cv citri* | Cankerous spots on leaves, twigs etc. | Copper oxychloride-3 g | Rainy season |

Chapter 3

# VEGETABLE CROPS

China is the largest producer of vegetables contributing 47.3% of the total world production (Table 3.1). Vegetables occupy an area of 22.518 million ha in China with a production of 434.332 million MT. Productivity of vegetable crops is highest (33.8 MT/ha) in Spain.

India is the second largest producer of vegetables next only to China contributing 11.9% of the total world production. Vegetables occupy an area of 7.164 million ha with a production of 109.050 million tonnes (Bijay Kumar, 2007) (Tables 3.1 and 3.2). Vegetable crops contribute 58.93% share to the total production of horticultural crops. India produces 14.7% of the world onions. In vegetables, India occupies the first position in peas; second in cabbage, cauliflower, onion, brinjal; third position in potato and fourth in tomato. The per capita availability of vegetables has increased from 76 g in 1951 to 256 g in 2001 as against 280 g/day recommended by ICMR for a balanced diet. Increase in production is mainly due to increase in productivity. Productivity is still lower in India (15 t/ha) than the world average productivity (15.7 t/ha). There is scope for increasing the productivity of vegetables in India further. The vegetable production target for 2020 has been fixed at 150 million tonnes.

**Table 3.1. Major vegetable producing countries of the world**

| Country | Area (in '000 ha) | Production (in '000 MT) | Productivity (in MT/ha) | % share of world production |
|---|---|---|---|---|
| China | 22518 | 434332 | 19.3 | 47.3 |
| India | 7164 | 109050 | 15.2 | 11.9 |
| USA | 1307 | 36845 | 28.2 | 4.0 |
| Turkey | 1060 | 26290 | 24.8 | 2.9 |
| Egypt | 600 | 16165 | 26.9 | 1.8 |
| Italy | 593 | 15994 | 27.0 | 1.7 |
| Iran | 631 | 15760 | 25.0 | 1.7 |
| Russia | 928 | 15404 | 16.6 | 1.7 |

| | | | | |
|---|---|---|---|---|
| Spain | 395 | 13356 | 33.8 | 1.5 |
| Japan | 424 | 11745 | 27.7 | 1.3 |
| Mexico | 687 | 11241 | 16.4 | 1.2 |
| Korea | 329 | 10965 | 33.3 | 1.2 |
| Others | 15984 | 200416 | 12.5 | 22.0 |
| Total | 52620 | 917563 | 15.7 | --- |

Source: FAO (except Indian data) (Indian data source: Indian Horticulture Database 2006)

**Table 3.2. Area, production and productivity of vegetable crops in India (2005-06)**

| Vegetable crops | Area ('000 ha) | Production ('000 MT) | Productivity (MT/ha) | % share of India production |
|---|---|---|---|---|
| Brinjal | 553.9 | 9136.3 | 16.5 | 8.4 |
| Cabbage | 267.4 | 6921.6 | 22.1 | 5.4 |
| Cauliflower | 291.8 | 5260.1 | 18.0 | 4.8 |
| Okra | 376.1 | 3684.0 | 9.8 | 3.4 |
| Onion | 695.1 | 9248.4 | 13.1 | 8.5 |
| Peas | 314.0 | 2298.7 | 7.3 | 2.1 |
| Potato | 1550.2 | 29093.7 | 18.8 | 26.7 |
| Sweet potato | 129.4 | 1148.6 | 8.9 | 1.1 |
| Tapioca | 252.7 | 7924.9 | 31.4 | 7.3 |
| Tomato | 534.5 | 9361.8 | 17.5 | 8.6 |
| Others | 2199.5 | 25971.4 | 11.8 | 23.8 |
| Total | 7163.9 | 109049.5 | 15.2 | |

(Source: Indian Horticulture Database 2006)

Potato, tomato, brinjal, chilli, cabbage, cauliflower, onion, peas, gourds, okra and melons are now grown on commercial scale with increasing demand, better output, change in production system and development of suitable cvs. These crops have high potential for foreign exchange earning. Today the emphasis on vegetable breeding is on value addition, disease resistance and better quality.

The important vegetables exported include onions-7.1%, cabbage-4%, cauliflower-4%, okra-3%, peas-3%. Traditional vegetables like onion, potato, okra, bottle gourd and chilli are exported to Malaysia, Singapore, Gulf, Sri Lanka, Bangladesh, Pakistan and Nepal. Non-traditional vegetables like asparagus, cabbage, capsicum, sweet corn, baby corn, green peas, French bean, cucumber, gherkin, cherry tomato, Brussels sprout, broccoli, processed and frozen French bean and lima bean are exported to European countries, Australia, Gulf and S.E. Asian countries. French bean, lima bean, sweet corn and capsicum are exported to Gulf. Some of the new products introduced in export market are tomato paste in bulk aseptic packs, processed button mushrooms, freeze dried and instant frozen vegetables. Dehydrated onion and tomato products have great potential as export commodities.

## Disease Scenario

Vegetable crops are more susceptible to diseases caused by bacteria, virus, viroid, mycoplasma and phytoplasma that result in reduction of yield and quality. Seed-borne diseases and their inoculum in vegetables have increased with adoption of OGL and GATT agreement. Diseases affecting from seedling stage to harvest and spoilage caused by pathogens during transit, storage and marketing are the main constraints in vegetable production. A total of 65 million tonnes as pre-harvest losses in vegetables due to diseases alone in the world have been estimated that accounts for 10.5%. Introduction of hybrids in vegetable crops has resulted in new disease problems. Disease intensity of enation leaf curl in okra, chilli leaf curl and viral diseases in cucurbits have increased throughout the country. Intensive and monoculture of vegetables also increased the disease pressure. The off-season cultivation of vegetables has prolonged the survival period of pathogens in the soil.

## 3.1. POTATO

Potato (*Solanum tuberosum*) is one of the important vegetable crops grown in India. It is being cultivated in 1.550 million hectares producing 29.094 million tonnes of potatoes with an average yield of 18.8 tonnes per hectare (Bijay Kumar, 2007) (Tables 3.3 and 3.4). It thrives best in cool climate. Therefore it is a summer crop in the hills and a winter crop in the plains.

**Table 3.3. Major potato producing countries of the world**

| Country | Area (in '000 ha) | Production (in '000 MT) | Productivity (in MT/ha) | % Share of world production |
|---|---|---|---|---|
| China | 4884 | 73460 | 15.0 | 22.4 |
| Russia | 3071 | 37280 | 12.1 | 11.4 |
| India | 1550 | 29094 | 18.8 | 8.9 |
| Ukraine | 1516 | 19462 | 12.8 | 5.9 |
| USA | 439 | 19091 | 43.5 | 5.8 |
| Germany | 277 | 11624 | 42.0 | 3.6 |
| Poland | 588 | 10369 | 17.6 | 3.2 |
| Belarus | 462 | 8185 | 17.7 | 2.5 |
| Netherlands | 156 | 6777 | 43.4 | 2.1 |
| France | 156 | 6605 | 42.2 | 2.0 |
| Others | 6102 | 105368 | 17.3 | 32.2 |
| Total | 19200 | 327314 | --- | --- |

Source: FAO (except Indian data) (Indian data source: Indian Horticulture Database 2006)

**Table 3.4. Area, production and productivity of leading potato growing states in India (2005-06)**

| State | Area (in'000 ha) | Production (in '000 MT) | Productivity (MT/ha) |
|---|---|---|---|
| Uttar Pradesh | 443.2 | 9986.8 | 22.5 |
| West Bengal | 320.6 | 7106.6 | 22.2 |
| Bihar | 308.9 | 5702.5 | 18.5 |
| Punjab | 75.0 | 1223.5 | 16.3 |
| Gujarat | 41.0 | 1088.5 | 26.6 |
| Madhya Pradesh | 53.3 | 799.7 | 15.0 |
| Assam | 90.1 | 629.1 | 7.0 |
| Uttarakhand | 22.2 | 450.6 | 20.3 |
| Haryana | 19.0 | 374.4 | 19.7 |
| Karnataka | 57.5 | 368.4 | 6.4 |
| Others | 119.5 | 1363.6 | 11.4 |
| Total | 1550.2 | 29093.7 | 18.8 |

(Source: Indian Horticulture Database 2006)

### 3.1.1. Bacterial Diseases

#### 3.1.1.1. Common Scab, *Streptomyces scabies*

Potato scab is caused by a soil-borne organism that is worldwide in distribution. The same organism also attacks many other root crops as well as the fleshy roots of weeds. It has been reported on beet, turnip, rutabaga, radish, carrot and parsnip. Beets are especially susceptible when soil conditions are favorable for the disease.

**i) Symptoms**: Scab spots are more or less circular brown, roughened areas, with irregular margins. Sometimes the ridged portions are in broken concentric rings. The spots may be raised or warty in appearance, level with the surface, or somewhat sunken. Superficial russetting to deep pitting may occur (Fig. 3.1). White grubs, wire worms and millipedes are attracted to this affected tissue so that scab spots may be further enlarged or deepened. Scab begins when tubers start forming. Initially the spots may be so small that they are not noticed. As the tuber continues to grow, the areas of these reddish-brown spots also enlarge. An older tuber has too thick a protective layer on its surface to be invaded readily.

**ii) Disease cycle**: *Streptomyces scabies* can live on decomposing material in the soil and does not require a potato or root crop to remain alive. However, it does become more aggressive on any given crop, such as potatoes, if the same crop is grown year after year without rotation. The organism infects the tuber by means of a tiny thread that pushes directly through the tender skin of the forming potato. It slowly continues to grow until the crop is harvested. The infected areas on the potato respond at once by laying down a corky layer. As each layer is invaded by the scab organism, new cork formation takes place until a mature scab spot is produced. Scab does not develop further after the potatoes are dug, but the organism remains alive all winter in storage.

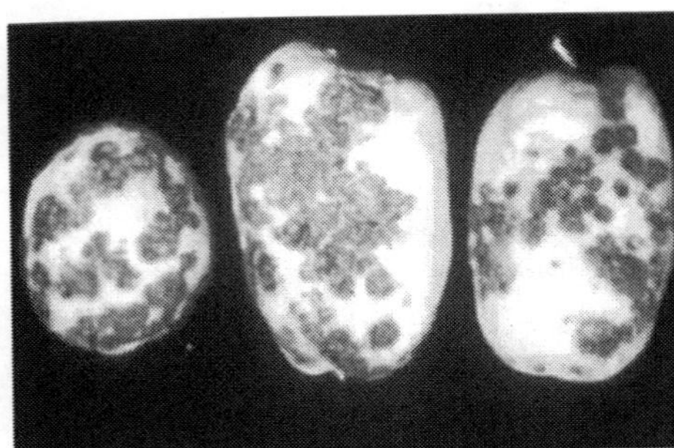

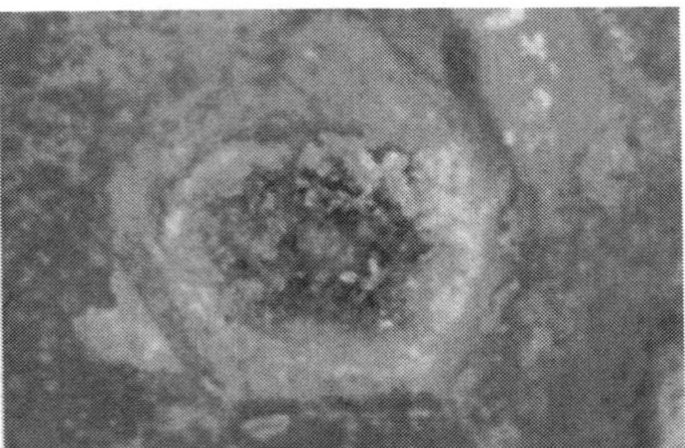

**Fig. 3.1.** Common scab symptoms on tubers and a close-up view of the scab.

**iii) Management :** Use disease-free seed tubers for planting. Irrigate the crop repeatedly to keep the moisture to the field capacity right from tuber initiation until the tubers measure 1 cm diameter. Follow crop rotation with wheat, peas, oat, barley, lupin, soybean, sorghum, bajra and adopt green manuring to keep the disease in check. Plough the potato fields in April and leave the soil exposed to high temperature during May-June. Grow disease resistant cvs. like Patna Red, Kufri Alankar, Kufri Sindhuri, Kufri Dewa and Kufri Lalima.

Management in the vegetable field involves correcting the conditions that are favourable for the scab organism. Some of the important points to be considered are given below:

- Rotations with other than root crops should be employed as the size of the field permits. This way the build up of aggressive scab strains can be avoided.
- Purchase certified potato seed pieces. Seed treatment will help to prevent the introduction of the organism into relatively scab-free soils but is not a replacement for using clean seed. Currently, thiophanate-methyl + mancozeb (Topsin MZ) is recommended for seed piece treatment against scab.
- Plant resistant varieties. If scab has been a problem before, test one of the following varieties under local conditions to determine whether it is suitable: Belchip, Centennial Russet, Keuka Gold, Norchip, Norgold Russet, Norland, Onaway, Pike, Russet Burbank, Salem or Superior. Of these, Norchip, Norland, Superior and Russet Burbank are probably the most available as seed tubers.
- In a garden where irrigation is available, scab can be reduced by keeping the soil reasonably wet for several weeks while the young tubers are beginning to form.

- Great care should be taken to avoid the application of fertilizer or other materials that tend to make the soil highly alkaline. At the head of the list are lime, ashes, and farm yard manure.
- One of the best methods for combating scab is the use of acid producing fertilizers, especially those that contain liberal amounts of sulphate of ammonia. If this practice is repeated for three or four successive years, the soil will finally become acid enough (pH 5.2) to keep the scab organism from developing.
- Test the pH of the soil and apply elemental sulphur to lower the soil pH to 5.2. Sulphur is usually slow in action so it may be more effective if it is applied a year ahead of the potato crop.

### 3.1.1.2. Bacterial Wilt or Brown Rot, *Ralstonia solanacearum*

Bacterial wilt of potato continues to be a major disease since 19$^{th}$ century. In India, the disease was first recorded on potatoes by Cappel (1892). This was the first bacterial disease recorded in India. The disease is more severe in tropical, subtropical and warm climates. In Ranchi district of Jharkhand (India), cultivation of rainfed potato was abandoned due to this disease (Gufran and Chakravarti, 1960) and similar instances are recorded in Indonesia (Machmud, 1986) and in Peru (French, 1986a). In India, the losses reported are up to 55% in Kumaon hills (Paharia, 1963), 0.33-40% in Maharashtra (Paharia, 1963), 20-25% in Hyderabad (Nath *et al.*, 1958) and over 75% in some localities of Karnataka (Gadewar *et al.*, 1991).

**i) Symptoms:** The disease affects both above and under ground parts of the plant. Infected plants show sudden wilting and complete collapse of plants. Vascular browning has been reported to be the characteristic symptom of the disease and the bacterial mass ooze out.

**ii) Epidemiology:** Latently infected material (tubers, rhizomes and corms) is the single most important carrier of the bacterium from one place to another. Bacteria spread from plant to plant by root contact or through the water films or flowing water. In general, increase in the temperature enhances wilting. High soil moisture favours survival of the pathogen in the soil, infection and exit of the bacterium from the host and spread in the soils.

**iii) Management:** The management practices ought to be directed towards reducing the inoculum in the field and avoiding introduction of fresh inoculum, growing the crop when the environmental conditions are not favourable for the pathogen and developing the resistant varieties or by biological means such as the use of avirulent strains of the pathogens and/or rhizobacteria.

**a) Cultural Methods:** Seed production may be undertaken only in disease-free areas (higher hills of Himachal Pradesh and Plains of Punjab). Selection of disease-free tubers for planting, soil application of bleaching powder, hot/cold weather cultivation, disinfection of cutting knives with mercurials, adopting crop rotation for 2-3 years with non-host crops (maize, sorghum, finger millet, cereals, garlic, onion, lupin, cabbage, green manuring crops like sunnhemp, etc.) and use of resistant cultivars have been recommended. Restrict post-emergence tillage to minimum. Practice full earthing up immediately after planting. Collect and burn all diseased plant material including tubers after harvest. The pathogen is susceptible to high temperature. Raising the soil temperature by covering the soil with polythene film has been found effective in reducing the wilt. Opening the deep furrows in infested fields at 50-60 cm intervals, spreading the straw and burning (Jhumming cultivation) reduced the wilt incidence by 100%.

In North Eastern and North Western hills, early planting of the summer crop in February and harvesting in June can avoid the wilt. In Shillong and Ooty, planting in last week of July reduced the bacterial wilt and the wilt was negligible to nil in crop planted in first week of September. These practices reduce the cropping period to only 90 days. Therefore short duration varieties can be used to minimize the losses.

Crop rotation with maize, wheat, barley, oat, sunnhemp, finger millet and vegetables like cabbage, onion, garlic reduced the wilt incidence to the extent of 94%. Studies undertaken in North Western hills using 6 cropping sequences revealed that the reduction in wilt incidence ranged between 25 to 81% by following 2 years rotations. Potato-finger millet-finger millet-potato gave maximum reduction in wilt incidence (81%) (Table 3.5). Roots of finger millet were free from *R. solanacearum* infection, whereas paddy, maize, wheat and beans carried the infection. However, rhizosphere population of the pathogen was found to be reduced.

**Table 3.5. Effect of crop rotation on bacterial wilt incidence and yield of potato in North Western Hills (Bhowali)**

| Crop rotation | % Wilt incidence | Yield loss | Popn. of *R. solanacearum* in cfu/root system of different crops |
|---|---|---|---|
| Potato- Potato- Potato- Potato | 51.7 ( + 92) | 56.9 | $9.0 \times 10^6$ ( Potato) |
| Potato-Bean-Bean- Potato | 19.7 ( - 25) | 18.3 | $9.2 \times 10^4$ ( Bean) |
| Potato-Paddy-Paddy- Potato | 12.0 ( - 53) | 29.7 | $1.0 \times 10^5$ (Paddy) |
| Potato-Wheat-Wheat- Potato | 9.2 ( - 63) | 18.8 | $4.2 \times 10^4$ (Wheat) |
| Potato-Maize-Maize- Potato | 8.7 ( - 66) | 13.1 | $1.0 \times 10^6$ (Maize) |
| Potato-Finger millet-Finger millet-Potato | 4.7 ( - 81) | 7.1 | 0.0 ( Finger millet) |

Intercropping of potato with maize and cowpea were the most effective in reducing the wilt incidence.

In non-endemic areas (Gujarat, Maharashtra, North Western and North Central Plains and North Western high hills), use of disease-free seeds alone is adequate to control the disease completely. In mid hills (North Western mid hills up to 2200 MSL, North Eastern hills and Nilgiris hills), use of disease-free seeds, cold weather cultivation and application of bleaching powder at 12 kg/ha will be adequate for the disease control. Application of bleaching powder can be substituted with 2 years crop rotation with wheat, barley, finger millet, cabbage, cauliflower, knol-khol, carrot, onion, garlic. In

Eastern Plains and Deccan Plateau, the disease can be reduced to considerable extent by following cultural practices like use of disease-free seeds, adjusting the date of planting, application of bleaching powder, straw burning in infested soil, blind earthing, hot weather cultivation and crop rotation (Table 3.6).

**Table 3.6. Control of bacterial wilt and brown rot**

| Treatment | % reduction | | % yield increase |
|---|---|---|---|
| | Wilt | Brown rot | |
| Healthy seed | 50 – 76 | 40 – 46 | 18 - 27 |
| Bleaching powder at 12 kg/ha + Blind earthing | 70 – 94 | 50 – 85 | 22 – 28 |
| Summer ploughing in Plains | 50 – 83 | 50 – 70 | 5 – 25 |
| Winter ploughing in Hills | 60 – 75 | 62 – 78 | 35 – 40 |
| Early planting in Hills | 77 - 80 | 75 - 85 | 44 - 68 |

**b) Biological Methods:** Promising biocontrol agents include *Bacillus* spp., *B. polymyxa* and *Pseudomonas fluorescens* which have been found to reduce the wilt development and incidence. Root colonization by *P. fluorescens* has been reported. *Bacillus* spp. reduced bacterial wilt incidence up to 79%, while *P. fluorescens* by 43 to 75%. Summer (during May-June in Plateau and Plains where very hot and dry climate prevails) and winter ploughing (during December-January in hills where cold temperature prevails) alone reduced wilt incidence by 66% and 50%, respectively. Dipping of seed tubers in 0.25% solution of the formulation of *Bacillus cereus* (strain B4) and *B. subtilis* (strain B5) for 20 min. was effective in reducing the wilt by 59-72% and enhancement of tuber yield by 16.7-25.6%, respectively. Tuber treatment with *P. fluorescens/B. subtilis* reduced wilt incidence by 76.31 and 53.15% and increased tuber yield by 220 and 129%, respectively.

Avirulent and bacteriophage producing strains of *R. solanacearum* induce resistance to bacterial wilt. Tuber bacterization followed by one soil application of avirulent strain of *Ralstonia solanacearum* significantly reduced the disease incidence to an extent of 80 to 82.5% and increased the tuber yield (15.08 to 15.83 t/ha) compared to control (6.04 to 6.94 t/ha).

**c) Host Resistance:** Some of the resistant/tolerant cvs. under cultivation are Ontario, Sebago, Kathadin and Green Mountain in USA; Prisca in Madagascar; Myrylin and Greta in Mauritius and Caxamarca and Molinera in South America.

### 3.1.1.3. Soft Rot, *Erwinia carotovora*

**i) Symptoms:** Fruits develop sub-epidermal soft-rot, which spreads rapidly reducing the tissue to a pulp in 4 to 6 days.

**ii) Management:** Avoid excess irrigation. Provide proper drainage. Adjust planting time to avoid hot weather during plant emergence. Harvest the crop before the temperature raises to 28°C. Harvest the crop only when the tuber skin is fully cured. Avoid injury to tubers and sort out injured tubers. Treat tubers before storage with 3% boric acid for 30 min. and dry under shade. Store the produce in well ventilated cool stores or cold stores.

### 3.1.2. Viral Diseases

#### 3.1.2.1. Potato Virus X (PVX)

**i) Symptoms:** The characteristic symptoms of PVX include interveinal mosaic, mild or barely perceptible mosaic, light and dark green patches, mottling with stunting, crinkling (Fig. 3.2) with severe strains and top necrosis in some varieties. An infection of PVX has the potential to reduce the yields up to 10-30%.

**Fig. 3.2.** PVX on potato leaves.

**ii) Transmission:** Use of infected tubers is the main reason for spread of PVX from one place to another. PVX is contagious and readily spread upon contact of plant foliage and roots in the field as well as upon injury/cutting of the tubers.

PVX is readily transmitted mechanically. It is spread by foliar or sprout contact, seed cutters, pick-type planters, cultivators, spray equipment, and possibly by insects with chewing mouth parts.

**iii) Management:** Strict sanitation should be practiced in the field and also in stores. Disinfect all field equipments with 3% Trisodium phosphate or 1% Calcium hypochlorite solution. Use disease-free seed tubers for planting. Rouging of diseased plants along with their tubers at least twice in growing season helps in reducing the incidence of the disease. Dehaulm the crop before the aphids cross the critical level. Follow thermotherapy by meristem culture for eliminating the virus from seed tubers. Potato cvs. having resistance are Kufri Ashoka, Kufri Badshah, Kufri Jawahar, Kufri Jyoti, Kufri Lalima, Kufri Pukhraj and Kufri Sindhuri.

#### 3.1.2.2. Potato Virus Y (PVY)

**i) Symptoms:** The characteristic symptoms of PVY include mild to severe mosaic, veinal necrosis, leaf drop streaking, stunting of plants, rugosity, bunching or twisting of leaves, down-turning of leaflet margins and necrosis, dropping and collapse of older leaves (Fig. 3.3). Symptoms of PVY may vary by cultivar. Necrosis and veinal necrosis of petioles often cause leaves to drop or cling to the stems. Infected plants have bunchy tops and leafless lower stems or clinging necrotic leaves. An infection of PVY has the potential to reduce the yields up to 60-80%. Affected plants give a poor crop of small tubers. PVY is also tuber-borne.

**ii) Transmission:** PVY is a potyvirus which is transmitted through aphids, *Myzus persicae* and *Aphis gossypii* in non-persistent manner.

**Fig. 3.3.** PVY on potato leaves and tuber.

PVY is a stylet-borne virus transmitted by at least 30 species of aphids, particularly the green peach aphid and potato aphids. Cereal aphids in high numbers become important vectors of PVY. Aphids landing on infected plants acquire PVY within a few seconds of feeding and can transmit the virus immediately in a non-persistent manner. While aphid transmission is the most important means of spread in the field, all PVY strains also are mechanically transmissible by leaf contact and may be carried in seed tubers. Current-season infection late in the growing season may not cause foliar symptoms.

**iii) Management:** A combination of thermotherapy coupled with meristem culture eliminated the virus from seed tubers. Potato cvs. Kufri Jyoti, Kufri Kuber, Kufri Lalima, Kufri Sindhuri and Kufri Ashoka have in built resistance/tolerance to PVX, PVY and PLRV.

### 3.1.2.3. Potato Leaf Roll Virus (PLRV)

Potato leaf roll is caused by the potato leaf roll virus. The severity of symptoms depends on the strain of virus, susceptibility of the cultivar, time of infection, and environment. Yield may be substantially reduced in plants grown from infected tubers. An infection of PLRV has the potential to reduce the yields up to 60-80%. Potato leaf roll severity can be reduced as a result of a comprehensive aphid suppression and monitoring programme.

**i) Symptoms:** Potatoes infected with PLRV show an upward and inward rolling of leaflet margins (Fig. 3.4) and plants are dwarfed. Affected leaflets become thick and dry, and the lower leaves may become completely brown. Infected plants usually are stunted and more upright than healthy plants.

Infected plants produce tubers smaller in size and numbers. In the secondary infection, leaves turning leathery, brittle, pink or brown pigmented, stunting of plants and making rattling noise when shaken. Current-season infections that develop late in the season may cause no visible symptoms, but tubers usually become infected. Early season infections may reduce yield. PLRV also infects other solanaceous crops and weeds (e.g. tomato, tobacco, jimson weed).

**Fig. 3.4.** Leaf roll virus infected plants and tuber of potato.

**ii) Transmission:** The virus is transmitted by aphids, *Myzus persicae* and *Aphis gossypii* in persistent manner. The green peach aphid, *Myzus persicae* is the primary vector. Aphids acquire the virus after several hours of feeding and require an incubation period before being able to transmit the virus to non-infected plants. Once infected, the aphid usually retains the virus for life.

**iii) Epidemiology:** Leaf roll is introduced by planting infected seed or by aphids carrying the virus to the field from an outside source.

**iv) Management:** Plant high quality, certified seed. Two hour daily treatment of tubers for 6 weeks at 40°C inactivates PLRV. Treatment of tubers in hot water at 55°C for 15-20 min has also been found effective in eradicating the virus. Control aphid populations in the field, and on other hosts. Rogue infected plants early in the season to control disease sources in the field. Consider early vine-kill when high aphid populations are present.

Potato cvs. Kufri Jyoti, Kufri Kuber, Kufri Lalima, Kufri Sindhuri and Kufri Ashoka have in built resistance/tolerance to PVX, PVY and PLRV.

### 3.1.2.4. Potato Spindle Tuber Viroid (PSTVd)

**i) Symptoms:** It is particularly destructive for seed production. Potato plants severely infected are upright, stunted and much thinner than normal plants. Leaves of infected plants are smaller and may be grey or distorted. The affected stems are often more branched and branches are at very sharp angles to the stem. Symptoms are more obvious on tubers. Affected tubers are small, narrow and spindle shaped (oblong).

In some varieties tubers develop knobs and swellings. In other varieties eyes on tubers are numerous, shallow and more prominent, and affected tubers are often cracked. This viroid is easily mechanically transmitted. An infection can be introduced by sowing infected seed tubers, by insects such as aphids, grass hoppers, flea beetles, and through pollen. Tomato and capsicum are additional hosts for PSTVd.

**ii) Management**

- Plant certified disease-free seed tubers.
- Plant whole seed tubers instead of cut pieces.
- Remove infected plants from the field.

**Integrated Management of Viral Diseases**

- Use virus-free seed tubers.
- Plant resistant varieties where available. In Kenya, the following potato varieties are claimed to possess some resistance to viral diseases: "Kenya Baraka", "Roslin Eburu", "Feldelslohn", "Annet" and "Dutch Robjn".
- Uproot infected plants to reduce the incidence of infection and spread of the disease within a field. For maximum effectiveness remove the diseased plant, the three plants on each side of the diseased plant in the same row, and the three closest plants in adjacent rows. This is most important in seed fields.
- Control insect vectors. This may reduce virus spread.
- Do not overlap potato crops.
- Practice good field sanitation.
- Control nightshades and volunteer potatoes because these plants are reservoirs for viruses.

## 3.2. TOMATO

Tomato (*Lycopersicon esculentum*) is the most important and remunerative vegetable crop in India. It is being cultivated in 0.535 million hectares producing 9.362 million tonnes of tomatoes with an average yield of 17.5 tonnes per hectare (Bijay Kumar, 2007) (Tables 3.7 and 3.8). A rich source of minerals, vitamins and organic acids, tomato fruits provides 3-4% total solids, 15-30 mg/100 g ascorbic acid, 7.5-10.0 mg/100 ml titratable acidity and 20-50 mg/100 g fruit weight of lycopene. Tomato fruits contain 0.4% protein, 2.5% carbohydrates, 1.5 mg iron and 2 mg vitamin C in 100g of fresh weight. Fruits are good for people suffering from constipation, jaundice and indigestion.

**Table 3.7. Major tomato producing countries of the world**

| Country | Area (in '000 ha) | Production (in '000 MT) | Productivity (in MT/ha) | % Share of world production |
|---|---|---|---|---|
| China | 1305 | 31626 | 24.2 | 24.7 |
| USA | 167 | 11043 | 66.3 | 8.6 |
| Turkey | 260 | 10050 | 38.7 | 7.9 |
| India | 535 | 9362 | 17.5 | 7.3 |
| Egypt | 195 | 7600 | 39.0 | 5.9 |
| Italy | 139 | 7187 | 51.8 | 5.6 |
| Spain | 72 | 4810 | 66.5 | 3.8 |
| Iran | 139 | 4781 | 34.4 | 3.7 |
| Brazil | 61 | 3453 | 57.0 | 2.7 |
| Mexico | 119 | 2800 | 23.6 | 2.2 |
| Others | 1624 | 35281 | 21.7 | 27.6 |
| Total | 4615 | 127993 | | |

Source: FAO (except Indian data) (Indian data source: Indian Horticulture Database 2006)

**Table 3.8. Area, production and productivity of leading tomato growing states in India (2005-06)**

| State | Area (in'000 ha) | Production (in '000 MT) | Productivity (MT/ha) |
|---|---|---|---|
| Andhra Pradesh | 76.5 | 1453.5 | 19.0 |
| Orissa | 100.4 | 1332.2 | 13.3 |
| Karnataka | 44.5 | 1188.1 | 26.7 |
| Maharashtra | 35.0 | 987.0 | 28.2 |
| West Bengal | 50.0 | 857.2 | 17.2 |
| Bihar | 46.0 | 727.2 | 15.8 |
| Gujarat | 29.3 | 650.0 | 22.2 |
| Chhattisgarh | 29.2 | 365.8 | 12.5 |
| Madhya Pradesh | 20.4 | 306.7 | 15.0 |
| Tamil Nadu | 22.0 | 277.7 | 12.6 |
| Haryana | 17.1 | 257.3 | 15.0 |
| Others | 64.1 | 959.1 | 15.0 |
| Total | 534.5 | 9361.8 | 17.5 |

(Source: Indian Horticulture Database 2006)

### 3.2.1. Bacterial Diseases

#### 3.2.1.1. Bacterial Wilt, *Ralstonia solanacearum*

This is the most serious bacterial disease of solanaceous crops. The disease is prevalent in warmer climate having high humidity under tropical and subtropical conditions.

**i) Symptoms:** Bacterial wilt has often been reported as the most serious handicap for tomato in the tropics. The disease causes rapid wilting and death of the entire plant without any yellowing or spotting of leaves. Total collapse of the plant usually occurs when temperature reach 32°C and above. Plant wilts while still green. The lower leaves may droop before wilting occurs. The vascular system becomes brown, root and rootlets rots leaving behind only corky portion of main root. When the stem of a wilting plant is cut across, the pith has a darkened water-soaked appearance, and on squeezing the cut stem, a white, yellow or grayish, slimy exudate may appear (Fig. 3.5). In later stages of the disease, decay of the pith may cause extensive hollowing of the stem.

**Fig. 3.5.** Bacterial wilt affected tomato plant.

The pathogen is soil inhabitant, persisting for long periods in some soils. It invades the plants through wounds, usually below ground. Relatively high soil moisture and soil temperature favours disease development.

**ii) Management**

**a) Cultural Methods:** Infested soil should be kept fallow for about a year and frequently disked during the hot and dry season to accelerate desiccation of plant debris and death of bacteria. Water stagnation should be avoided in the field. Light textured and well drained soil reduced the disease incidence. Avoid growing crop in acidic soil and reclaim the soil to neutral pH. If only a few wilted plants are found, immediately remove them from the field

A method called "bio-fumigation" is under development from the Austalian Centre for International Agricultural Research for the reduction and long-erm elimination of bacterial diseases from the soil by incorporating especially mustard or radish plants in large amounts into the soil immediately before planting tomatoes or potatoes. This practice is reported to reduce incidence of bacterial wilt by 50-70% in the Philippines.

Soil amendment with green manuring crops, transplanting disease-free seedlings and following 3 years rotation with cereals/crucifers are effective for the management of wilt. Crop rotation with cowpea-maize-cabbage, okra-cowpea-maize, maize-cowpea-maize and finger millet-brinjal (Pusa Purple Cluster)-French bean are reported effective in reducing bacterial wilt. Intercropping of tomato with sorghum, maize, onion, garlic and marigold resulted in better crop stand. Grafting tomato on brinjal rootstocks such as Pusa Purple Cluster and Dingra Multiple Purple helps in reducing the disease incidence.

**b) Chemical Methods:** Application of bleaching powder at 15 kg/ha and groundnut cake before planting was found effective.

**c) Biological Methods:** Bacterial wilt of tomato was successfully controlled by incorporation of *Pseudomonas fluorescens.* Prior inoculation of tomato roots with *T. viride* and *P. fluorescens* protected the plants. Application of paste formulation of *P. fluorescens* in vermicompost exhibited lower wilt incidence with maximum reduction of wilt pathogen in the rhizosphere and significant yield enhancement.

Dipping of seedling roots in avirulent strain of *Ralstonia solanacearum* (Av 10) before transplanting gave least wilt incidence (6.66%) as compared to control (75%). Seed treatment with *P. fluorescens, P. aeruginosa* and *B. subtilis* resulted in least seedling mortality (<1%, 2.66 and 4%, respectively) as compared to control (27.58%). Seedling root dip in *P. fluorescens/B. subtilis* followed by application of antagonists cell suspension to soil in root zone was found effective in reducing the disease incidence (16.66 and 25%, respectively) as compared to control (74.5%). Soil application of *Glomus mosseae* reduced the incidence of wilt.

**d) Host Resistance:** Tomato cvs. Arka Abha, Arka Alok, Arka Shreshta, Arka Abhijit, Megha, Shakthi, Sonali, Sun 7610, Sun 7611, BT-1, BT-10, LE 79-5 are resistant to bacterial wilt.

**e) Integrated Methods:** Integration of growing and incorporation of sunnhemp (green manure crop) into the soil and seed treatment with *Pseudomonas fluorescens* ($10^{10}$ cfu/ml) was highly effective in reducing the bacterial wilt incidence to 6.75% and significantly increasing the fruit yield to 25.5 MT/ha as compared to high wilt incidence (55.6%) and low yield (7.8 MT/ha) in control. Thus, the efficacy of *Pseudomonas fluorescens* is increased under green manure incorporated soil against bacterial wilt of tomato.

### 3.2.1.2. Bacterial Spot, *Xanthomonas campestris* pv. *vesicatoria*

**i) Symptoms:** It is a serious disease, occurring during the rainy season and continues up to fruit initiation stage. It is most noticeable on fruits, but also causes damage to the foliage and stems. On green fruits, the initial spot is very small and water soaked (Fig. 3.6). It eventually enlarges to about six mm. As the bacterial spot matures, it becomes brown and scabby without extending deep into the fruit. On foliage, irregular greasy, dark green spots, 2 to 3 mm wide, are observed (Fig. 3.6). The spots eventually dry and the tissue often tears. The disease is transmitted through the seed. The disease is prevalent in warm humid weather when relative humidity is more than 90% and temperature is 30-35°C.

**Fig. 3.6.** Bacterial spot on leaves and fruits.

**ii) Management:** Seeds should be collected from the disease-free plants. Use of disease-free seeds, removal and destruction of crop residues after harvest, clean cultivation and 3 sprays of 1% Bordeaux mixture reduced the disease incidence. One spray of 0.3% copper oxychloride or 150 ppm streptocycline should be given in the infected field to check the disease incidence. Use of tolerant varieties like LE-5 and spraying with 0.3% copper oxychloride provides satisfactory control (Pandey and Pandey, 2002).

### 3.2.1.3. Bacterial Stem and Fruit Canker, *Clavibacter michiganensis* pv. *michiganensis*

Bacterial canker (*Clavibacter michiganensis* pv *michiganensis*) is one of the most destructive and puzzling tomato diseases. This disease of tomato was observed first during 1909 in the USA In India, the disease was reported from Nagpur market during 1949 (Jain, 1951).

**i) Symptoms:** Temporary and later on permanent wilting of leaflets of tomato plants are the first symptoms of the disease. Light streaks appear at the juncture of petiole and stem extending down the internodes and up to the petiole. As these streaks become darker, epidermal tissues may rift and longitudinal canker like opening may appear in stem, petiole and midrib. When the stem of diseased plants cut longitudinally, a creamy white, yellow or brown line follows the phloem. Discolouration of vascular system can be traced up to the fruit, placenta and seeds. Secondary infection appears on green fruits as water-soaked spots with a white halo. The spots remain less than 3 mm in diameter exclusive of halo. Halo is the distinguishing character of bacterial canker of tomato (Fig. 3.7). The pith is easily separated from the wood along this line.

The vascular bundles within the pith are destroyed and cavities are formed in the pith. Young affected fruits are small, deformed and seeds are aborted. Spots may also develop on older fruits. These spots are circular, up to three mm wide, with slightly raised brown centre surrounded by a pronounced white halo resembling a birds' eye. The halo is flat. The bacterium is seed-borne and can survive up to three years in soil in crop debris. Spread within a crop is by rain splash, sprinklers and pruning knives.

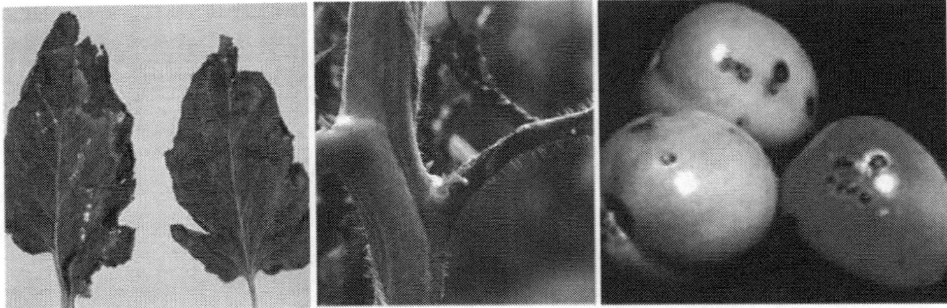

**Fig. 3.7.** Bacterial canker on leaves, stem and fruits of tomato.

**ii) Epidemiology:** The bacterial canker pathogen can be seed-borne, both on the surface of the seed and under the seed coat. The pathogen can move readily from infected plants onto clean plants in the course of regular activities, and carryover in the field from one season to the next is important.

Bacteria can enter from the surface of the leaf through natural leaf openings or wounds, causes the "marginal scorch" symptoms. Once the bacteria enters the plant which could occur in the field it can take anywhere from 7 to 84 days for symptoms to appear. If infected plants are present, the movement of bacteria from one plant to another during normal watering and handling, activities occurs readily.

**iii) Disease Cycle:** The disease is seed-borne and up to 70% seed-borne infection has been reported (Nedumaran and Vidyasekaran, 1982). Bacterium also survives in soil and the disease spreads to new area through infected seeds.

**iv) Management:** Three key principles for preventing losses to bacterial canker can be summed up as follows:

- Use disease-free seed.
- Control bacterial populations that may be present on the leaf surface of transplants in the greenhouse.

- Plant into a clean field.

The following practices may be followed to manage the disease:

- Seed treatment with hot water or organomercurials or streptomycin is beneficial.
- Three years crop rotation with non-host crops is effective.
- Bacteria on leaf surfaces were effectively suppressed by spraying copper hydroxide, alone or mixed with mancozeb, or streptomycin sulphate on a five-day schedule from the first true leaf stage until transplanting. Unsprayed, infected plants had lower yields. Weekly sprays of streptomycin on greenhouse-inoculated plants eliminated all bacteria on sprayed plants.
- Rotate tomatoes to a different field. Keep each field out of tomato (and related crops such as potato and eggplant) for two to three years.
- Bacterial canker survives in the field as long as there is any infected crop debris. It lasts longer in debris on the surface than it does in buried debris. Ploughing after harvest will help to speed up the decomposition.

#### 3.2.1.4. Bacterial Speck, *Pseudomonas syringae* pv. *tomato*

Bacterial speck of tomato was first reported from Formosa in 1933. In India, the disease was observed during November 1960 to February 1961 at South Arcot, Tamil Nadu (Rangaswami and Gowda, 1963) and in August 1978 at Kanpur, Uttar Pradesh (Singh *et al.*, 1982).

**i) Symptoms:** The disease starts as small, round, water-soaked spots, which gradually become yellowish brown and finally blackish brown with sunken ashy centre and pale yellowish green halo. Later on, spots become irregular to circular, measuring 1 mm in diameter, distributed over the entire leaf lamina and in severe cases forming large blotches. Spots on petioles are more or less oval.

**ii) Management:** Seed treatment with 100 ppm Agrimycin-100 or 200 ppm Sandox or 100 ppm streptocycline is effective.

### 3.2.2. Viral Diseases

#### 3.2.2.1. Tomato Leaf Curl Virus (TLCV)

This is one of the most important and devastating diseases of tomato which causes yield losses up to 92.4% if the plants get infection within 20 days after transplanting. However, if the plants are infected after 35 and 50 days after transplanting, the loss in yield would be reduced to 74 and 28%, respectively (Sastry and Singh, 1973b).

**i) Symptoms:** Disease appears in severe form during the months of September to November. Tomato plants infected early in the season are normally stunted and excessively branched. Leaves show downward rolling, curling, twisting and chlorosis (Fig. 3.8). Old leaves turn leathery and brittle. Plants exhibit stunting, shortening of internodes, bushy appearance, non-flowering and fruiting at later stage of virus infection. Flower drop is common, and therefore, infected plants have a reduced number of flowers and fruits. The diseased plants induce partial or complete sterility. In case of late infection, plants do bear fruits but they are very small, deformed and unmarketable.

**Fig. 3.8.** Leaf curl virus on tomato leaves.

**ii) Transmission:** TLCV is caused by Gemini virus which is transmitted by the white fly, *Bemisia tabaci*. Even a single white fly can transmit the disease. It is not seed-borne.

**iii) Epidemiology:** High temperature and low humidity favours the disease incidence. The disease progresses during February-June with the prevalence of dry hot season with low humidity which favour white fly multiplication and disease spread. The incidence may reach even up to 100% during summer months (Sastry *et al.*, 1978). The virus infects a number of cultivated hosts viz., papaya, tobacco, bell pepper, sunnhemp, hibiscus, zinnia, petunia and weeds which serve as a source of inoculum.

**iv) Management:** Seed treatment with hot water maintained at 50°C or 2% trisodium phosphate solution for 25 min is recommended.

The nursery should be raised under nylon net to prevent entry of whiteflies which transmits leaf curl (Fig. 3.9). Field sanitation should be given top priority as many weeds are alternate hosts for the white fly vector. Crop should be frequently irrigated which invites less oviposition by white flies. Leaf curl affected plants should be uprooted and destroyed.

When tomato is grown along with French bean and brinjal, the incidence of TLCV is low as they act as trap crops for white flies.

The disease spread can be minimized by growing border or barrier crops like maize, sorghum and bajra (5 to 6 rows of these crops should be sown all-round the tomato plot 50 to 60 days before transplanting tomato) which prevents incoming viruliferous white flies from entering into the tomato crop (Sastry *et al.*, 1977). Sowing of 5 rows of maize (border crop) about 5 weeks before transplanting tomato resulted in low incidence of leaf curl disease (14.42%) followed by sunnhemp (18.27%) and sorghum (23.94%) as compared to 48.09% in control. Hence growing maize as barrier crop around tomato crop is effective in preventing further spread of the leaf curl disease.

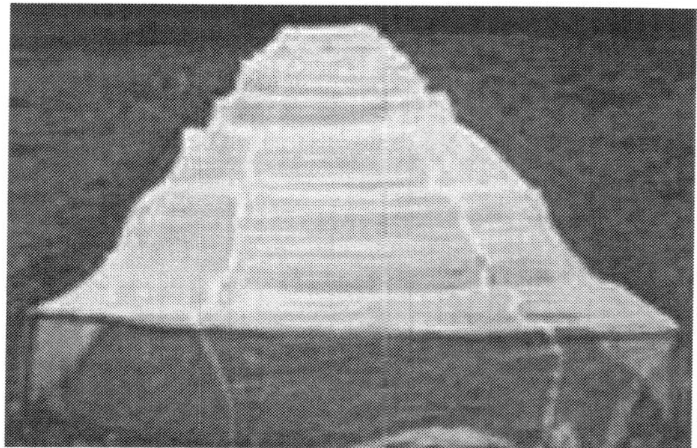

**Fig. 3.9.** Tomato nursery covered with nylon net to prevent entry of white flies.

Mulching tomato fields with saw dust, straw or yellow polythene sheets during early crop stages shall attract white flies which stay there long enough to be killed by reflected heat. Yellow sticky traps can also be used for controlling white fly population (Smith *et al.*, 1979). Use of biocontrol agent like *Encarsia* has shown great promise in reducing white fly population. The greenhouse white fly, *Trialeurodes vaporarium* gave effective control of *Bemisia tabaci*.

Soil application of carbofuran at 1.5 kg a.i./ha in nursery bed and also during transplanting of the seedlings is effective. Two to three foliar sprays of monocrotophos or dimethoate both at 0.1% at 10 days interval are also effective (Singh and Sastry, 1976). Singh *et al.* (1973) found that krishi oil (power oil) at 2% was very effective in reducing the spread of TLCV. The use of oil prevents the acquisition and inoculation of the virus by white flies. It is advisable to spray krishi oil after fruit formation and spraying of insectcides should be discontinued later (Singh and Sastry, 1976).

Tomato cvs. Avinash-2 (T), Hisar Anmol (H-24), Hisar Gourav, JK Asha, Nandi, Abhinav, Vaibhav, Sankranti and Mruthyunjaya-3 are reported to be resistant/tolerant to leaf curl.

### Integrated Methods

- Use resistant/tolerant varieties such as "Tengeru 97", "Amareto", "Peto 86", "Fiona F1", "Perlina", "Denise", "Cheyenne" ("E448"), "Rover". "Roma" and "Marglobe" are highly susceptible and should not be used in areas where the disease is common.
- Protect seed beds with a white nylon net (40 mesh), as the later the plants are infected with the virus the less impact the disease will have on yield reductions.
- Eradicate weeds.

- Rogue out diseased plants (in the seedbed and in the field) and destroy them.
- Plant barrier crops.
- Avoid continuous growing of tomato.
- Control white flies.

### 3.2.2.2. Tomato Spotted Wilt Virus (TSWV)

This disease was first reported from India by Todd *et al.* (1975) from Nilgiri Hills.

**i) Symptoms:** The most characteristic symptoms are sudden browning of leaves followed by cessation of growth of the plants. Later on, the leaves become distorted and areas of necrosis are common. Leaves exhibit bronze colouring. The intensity of bronzing varies from inconspicuous green brown glaze to a distinct dark brown to an almost black glazed area. Bronzed areas usually roll inward and affected areas often become necrotic. The fruits on infected plants show numerous spots of about 1.2 cm in diameter with concentric circular markings. On ripe fruits, these markings consist of alternate bands of red and yellow which are diagnostic for the virus (Fig. 3.10).

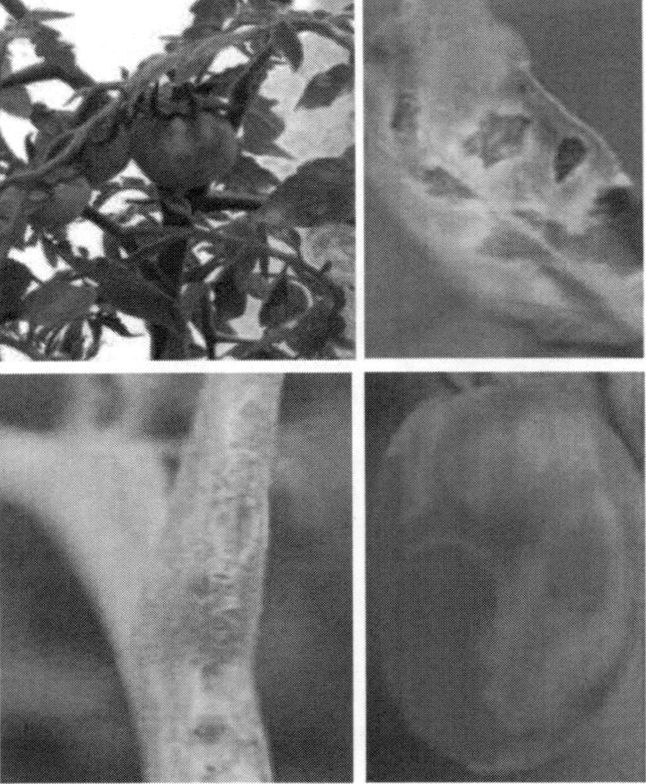

**Fig. 3.10.** Spotted wilt virus on leaf, stem and fruit.

**ii) Transmission:** A tospo virus (TSWV) causes the disease. TSWV is transmitted by several species of thrips and the main vector is *Frankliniella occidentalis*. The virus is transmitted by mechanical means. The important means of natural transmission is through thrips, *Thrips tabaci, Frankliniella schultzi* and *F. occidentalis*. The virus is acquired first by the larval form, and then subsequent adult can then transmit the virus. The minimum acquisition period is 15 minutes.

**iii) Host Range:** The virus has a wide host range including many ornamental plants, vegetable crops and weeds. In addition to tomato, this virus infects potato, brinjal, bell pepper, groundnut, pea, French bean, cowpea, lettuce, pineapple, marigold, Chrysanthemum, *Zinnia elegans, Aster sp., Gerbera sp.* and *Dahlia variablis*. These hosts are important source from which the disease can spread each year.

**iv) Management**

- Raise the seedlings under nylon net cover (50 mesh) to prevent vectors (thrips).
- Uproot and burn the affected plants as soon as the disease is noticed.
- The disease spread can be controlled by spraying systemic insecticides.
- Three to four foliar sprays of 4% NSKE or 0.1% monocrotophos or dimethoate at 10 days interval are effective.
- Efficient weed control and avoiding cultivation of other hosts of the virus in the vicinity of susceptible crops may help in disease control.
- Physical barriers that limit the movement of thrips have been shown to reduce the disease incidence.
- Protected cultivation of tomato crop in greenhouses.
- Mulching of soil with aluminium-surfaced plastic mulch significantly reduced the disease incidence by repelling thrip vectors (Black *et al.*, 1986).
- Apply carbofuran at 1 kg a.i./ha to the nursery beds at the time of sowing.
- Spray the crop with 0.2% acephate or 0.03% imidachloprid or 0.1% fipronil.
- Crop rotation with non-host crops of thrips.
- Plant resistant cultivars, (e.g. "Star 9006", "Star 9008").

Spraying of Adathoda leaf extract (10%) was found to be equally effective as that of the existing practice of spraying monocrotophos (0.15%) for the management of TSWV. Both these treatments recorded the least incidence of 10% and maximum yield of 3.2 and 3.9 kg/plot, respectively.

### 3.2.2.3. Tomato Mosaic Virus (TMV)

The disease is caused by a strain of tobacco mosaic virus. The virus infects many solanaceous hosts. The loss in yield may vary from 20-80% depending upon the time of infection and age of the plant. TMV was first reported from India by Das and Raychaudhuri (1953).

**i) Symptoms:** Affected plants show light and dark green mottling and distortion of the youngest leaves. The latter may be stunted or elongated resembling fern leaves. Under high temperature and light intensity, mottling can be severe (Fig. 3.11). Under low temperature and low light intensity, stunting and leaf distortion are severe. Internal browning of the fruit sometimes occurs; this symptom is most common when fruits become infected at mature green or pink stage. The disease can be seed-borne but can also survive on plant debris in the soil and so re-infect newly planted crops. The virus is easily mechanically transmissible by contact between plants or through human activities (transplanting seedlings or pruning).

**Fig. 3.11.** Mosaic virus on tomato.

**ii) Transmission:** The virus is easily sap transmitted. TMV is a highly contagious disease and can be transmitted by clothes, tools, farm machinery, soil debris, etc. The virus is also seed-borne up to 80% in tomato.

**iii) Management:** The following practices may be adopted for effective management of the disease:

- Select seeds from healthy plants and as an added precaution, soak them in 0.3% solution of trisodium phosphate for 48 hours before sowing.
- Dip the seedlings in solution of trisodium phosphate or monocrotophos both at 0.2% concentration for 10 minutes before transplanting.

- Carefully remove and destroy all infected plants in the nursery or in main field. Do not use TMV infected seedlings for transplanting.
- Crop rotation with non-solanaceous crops helps in managing the disease.
- Use resistant varieties (e.g. "Alboran", "Kentom", "Meru", "Shen-gena", "Tengeru 97")
- Remove crop refuse and roots from fields.
- Eradicate weeds.
- Workers should not smoke or take snuff when working in tomato fields since the virus can be transmitted from tobacco

### 3.2.2.4. Tomato Big Bud

The disease is caused by phytoplasma. This disease causes heavy losses in yield of tomato fruits.

**i) Symptoms:** The important symptoms of the disease are phyllody and virescence of floral parts, proliferation of auxiliary buds and shortening of leaves. The whole calyx is enlarged and forms a bladder-like structure with a toothed opening at the tip (known as big bud). Thickening of leaves, stem, petiole and peduncle is also common.

**ii) Transmission:** the disease is transmitted by leaf hopper vector, *Orosius argentatus* and by grafting.

**iii) Management**

- Strict rouging of infected plants from the field.
- Application of carbofuran at 1.5 kg a.i./ha at transplanting followed by 3 to 4 foliar sprays of 0.1% monocrotophos at an interval of 10 days.
- Spraying of 200 ppm oxytetracycline hydrochloride solution at weekly interval.

### 3.2.2.5. Marginal Flavescence of Tomato

This is a new disease reported for the first time from Bangalore by Singh and Sastry (1980b). The disease is caused by mycoplasma-like organism (Singh, 1987).

**i) Symptoms:** The disease is characterized by slight chlorosis along the leaf margins. The infected leaves are small and plant show stunted growth and short internodes. The top infected leaves become smaller manifold in comparison to healthy ones. The infected plants produce numerous axillary branches giving bushy appearance to the plant. The flower buds take upright position. As the plant ages, it loses its resemblance to a tomato plant because of excessive crowding of short branches and production of smaller leaves which give a rosette appearance to the plant. The infected plants either do not bear fruits or produce very few fruits which are often very small and unmarketable.

**ii) Transmission:** In nature, the disease is transmitted by leaf hopper vector, *Orosius albicinctus*. Transmission has also been achieved by grafting (Singh and Sastry, 1980b).

**iii) Host Range:** Besides tomato, the pathogen also infects potato and *Datura stramonium*.

**iv) Management**

- Spraying with systemic insecticides like 0.05% monocrotophos or 0.05% dimethoate at 10 days interval to control leaf hopper population.
- Removal and destruction of infected plants as soon as they are noticed.
- Spraying of 500 ppm of oxytetracycline hydrochloride at weekly intervals helps in suppression of symptoms.

### 3.2.2.6. Tomato Fern Leaf

**i) Symptoms:** The first characteristic symptom of the disease consists of spindling appearance of the young leaves in the terminal bud. The leaves twist round in a cork screw fashion, remain folded, curve downwards or curl up in spirals. Chlorosis of older leaves, especially along the veins, is also an early symptom. About 3 weeks later, the spindling leaves unfold and show typical distortion resulting in the formation of filiform leaves. About 5 weeks later, excessive number of lateral leaflets are produced.

**ii) Etiology:** The disease is caused by a strain of the CMV.

**iii) Transmission:** The virus is easily sap transmitted. In nature, the disease is transmitted by aphid vectors such as *Aphis craccivora, A. gossypii, Macrosiphum euphorbiae* and *Myzus persicae* in a non-persistent manner.

**iv) Management**

- Rouging of infected plants.
- Control of weed hosts.
- Spraying of systemic insecticides like monocrotophos or phosphamidon each at 0.05% or oxydemeton methyl at 0.02% at 10 days interval to control aphid vectors.

## 3.3. BRINJAL

Brinjal (*Solanum melongena*) is a widely grown vegetable crop in Asian countries. It is being cultivated in 0.553 million hectares producing 9.136 million tonnes of brinjal with an average yield of 16.5 tonnes per hectare (Bijay Kumar, 2007) (Tables 3.9 and 3.10). In India, it is adapted to a wide range of climatic conditions from north to south and east to west. In hilly regions, it is grown only in summer. Brinjal is used in a variety of culinary preparations. Pickles and industrially processed foods are also produced. It is rich in Vitamins A and B and good for diabetic patients.

**Table 3.9. Major brinjal producing countries of the world**

| Country | Area (in '000 ha) | Production (in '000 MT) | Productivity (in MT/ha) | % Share of world production |
|---|---|---|---|---|
| China | 952 | 17028 | 17.9 | 53.5 |
| India | 553 | 9136 | 16.5 | 28.7 |
| Egypt | 43 | 1000 | 23.3 | 3.1 |
| Turkey | 35 | 930 | 26.6 | 2.9 |
| Iraq | 23 | 439 | 19.1 | 1.4 |
| Japan | 11 | 396 | 34.7 | 1.2 |
| Italy | 12 | 339 | 27.8 | 1.1 |
| Sudan | 15 | 272 | 18.7 | 0.9 |
| Indonesia | 43 | 252 | 5.9 | 0.8 |
| Philippines | 21 | 188 | 8.8 | 0.6 |
| Others | 120 | 1856 | 15.4 | 5.8 |
| Total | 1829 | 31836 | | |

Source: FAO (except Indian data) (Indian data source: Indian Horticulture Database 2006)

**Table 3.10. Area, production and productivity of leading brinjal growing states in India (2005-06)**

| State | Area (in'000 ha) | Production (in '000 MT) | Productivity (MT/ha) |
|---|---|---|---|
| West Bengal | 152.9 | 2757.4 | 18.0 |
| Orissa | 127.9 | 1853.9 | 14.5 |
| Bihar | 53.7 | 1031.0 | 19.2 |
| Gujarat | 62.7 | 982.6 | 15.7 |
| Andhra Pradesh | 26.2 | 549.4 | 21.0 |
| Maharashtra | 32.0 | 521.6 | 16.3 |
| Karnataka | 14.7 | 326.0 | 22.1 |
| Chhattisgarh | 19.4 | 266.4 | 13.7 |
| Madhya Pradesh | 14.8 | 221.9 | 15.0 |
| Assam | 13.0 | 171.5 | 13.2 |
| Haryana | 11.5 | 148.7 | 12.9 |
| Others | 24.5 | 305.8 | 12.5 |
| Total | 553.3 | 9136.4 | 16.5 |

(Source: Indian Horticulture Database 2006)

## 3.3.1. Bacterial Diseases

### 3.3.1.1. Bacterial Wilt, *Ralstonia solanacearum*

**i) Symptoms:** Characteristic symptoms of bacterial wilt are rapid and complete wilting of normal grown up plants (Fig. 3.12). Pathogen is mostly confined to vascular region. In advanced cases, it may invade the cortex and pith and cause yellowish-brown discolouration of tissues. Infected plant parts when cut and immersed in clear water, a white streak of bacterial ooze coming out from cut ends is visible.

**Fig. 3.12.** Bacterial wilt affected plants of brinjal.

This disease occurs with bad drainage, especially in the hot, wet season and is often combined with symptoms of root-knot nematodes. The pathogen is soil-borne with a wide host range. Root-knot nematode infestation aggravates the disease development.

**ii) Management:** Crop rotation with cowpea-maize-cabbage, okra-cowpea-maize, maize-cowpea-maize and finger millet-brinjal (Pusa Purple Cluster)-French bean are reported effective in reducing bacterial wilt. Grafting brinjal on *Solanum torvum* rootstock was found effective against wilt. Removal and destruction of infected plants from fields helps to reduce the incidence of the disease.

Bacterial wilt of brinjal was successfully controlled by incorporation of *Pseudomonas fluorescens*.

Brinjal cvs. Pusa Purple Cluster, Dingra multiple purple, Arka Keshav, Arka Nidhi, Arka Neelkanth, Gulla, Vijay hybrid and Banaras Giant Green are found resistant.

### 3.3.1.2. Bacterial Wilt, *Ralstonia solanacearum* and Root-knot Nematode, *Meloidogyne incognita* Disease Complex

Eggplant is prone to many soil borne diseases among which the bacterial wilt (*R. solanacearum*) in combination with root-knot nematode (*M. incognita*) takes heavy toll every year all over the world (Naik, 2004).

**i) Symptoms:** The root-knot nematode, *M. incognita*, present along with *R. solanacearum*, greatly increases the incidence of bacterial wilt of brinjal (Fig. 3.13). The root-knot nematode is responsible for breaking bacterial wilt resistance in "Pusa Purple Cluster" cultivar of brinjal (Parvatha Reddy *et al.*, 1979). *M. incognita* interacts with *R. solanacearum* in increasing the speed of development and severity of wilt disease of eggplant. Maximum disease occurred when the bacterium and nematode were inoculated simultaneously. Inoculation with *R. solanacearum* alone resulted in less severe disease. It was concluded that root-knot nematodes modify the plant tissues facilitating bacterial colonization (Nayar *et al.*, 1988).

**Fig. 3.13.** Bacterial wilt and root-knot nematode disease complex in brinjal.

**ii) Management**

**a) Integrated Methods:** Combined application of *T. harzianum* and *P. chlamydosporia* was effective for management of the disease complex and gave least root gall index (RGI), least wilt index (WI) and maximum fruit yield (1.8 RGI, 22.18% WI and 2.024 kg/3 sq.m. yield) followed by *P. fluorescens* + *P. chlamydosporia* (1.81 RGI, 25.13 WI and 1.782 kg/3 sq.m. yield) and *T. harzianum* + *P. fluorescens* ( 1.93 RGI, 27.28 WI and 1.680 kg/3 sq.m. yield) (Naik, 2004).

### 3.3.2. Viral Diseases

#### 3.3.2.1. Brinjal Mosaic Virus

The disease is caused by potato virus Y (Sastry *et al.* (1974b).

**i) Symptoms:** The characteristic symptoms of the disease are mosaic mottling of the leaves and stunting of plants. The mosaic symptoms in early stages are mild but later it takes severe forms. The infected leaves are deformed, become small and leathery. The plants infected in early stages remain stunted in growth. Very few fruits are produced on infected plants.

**ii) Transmission:** The virus is easily sap transmissible. In nature, the virus is transmitted by aphids such as *Aphis gossypii* and *Myzus persicae* in a non-persistent manner.

**iii) Host Range:** In addition to brinjal, the virus can infect *Solanum nigrum, S. xanthocarpum, S. nodoflorus, Datura metal, Petunia hybrida, Physalis floridana, P. peruviana, P. minima* and tomato and produce mosaic mottling symptoms.

**iv) Survival:** The virus perpetuates on weed hosts like *Solanum nigrum, S. xanthocarpum* and *Nicandra physaloides*.

**v) Management**

- Removal and destruction of infected plants from the field.
- Removal of all the weeds in and around the brinjal field.
- Spraying 0.05% nuvacron or 0.02% metasystox at 10 days interval to control aphid vectors.
- Soil application of carbofuran at 1.5 kg a.i./ha at the time of sowing seeds in the seed bed followed by 1-2 foliar sprays of 0.05% nuvacron.
- Spraying of 1.5 to 1.8% mineral oil at 5-6 days interval has been found effective in reducing the disease incidence.

### 3.3.2.2. Little Leaf (Phytoplasma)

Brinjal little leaf disease was first reported by Thomas and Krishnaswamy (1939) from Coimbatore, India. The disease is caused by phytoplasma-like organism.

**i) Symptoms:** The characteristic symptoms include drastic reduction in leaf size, aggregation of numerous leaves in each branch giving bushy appearance and shortening of internodes (Fig. 3.14). Auxiliary buds enlarge but petioles and leaves remain shortened, narrow, pale green, soft and smooth and fruit size is reduced. Purple coloured fruits turn white after infection. At later stage, infection leads to complete sterility of the plant.

**Fig. 3.14.** Little leaf affected plant of brinjal.

Infected plants normally do not bear fruits. However, if any fruit is formed, it becomes hard and tough, and fail to mature. The very young fruit become necrotic, get mummified and cling to the plant (Anjaneyulu and Ramakrishnan, 1972b).

**ii) Transmission:** The disease is transmitted in nature by leaf hopper, *Hishimonas phycitis*. It is also graft transmissible.

**iii) Host Range:** The disease has got a very wide host range (Anjaneyulu and Ramakrishnan, 1973).

**iv) Management:** Early planted crop is severely affected by the disease. Hence avoid early transplanting of the crop. Uproot the severely infected plants and burn before spraying with insecticides (4-5 sprays of dimethoate or monocrotophos each at 0.1% at 10 days interval). Removal of weed hosts like *Datura fistulosa* and *Catharanthus roseus* and diseased plants from the field and nearby areas and use of disease resistant/tolerant varieties like Pusa Purple Cluster and Arka Sheel shall keep the disease under check. Periodical spraying of systemic insecticides is recommended to control the vector. During the initial stage of infection, spray 100 ppm streptocycline to check the growth of phytoplasma. The spray should be repeated twice in early sown crop (Anjaneyulu and Ramakrishnan, 1972a). Maximum disease control could be achieved by seedling dip in tetracycline or monocrotophos followed by soil application of phorate (Maheshwari *et al.*, 1988).

### 3.3.2.3. Tobacco Mosaic Virus

**i) Symptoms:** The characteristic symptom of this disease is conspicuous mosaic mottling of leaves. The leaves also develop blisters in advanced cases. The severely infected leaves become small and misshapen. Plants infected in early stages of their growth do not grow much and remain stunted. Vigour of the plant is reduced and severely infected plants wither, do not bear fruits or bear only one or two fruits. Such fruits also show some patches of light and dark areas on the surface and remain small and deformed.

**ii) Transmission:** The virus is highly contagious and transmitted by sap, contamination through implements, soil debris, hands of the working labourers and contaminated clothes. No insect vectors are known.

**iii) Host Range:** The virus has got a very wide host range since this disease is caused by a strain of TMV.

**iv) Survival and Spread:** The virus can perpetuate on many cultivated plants like tomato, tobacco, bell pepper, eggplant, cucurbits, legumes and many weed hosts. The virus can survive in the plant debris in the soil which act as a source of inoculum.

#### v) Management

- Destroy all perennial weeds in and near the seed beds and the field.
- Wash the hands well with soap and water before working in the seed beds.

- Prohibit smoking or chewing of tobacco by those handling brinjal seedlings.
- Avoid planting bell pepper, tobacco, cucumbers, etc. near brinjal seed beds.

## 3.4. CHILLI

Chilli (*Capsicum annuum*) is valued for its diverse commercial uses. India is a major producer, exporter and consumer of chilli. Indian chillies reach over 90 countries in the world. Bangladesh, Bahrain, Canada, Sri Lanka, Saudi Arabia, USA and UAE are the leading importing countries. It is being cultivated in 0.834 million hectares producing 0.847 million tonnes of chillies with an average yield of 1.0 tonne per hectare. Andhra Pradesh has been the largest chilli-growing state followed by Karnataka and Maharashtra. Productivity of chilli is highest in Andhra Pradesh, followed by Arunachal Pradesh and Punjab. The productivity of chilli in Andhra Pradesh and Punjab is higher since the crop is raised under irrigated conditions compared with Maharashtra and Karnataka. Chilli is rich in vitamins A and C.

### 3.4.1. Bacterial Diseases

#### 3.4.1.1. Bacterial Wilt, *Ralstonia solanacearum*

**i) Symptoms:** The disease occurs in scattered plants or groups of plants in the field. Characteristic symptom is wilting of the entire plant with no leaf yellowing (Fig. 3.15). Cross sections cut from roots and lower stems of diseased plants exude milky streams of bacteria from the vascular system when suspended in water. The bacteria have a wide host range and can survive in the soil for long periods. The disease is favoured by wet, warm conditions. Peppers are not as susceptible as eggplants, potatoes, tobacco or tomatoes. Rotation is not effective as the pathogen can survive for a long period - several years - in the soil and also attack a wide range of crops and solanaceous weeds.

**Fig. 3.15.** Bacterial wilt on chilli.

**ii) Management:** Chilli cvs. Kandhari, Pungent Pride, Cherry Red, Vattal, Dark Purple, Long Red and Pant C-1 were reported resistant. Manjari and Ujwala developed by Kerala Agricultural University, Vellanikkara, are also highly resistant to bacterial wilt.

### 3.4.1.2. Bacterial Leaf Spot, *Xanthomonas campestris* pv. *vesicatoria*

Bacterial leaf spot, also called as 'Scab spot' was first observed in the USA on chillies in 1912. At present, it is prevalent in many countries (Patel, 1972) including India where it is present in almost all the chilli growing areas. Loss in chilli yield due to disease has been reported up to 16.6% (Shekhawat and Chakravarti, 1977a).

**i) Symptoms:** Small brown spots appear on leaves, petiole, stem and fruit of chillies are distinctly raised (Fig. 3.16). These spots turn into greasy or black spots.

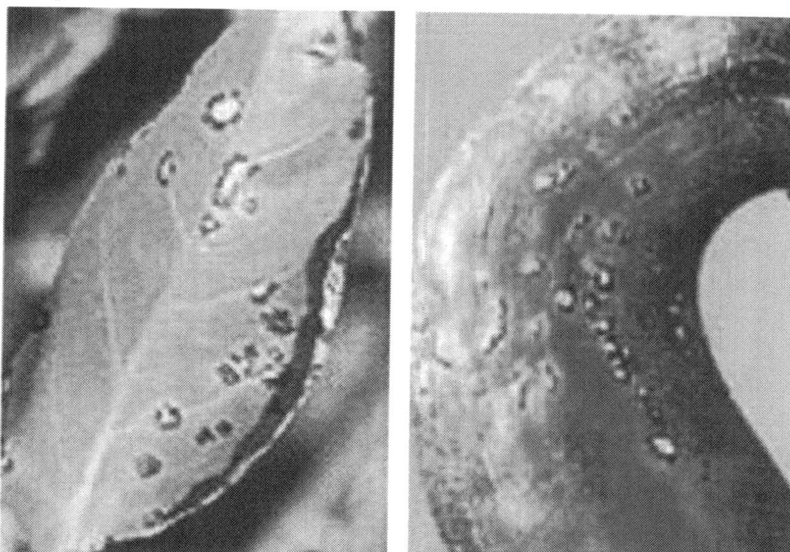

**Fig. 3.16.** Bacterial leaf spot on chilli leaf and fruit.

**ii) Epidemiology:** Pathogen is seed-borne (Kore and Dhaliwal, 1981) and also survives on stray chilli plants (Shekhawat and Chakravarti, 1978b). The bacterium survives for 14 and 10 months in seed at 8 and 10-38°C, respectively. It survives for 150 days in plant debris. Disease spreads to healthy plants through wind splashed rains and to new area through infected seed and planting material.

Maximum disease development has been found during July-October (Shekhawat and Chakravarti, 1976) and chilli plants at the age of 42-49 days are found most susceptible (Kore and Dhaliwal, 1981).

**iii) Management:** Hot water treatment of seeds at 50°C for 25 min is reported to be effective. Seed treatment with mercuric chloride (1: 1000) is also recommended for the control of the disease. Spraying of 3: 3: 50 Bordeaux mixture or 0.2% zineb or 20 ppm streptocycline or 100 ppm agrimycin + copper oxychloride or copper oxychloride + streptocycline or 200 ppm paushamycin + copper oxychloride is beneficial to check the disease (Raju and Rao, 1984). Chilli cv. Jwala is reported to be field resistant.

### 3.4.1.3. Bacterial Soft Rot, *Erwinia carotovora* pv. *carotovora*

**i) Symptoms:** Soft rot often begins in the peduncle and calyx tissues of harvested fruit. Infection can occur through wounds anywhere on the fruit. Fruit infected on the plant collapses and hangs on the plant like a water-filled bag. When the contents leak out, a dry shell of the fruit remains. The bacteria are soil-borne. Soft rot is primarily a post-harvest problem although it can occur in the field being facilitated by injuries to the fruit by insects and water splash. The disease is serious during rainy periods because the bacteria are splashed from the soil onto the fruit, which are more susceptible due to their high moisture content.

**ii) Management**

- Rotation with beans or maize
- Control of insects that cause injury to fruits.
- Post-harvest decay can be reduced by harvesting fruits when dry, minimizing injury during handling, and storing at cool temperatures.
- If fruits must be washed, add free chlorine in the wash water to eliminate soft-rot bacteria. However, this treatment prevents infection during washing but does not prevent soft rot development in fruits that were already infected before washing.

### 3.4.2. Viral Diseases

### 3.4.2.1. Indian Chilli Mosaic Virus

About 21-60% yield loss has been reported due to this disease (Mishra *et al.*, 1977).

**i) Symptoms:** The virus induces clearing of veins of younger leaves of chilli plants. Later yellowing of whole leaf occurs followed by irregular pattern of light (yellow) and dark (green) patches. The general stunting of diseased plants is accompanied by defective development of reproductive parts. Distortion of veins has also been noticed. Infected plant produce very few fruits.

**ii) Transmission:** The virus is easily mechanically sap transmitted. It is also transmitted by grafting. The insect vectors include species of aphids like *Aphis gossypii, A. evonymii* and *Myzus persicae* (Jha and Raychaudhuri, 1956) in a non-persistent manner (Nariani and Sastry, 1964).

The virus is transmitted through the pollen of infected flowers and also by root contact (Mishra and Jha, 1973).

**iii) Host Range:** The virus infects chilli, bell pepper, tobacco, tomato, eggplant, okra, peri winkle, *Nicotiana rustica, N. glutinosa, Petunia hybrida, Datura alba, Cucumis* spp., *Gomphrena globosa* and *Chrysanthemum* spp.

**iv) Management**

- Removal and destruction of infected plants from the field as soon as they are noticed.

- Eradication of weeds from the field.
- Spraying of 0.02% methyl parathion at weekly intervals to reduce the aphid vector population.
- Chilli cvs. NP 20 and NP 23 have been found fairly resistant to this virus (Jha *et al.*, 1953).

### 3.4.2.2. Tobacco Mosaic Virus (TMV)

**i) Symptoms:** The leaves of the plant affected with TMV show a greenish yellow mottling and may be slightly curled or irregular in shape (Fig. 3.17). TMV often causes a brown streaking of some of the branches. This causes yellowing and dropping of the leaves. At times, plants affected by this virus may show a general yellowing of many of the leaves with little streaking on the stem. The common form (green mottle) of TMV often causes little, if any, mottled symptoms on the fruit. However, fruits on streaked branches that have dropped many leaves often are yellow, wrinkled and withered. One strain of this virus causes a brilliant yellow mottling of the leaves. It may also cause a yellow spotting on some fruits. The spots are of irregular shape and not noticeably sunken or wrinkled.

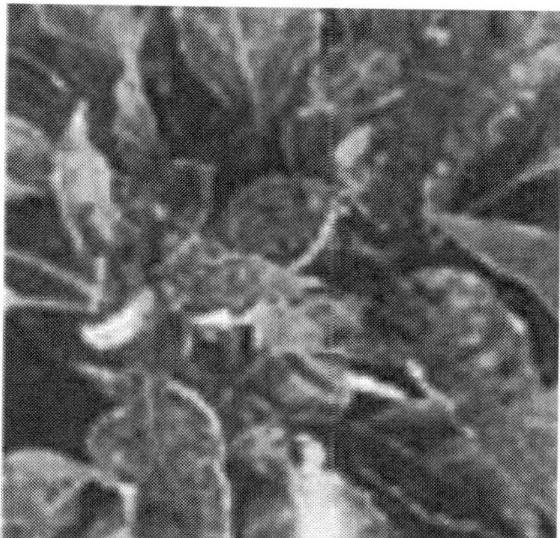

**Fig. 3.17.** TMV on chilli leaves.

**ii) Transmission:** The virus is highly contagious and transmitted by mechanical means by contamination through implements, soil debris, clothes, hands of workers and touching of one plant with another. The TMV is also transmitted through infected seeds of chilli up to 21%.

**iii) Host Range:** The TMV has got a very wide host range (Horvath, 1978).

**iv) Management:** Chilli cv. Florida FR-2 is resistant to common strains of TMV.

### 3.4.2.3. Cucumber Mosaic Virus

The virus has been reported from India by Anjaneyulu and Appa Rao (1967).

**i) Symptoms:** Plants infected with the common form of CMV develop a mottling somewhat like that of TMV but with more contrast between the light and dark portions of the leaf. The leaves curl upwards at the edge and often are abnormally narrow and pointed. Few fruits, those produced sometimes, are misshapen and have dark green raised spots on the surface. Infected plants often are abnormally short and compact in appearance. At times, yellow rings are produced on the leaves and fruits. These rings are a quarter to half inch in diameter. On fruits, they are slightly sunken and have concentric markings. When both TMV and CMV infect the same plant, there is more yellow mottling of the foliage than occurs with either virus alone. The fruits also show more mottling than those plants only infected with CMV.

**ii) Transmission:** The virus is very readily transmitted by mechanical means. In nature, it is transmitted by several species of aphid vectors like *Aphis craccivora, A. gossypii* and *Myzus persicae.*

**iii) Management**

- Removal and destruction of infected plants from the field.
- Removal of all the weeds in and around the chilli field.
- Spraying 0.05% nuvacron or 0.02% metasystox at 10 days interval to control aphid vectors.
- Soil application of carbofuran at 1.5 kg a.i./ha at the time of sowing seeds in the seed bed followed by 1-2 foliar sprays of 0.05% nuvacron.
- Spraying of 1.5 to 1.8% mineral oil at 5-6 days interval has been found effective in reducing the disease incidence.

### 3.4.2.4. Tobacco Etch Virus (TEV)

**i) Symptoms:** Chilli infected with TEV show a yellow and green mottling somewhat like that of CMV, although the symptoms are less pronounced at times. The yield is somewhat reduced.

When plants are infected with TMV and TEV, there is an intense yellow mottling of young leaves. Some of the older leaves become yellow and then drop. On such plants, yellow-green circular spots of about ¼ inch in diameter often appear below the smooth surface of the green fruits. As the fruits ripen, the spotted parts become more yellow and they shrivel slightly. The spots are raised in a way that gives the fruits a characteristic roughened appearance.

**ii) Transmission:** The virus is transmitted by mechanical sap inoculation and also by some species of aphids like *Aphis gossypii, A. spiraecola, Macrosiphum solanifolii, M. pisi* and *Myzus persicae.*

**iii) Host Range:** The virus has very wide host range. It infects many Solanaceous plants including tobacco, petunia, potato and *Datura stramonium*. Local lesion hosts include *Chenopodium amaranticolor* and *Physalis peruvianum*.

**iv) Management:** Spraying 0.05% nuvacron or 0.02% metasystox at 10 days interval to control aphid vectors.

### 3.4.2.5. Potato Virus Y (PVY)

**i) Symptoms:** The disease is characterizes by severe mosaic mottling, vein banding and interveinal hypertrophy of the infected leaves. The midrib of the leaf is distorted and the leaf area is reduced considerably. The plants infected in the early stage show stunted growth. Fruits from diseased plant show chlorotic streaks and rough skin. The virus was reported from India by Jayarajan and Ramakrishnan (1969).

**ii) Transmission:** In nature, the virus is easily transmitted to a number of herbaceous plants by several aphid vectors like *Aphis craccivora, A. gossypii, A. spiraecola, A. evonymii, Myzus persicae, Macrosiphum solanifolii,* and *M. pisi* (Jayarajan and Ramakrishnan 1969; Khatri and Sekhon, 1974).

**iii) Host Range:** The virus has got very wide host range. The important hosts include potato, tomato, tobacco, *Solanum dulcamera, S. nigrum, Hyoscymus niger, Petunia* spp., *Cyphomandra betacea, Dahlia* spp., and *Datura metal.*

**iv) Management**

- Remove weed hosts around chilli crop.
- Soil application of carbofuran at 1.5 kg a.i./ha in the nursery beds at the time of sowing seeds, followed by 2-3 foliar sprays of 0.05% dimecron at 10 days interval.
- Spraying of 1.5 to 2.0% mineral oil reduce the spread of the virus.
- Chilli cvs. Florida VR-2 and Delray Bell are resistant to all isolates of PVY (Zitter, 1979).

### 3.4.2.6. Leaf Curl Virus

The disease causes heavy losses in yield and quality of fruits. If the plants are infected within 20-25 days after transplanting, the loss in yield goes up to 80-90%. However, in case of later infections, the loss is comparatively less (Singh *et al.*, 1979b).

**i) Symptoms:** Symptoms include abaxial and adaxial curling of the leaves accompanied by puckering and blistering of interveinal areas and thickening and swelling of the veins. In advanced stages, Maxillary buds are stimulated to produce clusters of leaves which are reduced in size (Fig. 3.18). The whole plant assumes bushy appearance with stunted growth. Flowers and fruits are few and small, being curled at the end. The virus has a wide host range. However, important hosts are tobacco, bell pepper and several weed hosts.

**ii) Transmission:** The virus is transmitted by the white fly, *Bemisia tabaci.*

**iii) Host Range:** The virus has very wide host range. The important hosts include tobacco, bell pepper and several weed hosts.

**iv) Epidemiology:** The disease spread is more during summer months which favour white fly multiplication and disease spread. The virus persists on several cultivated plants and weed hosts (Singh *et al.*, 1979b).

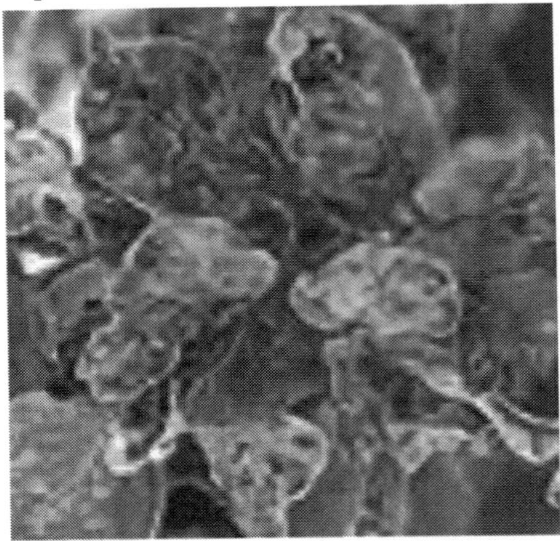

**Fig. 3.18.** Leaf curl virus on chilli leaves.

**v) Management**

- Rouging of infected plants from the field.
- Removal of all the weeds from and around the field.
- Soil application of carbofuran at 1.5 kg a.i./ha in the nursery and at the time of transplanting followed by 2-3 sprays of dimethoate or monocrotophos both at 0.05% at 10 days interval prevents whitefly multiplication and spread of the disease (Singh *et al.*, 1979a).

## Integrated Management of Insect-borne Viruses

Use of nylon net cover in nursery beds, soil application of carbofuran at 1 kg a.i./ha in combination with foliar application of 0.1% imidachloprid at 10 days interval effectively reduced the vector population as well as the disease incidence.

At Coimbatore, Soil application of carbofuran at 1 kg a.i./ha + covering the nursery bed with nylon net of 400 mesh followed by 2 sprays of 2% neem oil at 15 and 35 DAP recorded lesser virus incidence and highest yield with cost: benefit ratio of 1: 3.3.

At Sabour, Use of nylon net and soil application of carbofuran at 1 kg a.i./ha in seed bed and application of carbofuran at 1.5 kg a.i./ha in the main field a week after transplanting + 3 sprays of nuvacron at 1 ml/litre was most effective in managing the disease with cost: benefit ratio of 1: 5.64.

## 3.5. BELL PEPPER

Bell pepper (*Capsicum annuum*), also known as capsicum or sweet pepper, is a popular vegetable in India. It is relatively a new entrant into our country. It is mainly cultivated in Himachal Pradesh, Uttar Pradesh, parts of Gujarat, Maharashtra, Karnataka, Ranchi region of Bihar and hilly regions of Tamil Nadu. It grows well in summer season in hills and cooler season in the plains.

### 3.5.1. Bacterial Diseases

#### 3.5.1.1. Bacterial Spot, *Xanthomonas campestris* pv. *vesicatora*

**i) Symptoms:** Bacterial spot caused by *Xanthomonas campestris* pv. *vesicatora* (Xcv) is present wherever peppers are grown. Leaves, fruit and stems can be attacked by the disease. Leaf spots begin as circular, water-soaked that becomes necrotic with brown centres with yellowish borders. The spots are sunken on the upper leaf surface and slightly raised on the lower surface (Fig. 3.19). On stems, spots are elongated. Affected leaves turn yellow and drop. Affected fruits have raised brown spots that are wart-like in appearance (Fig. 3.19). It spreads rapidly during warm, rainy conditions. High relative humidity with free moisture on leaves for long periods favour infection. Ideal temperatures for infection are above 20°C. The bacterium is seed-borne and survives in crop debris.

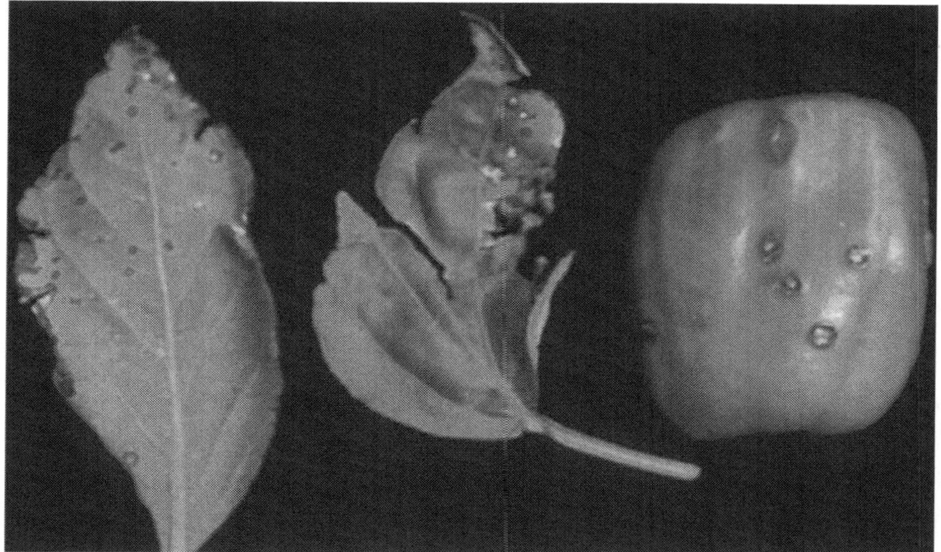

**Fig. 3.19.** Bacterial spot on capsicum leaves and fruit.

Bacterial spot lesions do not have concentric zones or a prominent halo. When conditions are optimal for disease development, spots can coalesce to form long, dark streaks. A general yellowing may appear on foliage with many lesions giving the plants a scorched appearance, and the plants may exhibit severe epinasty. Only green fruit is susceptible to infection and lesions are quite distinct, beginning as minute, slightly raised blisters with a halo that resemble the bird's eye spot caused by *Clavibacter michiganense* (bacterial canker). As lesions enlarge, they loose their halo and become brown, raised, and scab-like. Lesions on ripe pepper fruit may be scab-like or sunken.

**ii) Survival and Spread:** The bacterium is able to survive on volunteer plants and can over winter in diseased plant debris. Seed is an important mechanism for survival and dissemination of Xcv. Disease development is favoured by temperatures between 80° and 90°F and by heavy rainfall. The bacterium is spread by wind driven rain, workers, farm machinery, and aerosols. It penetrates through stomatoes and wounds created by insects, wind driven sand, and tools. Xcv affects all above ground plant parts. On leaves, the spots are generally brown, circular, and water-soaked.

**iii) Management**

- Buy certified seed from a reputable source or use seed treatments to reduce transmission.
- Produce disease-free plants by raising transplants in an area where production does not occur. Inspect all purchased transplants carefully and if transplants originate in southern states they should be certified.
- Rotate fields to avoid carryover on volunteer plants or crop residue.
- Keep fields free from volunteer plants, weeds, and cull piles.
- Crop rotation with non-host crops.
- Avoid working in fields when bacterial diseases are present and the fields are wet.
- Apply appropriate bactericides or combination pesticides.

**Chemical recommendations**

- Copper hydroxide: Begin applications when disease first threatens and repeat at 7-10 day intervals as needed. Do not apply spray solution with pH less than 6.0 to prevent phytotoxicity.
- Mancozeb plus copper hydroxide: Begin applications when disease threatens and repeat at 7-10 day intervals as needed. Use higher rates and 3-7 days interval when disease pressure is severe.
- Streptomycin sulfate (Agrimycin): 200 ppm (REI 12 h). Use only up to transplant.

### 3.5.1.2. Soft Rot, *Erwinia carotovora* pv. *carotovora*

**i) Symptoms:** Soft rot often begins in the peduncle and calyx tissues of harvested fruit. Infection can occur through wounds anywhere on the fruit.

Fruit infected on the plant collapses and hangs on the plant like a water-filled bag. When the contents leak out, a dry shell of the fruit remains. The bacteria are soil-borne. Soft rot is primarily a post-harvest problem although it can occur in the field being facilitated by injuries to the fruit by insects and water splash. The disease is serious during rainy periods because the bacteria are splashed from the soil onto the fruit, which are more susceptible due to their high moisture content.

**ii) Management**

- Rotation with beans or maize
- Control of insects that cause injury to fruits.
- Post-harvest decay can be reduced by harvesting fruits when dry, minimizing injury during handling, and store at cool temperatures.
- If fruits must be washed, add free chlorine in the wash water to eliminate soft-rot bacteria. However, this treatment prevents infection during washing but does not prevent soft rot development in fruits that were already infected before washing.

### 3.5.1.3. Bacterial Wilt, *Ralstonia solanacearum*

**i) Symptoms:** The disease occurs in scattered plants or groups of plants in the field. Characteristic symptom is wilting of the entire plant with no leaf yellowing (Fig. 3.20). Cross sections cut from roots and lower stems of diseased plants exude milky streams of bacteria from the vascular system when suspended in water. The bacteria have a wide host range and can survive in the soil for long periods. The disease is favoured by wet, warm conditions. Peppers are not as susceptible as eggplants, potatoes, tobacco or tomatoes. Rotation is not effective as the pathogen can survive for a long period - several years - in the soil and also attack a wide range of crops and solanaceous weeds.

**Fig. 3.20.** Bacterial wilt on sweet pepper (Courtesy: A.M. Varela, ICIPE).

### ii) Management

- Plant varieties that are tolerant / resistant, if available.
- Do not grow crops in soil where bacterial wilt has occurred.
- Remove wilted plants from the field to reduce spread of the disease from plant to plant.
- Control root-knot nematodes since they could facilitate infection and spread of bacterial wilt.
- Soil amendments (organic manures) can suppress bacterial wilt pathogen in the soil.
- Rotation is of limited value since the disease has a wide host range, but still it is recommended to avoid continuous planting of solanaceous crops.

## 3.5.1.4. Bacterial Wilt, *Ralstonia solanacearum* and Root-knot Nematode, *Meloidogyne incognita* Disease Complex

Capsicum is prone to many soil borne diseases among which the bacterial wilt (*R. solanacearum*) in combination with root-knot nematode (*M. incognita*) takes heavy toll every year all over the world (Naik, 2004).

### i) Management Methods

**a) Integrated Methods:** Combined application of neem based formulations of *P. fluorescens* and *P. chlamydosporia*/*T. harzianum* at 40 g/m$^2$ in nursery beds and transplanting these seedlings in the main field resulted in significant reduction in disease index and root-knot index in capsicum to the tune of 70% and increased the crop yield by 37% (Rao *et al.*, 2002) (Table 3.11). Combinations of the bioagents did not affect the colonization of the individual bioagents on the roots and hence the transplants carried the bioagents to the main field. This has resulted in the effective management of the pathogens involved in the disease complex.

**Table 3.11. Effect of integration of neem-based bioagents on the growth of transplants and management of disease complex and yield of capsicum**

| Treatment | Seedling weight (g) | Root-knot index (1-10) | Disease index (1-9) | Yield in kg/4 sq.m. |
|---|---|---|---|---|
| Seed treat. with *P. fluorescens* | 421 | 5.6 | 6.4 | 4.3 |
| Seed treat. with neem-based *P. fluorescens* | 428 | 5.2 | 6.7 | 4.7 |
| Nursery treat. with *P. fluorescens* | 435 | 4.6 | 5.4 | 4.8 |
| Nursery treat. with *P. chlamydosporia* | 364 | 4.4 | 7.3 | 3.2 |
| Nursery treat. with *T. harzianum* | 374 | 4.8 | 7.0 | 3.4 |

| | | | | |
|---|---|---|---|---|
| Nursery treat. with *P. fluorescens* + *P. chlamydosporia* | 463 | 4.1 | 5.2 | 4.0 |
| Nursery treat. with *P. fluorescens* + *T. harzianum* | 493 | 3.8 | 3.5 | 5.1 |
| Control | 340 | 8.7 | 8.2 | 2.6 |
| CD at 5% | 27.20 | 0.49 | 0.38 | 0.25 |

Naik (2004) reported that combination of *T. harzianum* along with *P. fluorescens* increased the yield (3.320 kg/3m$^2$ plot) followed by combination of *T. harzianum* and *P. chlamydosporia* (2.960 kg/3m$^2$). The least gall index was present where the combination of *T. harzianum* and *P. fluorescens* was used (1.5) followed by *T. harzianum* + *P. chlamydosporia* (1.7) and *P. fluorescens* + *P. chlamydosporia* (1.7). But among combination treatment of bio-agents, all the treatments were on par (*P. fluorescens* + *T. harzianum*, *P. fluorescens* + *P. chlamydosporia*, *T. harzianum* + *P. chlamydosporia*). Results related to the percent bacterial wilt disease (*R. solanacearum)* incidence also showed similar trend in which *T. harzianum* + *P. fluorescens* performed very well and the percent incidence was 16.90 followed by *T. harzianum* + *P. chlamydosporia* (17.20).

### 3.5.2. Viral Diseases

#### 3.5.2.1. Mosaic Virus

**i) Symptoms:** The important symptoms of this disease are mottling, distortion and reduction in leaf lamina. Sometimes the infected plants show vein banding, yellowing and leaf curling symptoms. In case of early infection, stunted growth and reduction in fruit size and yield are seen. In later stages of infection, the fruits become malformed and unfit for consumption. Yellow and green streaks may be produced on the fruits.

**ii) Management**

Effective management of thrips and mite population in the polyhouse reduces the disease incidence. Application of 0.15% acephate or 0.1% fipronil or 0.1% dicofol or 1% emulsified neem oil is recommended for insect vector management.

#### 3.5.2.2. Yellow Mosaic

**i) Symptoms:** The diseased plants exhibit vein clearing of infected leaves followed by yellow mosaic consisting of yellow and dark green irregular patches scattered over leaf lamina (Capoor and Sawant, 1987).

**ii) Etiology:** The causal virus is a strain of TMV.

**iii) Transmission:** The virus is readily transmitted by mechanical sap inoculation. Seed transmission up to 5.8% has been reported (Capoor and Sawant, 1987).

**iv) Management**

- Select seeds from healthy plants and as an added precaution, soak them in 0.3% solution of trisodium phosphate for 48 hours before sowing.
- Dip the seedlings in solution of trisodium phosphate or monocrotophos both at 0.2% concentration for 10 minutes before transplanting.
- Carefully remove and destroy all infected plants in the nursery or in main field. Do not use TMV infected seedlings for transplanting.
- Crop rotation with non-solanaceous crops helps in managing the disease.
- Use resistant varieties.
- Remove crop refuse and roots from fields.
- Eradicate weeds.
- Workers should not smoke or take snuff when working in bell pepper fields since the virus can be transmitted from tobacco.

## 3.6. ONION AND GARLIC

Onion (*Allium cepa*) is one of the most important commercial vegetables. The bulbs are consumed both in mature and tender stages. Even the leaves and stalks of inflorescence are used as vegetables. It is being cultivated in 0.695 million hectares producing 9.248 million tonnes of onions with an average yield of 13.3 tonnes per hectare (Bijay Kumar, 2007) (Tables 3.12 and 3.13).

**Table 3.12. Major onion producing countries of the world**

| Country | Area (in '000 ha) | Production (in '000 MT) | Productivity (in MT/ha) | % Share of world production |
|---|---|---|---|---|
| China | 901 | 19040 | 21.1 | 30.2 |
| India | 695 | 9248 | 13.3 | 14.7 |
| USA | 65 | 3346 | 51.2 | 5.3 |
| Turkey | 77 | 2070 | 26.8 | 3.3 |
| Pakistan | 128 | 1765 | 13.8 | 2.8 |
| Russia | 123 | 1759 | 14.3 | 2.8 |
| Iran | 50 | 1685 | 33.7 | 2.7 |
| Egypt | 43 | 1302 | 30.6 | 2.1 |
| Brazil | 58 | 1138 | 19.5 | 1.8 |
| Mexico | 40 | 1115 | 28.1 | 1.8 |
| Others | 1179 | 20500 | 17.4 | 32.6 |
| Total | 3359 | 62968 | | |

Source: FAO (except Indian data) (Indian data source: Indian Horticulture Database 2006)

**Table 3.13. Area, production and productivity of leading onion growing states in India (2005-06)**

| State | Area (in'000 ha) | Production (in '000 MT) | Productivity (MT/ha) |
|---|---|---|---|
| Maharashtra | 154.0 | 1878.8 | 12.2 |
| Gujarat | 84.9 | 1876.6 | 22.1 |
| Bihar | 49.8 | 1011.7 | 20.3 |
| Karnataka | 139.0 | 1008.5 | 7.3 |
| Andhra Pradesh | 39.4 | 670.0 | 17.0 |
| Madhya Pradesh | 40.0 | 639.8 | 16.0 |
| Rajasthan | 44.5 | 427.2 | 9.6 |
| Haryana | 15.5 | 317.7 | 20.5 |
| Uttar Pradesh | 21.2 | 283.1 | 13.3 |
| Orissa | 26.8 | 243.9 | 9.1 |
| Tamil Nadu | 29.2 | 233.8 | 8.0 |
| Others | 50.8 | 657.3 | 12.9 |
| Total | 695.1 | 9248.4 | 13.3 |

(Source: Indian Horticulture Database 2006)

Garlic (*Allium sativum*) is mainly used for flavouring vegetables and meat dishes. It is carminative, stimulant and cures digestive disorders. It has been found to be very effective in curing diabetes and high blood pressure. It is being cultivated in 0.107 million hectares producing 0.436 million tonnes of garlic with an average yield of 4.1 tonnes per hectare. Madhya Pradesh, Gujarat, Orissa, Maharashtra, Uttar Pradesh and Rajasthan are major garlic-growing states. More than 50% production comes from Madhya Pradesh and Gujarat only.

### 3.6.1. Bacterial Diseases

#### 3.6.1.1. Bacterial Stalk Rot, *Pseudomonas gladioli* pv. *allilicola*

The disease was observed for the first time on onion at Kanpur, Uttar Pradesh during 1971 (Swarup *et al.*, 1973). The disease incidence varies from 12.8 to 35.6% in seed crop and 8.5 to 30.6% in bulb crop (Kishun and Swarup, 1979b).

**i) Symptoms:** The disease is characterized by internal brown discolouration and dry rotting in the basal portion of the flower stalk resulting in partial to complete withering of inflorescence prior to seed formation. Such rotting may extend internally up to the seed bulb. Outwardly, the plants appear normal till either inflorescence gets dried up or stalks topple down. Yellowing of outer leaves from tip downwards followed by drying is seen in bulb crop. During storage, the inner scales of affected bulbs become water-soaked followed by browning and drying which results in the hollowing of the bulbs.

**ii) Host Range:** Besides onion, the bacterium is pathogenic to garlic, tuberose, rice, pearl millet (bajra), sorghum, cowpea and other plants (Swarup, 1983).

**iii) Epidemiology:** The pathogen survives in diseased plant debris in soil. Onion seedlings get infected in the field, and then it invades the bulbs and survives along with them, causing dry rot in the storage. Partially rotten bulbs while sprouting produce diseased plants in the next season (Kishun and Swarup, 1979b). The pathogen also survives in diseased plant debris at 6-8°C up to 12 months and at room temperature (25-30°C) for 6 months (Kishun and Swarup, 1979b). The bacterium enters the host through wounds and spread in storage by contact of diseased and healthy bulbs.

**iv) Management**

Soaking of onion bulbs in ceresan wet, sandoz seed dressing-6334, 2000 ppm aretan, agrimycin-100, 200 ppm streptocycline for 2 hours reduced the disease incidence in seed crop (Kishun and Swarup, 1982).

Onion lines 5001-12, 5308 (bulb crop), 5223-2, 5404-3, 6106, 6621 and VL-14 (seed crop) remain free from infection under natural conditions (Kishun and Swarup, 1982).

### 3.6.1.2. Storage Soft Rot, *Erwinia carotovora* pv. *carotovora*

**i) Symptoms:** Soft rot is one of the more prevalent causes of loss in storage onions (Fig. 3.21). Bacteria can enter the neck tissue as plants mature and then invade one or more scales. At this stage, the affected tissues are water-soaked and pale yellow to light brown. As the rot progresses, the invaded fleshy scales become soft. Diseased bulbs can be diagnosed by pressing on the bulb: a watery, foul-smelling fluid often can be squeezed from the neck of diseased bulbs. *E. carotovora* pv. *carotovora* enter only through wounds. Onion maggots (*Delia antiqua*) may carry the bacteria and introduce them while feeding. Onions with mechanical injuries, bruises or sunscald under warm, humid conditions are particularly susceptible to bacterial soft rot.

**ii) Host Range:** *E. carotovora* pv. *carotovora* can affect many vegetables including carrots, celery, potato and parsnip.

**iii) Management**

- Eliminate the rotted bulbs by periodical turning of the storage heaps for minimizing the losses during storage.
- Onion tops should be allowed to mature well before harvesting.
- Care should be taken to avoid bruising during harvesting and packing.
- Storage places should be well ventilated to avoid accumulation of moisture on the surfaces of bulbs.
- Onions should be stored at 0°C and a relative humidity of 65-70%.

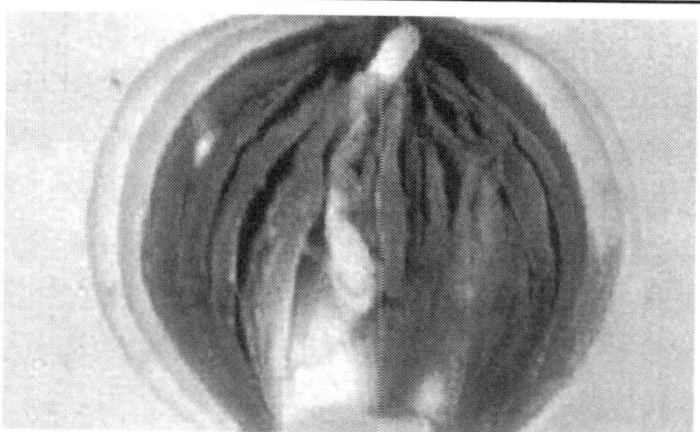

**Fig. 3.21.** Soft rot of onion.

### 3.6.1.3. Slippery Skin, *Pseudomonas allicola* pv. *allicola*

**i) Symptoms:** There are no symptoms on the outer surface of bulbs during early stages of disease development. When the bulb is cut open length-wise, one or more of the inner scales can be found water-soaked or appears as if it has been cooked. The rot does not progress crosswise in the bulb. After the decay has progressed, the tissue begins to dry, the onion shrivels and secondary organisms can enter and cause a wet rot. The base of the bulb can be pressed hard enough to cause the centre core to slip out at the top, and for this reason the disease is known as slippery skin. The disease is favoured by high moisture.

**ii) Management**

- Proper maturing of the crop.
- Quick drying after harvest.
- Proper storage as in the case of bacterial soft rot.

### 3.6.1.4. Sour Skin, *Pseudomonas cepacia*

**i) Symptoms:** In contrast to soft rot and slippery skin, infected scales are not water-soaked but are slimy and yellow. Symptoms usually visible only after onions are dug. The upper portion of the bulb shrinks, and in advanced stages of the disease, the outer dry skin readily slips off during handling while the centre of the bulb still remains firm. Outer layer of scales often becomes darkened and almost orange. Decay of inner scales leads to a soft rot that has a sour, vinegar-like odour. The disease is favoured by wet warm conditions.

## 3.6.2. Viral Diseases

### 3.6.2.1. Onion Yellow Dwarf Virus

**i) Symptoms:** The first sign of infection in plants grown from naturally infected onion bulbs is a series of short yellow streaks which appear at the base of the first leaf emerging through the neck of the bulb. In general, all the leaves emerging after the appearance of first symptoms show signs of disease and those previously developed remain apparently healthy. Under favourable conditions, the leaves showing short yellow streaks at the base become yellow throughout and also crinkled, somewhat flat and fall off. Flower stalks of infected plants producing seed show yellow streaks extending upwards from the base. Later, the streaks coalesce, the stalks become yellow throughout, twists and curls in a characteristic manner. The infected plants are stunted in growth. The flower clusters of infected mother plants are smaller and have fewer flowers than normal plants.

**ii) Transmission:** The virus is mechanically transmitted. The most important aphid vectors include *Aphis rumicis, A. maidis* and *Rhophalosiphum prunifoliae*. Other species of aphids which transmits virus include *Amphorophora rubi, Aphis gossypii, A. pomi, Brevocoryne brassicae, Hystoneura setariae, Macrosiphum euphorbiae, M. pisi, Myzus persicae* and *Rhophalosiphum pseudobrassicae*.

**iii) Host Range:** Besides onion, the virus also infects garlic (*Allium sativum*), Chinese sacred lily (*Narcissus tazetta*) and jonquil.

**iv) Management**

- Obtain virus-free planting material by indexing.
- Virus-free seed stocks of bulbs can be produced in disease-free areas.
- Rouge virus infected plants in the field from time to time.

### 3.6.2.2. Irish Yellow Spot Virus (IYSV)

**i) Distribution:** The thrips-IYSV complex is causing problems worldwide. It is present in Australia, Brazil, Chile, Egypt, Guatemala, Peru, India, Iran, Israel, Japan, The Netherlands, Reunion Island, Spain and the U.S. The economic impact associated with IYSV infections is increasing. Severe seed and bulb damage is occurring resulting in huge economic losses

**ii) Symptoms:** Symptoms include diamond shaped lesions on onion plant stems (Fig. 3.22). Lesion size increases with time of infection and age of the plant when the sampling was done.

**iii) Transmission:** Thrips are one of the most important contributing factors to virus spread and epidemics. There are about 5,000 species of thrips, but only one species, onion thrips, is the known IYSV vector.

The virus probably multiplies in the thrips vector, there is no evidence of seed transmission.

The overlap between seed and bulb crops in Pacific Northwest provides the "green bridge". Transplants, for example, can carry viruliferous thrips.

Both western flower thrips and onion thrips are prevalent in the Pacific Northwest. In Georgia, however, very few onion thrips are seen. Most present are either western flower thrips or tobacco thrips.

**Fig. 3.22.** Irish yellow spot virus on onion.

**iv) Management**

- **Field sanitation**: Clean up onions left over from fall harvest. Virus and/or viruliferous thrips can over winter on the culls or other plants. Come spring, they can act as a source of the virus and other pests for the new onion crop.
- **Reduce plant stressors**: Any thing that stresses the onion plant can compound the effects of IYSV. Maintain proper soil and plant fertility, uniform plant stand, and avoid soil compaction and irrigation stress.
- **Practice sound thrips management**: Take steps to reduce thrips populations. However, remember that thrips control alone may not reduce disease incidence. Examine crop rotation and nearby crops-Isolating seed and bulb crops to break the 'green bridge' may be helpful but often is not practical.

## 3.7. OKRA

Okra (*Abelmoschus esculentus*) is an annual vegetable crop grown in tropical and sub-tropical regions. It is being cultivated in 0.376 million hectares producing 3.684 million tonnes of okra with an average yield of 9.8 tonnes per hectare (Bijay Kumar, 2007) (Table 3.14). Tender green fruits are cooked in curry and soup. The root and stem are used for clearing cane juice in preparation of 'gur'. High iodine content of fruits helps to control goitre while leaves are used in inflammation and dysentery. It is a good source of Vitamins A, B and C. The fruits are also used in cases of renal colic, leucorrhoea and general weakness. The dry seed contains 13-22% good edible oil and 20-24% protein. The oil is used in soap, cosmetic industry and as vanaspati while protein is used for fortified preparations. The crushed seed is fed to cattle for more milk production and the fibre is utilized in jute, textile and paper industry.

**Table 3.14. Area, production and productivity of leading okra growing states in India (2005-06)**

| State | Area (in'000 ha) | Production (in '000 MT) | Productivity (MT/ha) |
|---|---|---|---|
| West Bengal | 65.8 | 718.9 | 10.9 |
| Bihar | 56.2 | 714.1 | 12.7 |
| Orissa | 71.4 | 618.8 | 8.7 |
| Gujarat | 41.0 | 350.3 | 8.5 |
| Andhra Pradesh | 25.6 | 332.7 | 13.0 |
| Chattishgarh | 18.1 | 184.7 | 10.2 |
| Maharashtra | 29.0 | 182.7 | 6.3 |
| Assam | 9.6 | 119.7 | 12.5 |
| Uttar Pradesh | 8.8 | 110.2 | 12.5 |
| Haryana | 14.0 | 93.7 | 6.7 |
| Karnataka | 7.7 | 63.2 | 8.2 |
| Others | 28.9 | 194.9 | 6.7 |
| Total | 376.1 | 3684.0 | 9.8 |

(Source: Indian Horticulture Database 2006)

### 3.7.1. Bacterial Diseases

#### 3.7.1.1. Bacterial Blight, *Xanthomonas campestris* pv. *malvacearum*

**i) Symptoms:** The bacterium is transmitted through seed and water splash. Symptoms consist of spots on leaves and pods. Leaf spots are initially soaked with water and are more obvious on the underside of leaves. They are restricted to the veins and are angular in shape. Spots on petioles are elongated, sometimes slightly sunken and blackish. Spots on pods are initially water-soaked and round. Later, they join together and turn oily black. All spots appear waxy and shiny (Fig. 3.23).

**ii) Management**

- Use certified disease-free seeds.
- Avoid dense planting.
- Avoid overhead irrigation.
- Avoid working in the field when wet.
- Remove crop debris after harvest.
- Spray with copper-based fungicides.

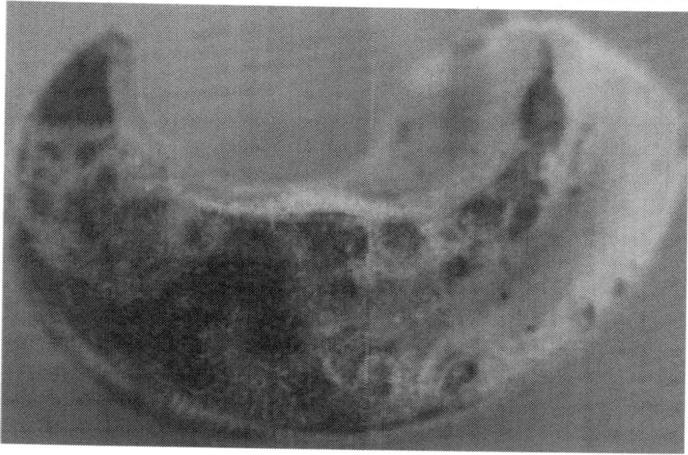

**Fig. 3.23.** Bacterial blight on young okra pod

### 3.7.2. Viral Diseases

#### 3.7.2.1. Yellow Vein Mosaic Virus (YVMV)

This is the most important and devastating virus disease of okra. Sastry and Singh (1974) estimated that if the plants are infected within 20 days after germination, the loss in yield was recorded up to 98%. The plants that infected at 35 and 50 days after germination, the loss in yield was estimated to be 83 and 49%, respectively.

**i) Symptoms:** YVMV is caused by Gemini virus and transmitted by white fly. Characteristic symptoms appear as yellow vein and vein lets leaving green tissue in interveinal area (Fig. 3.24). Severely infected leaves sometimes become completely yellow. Fruits also change in colour to yellow and become hard in early stage of development. The infected plants are stunted and bear very few yellow coloured fruits.

**Fig. 3.24.** YVMV infection on okra leaves.

**ii) Transmission:** In nature, the virus is transmitted by white fly, *Bemisia tabaci*. Even a single white fly can transmit the disease. White flies require 15-20 min to acquire the virus from infected plants. They can inoculate the plant within 15-20 min of inoculation feeding. Incubation of virus in the vector has been found to be 6-8 hr. Whiteflies once fed for 12 hr on infection sources, can remain viruliferous for the entire life.

**iii) Epidemiology:** Vector population and virus incidence were more during March to June when the atmospheric temperature remained high and humidity is less which is favourable for white fly multiplication and spread of the disease. On the contrary, low temperature and high humidity are detrimental for white fly multiplication and spread of the disease. There is a direct correlation between white fly population and virus spread. More the white fly population, more the disease incidence and *vice versa* (Singh, 1980, 1990).

**iv) Host Range:** Besides okra, the virus infects several species of *Hibiscus* and *Abelmoschus*. It commonly affects *Hibiscus tetraphyllus*, a weed in Deccan, and hollyhock, *Althea rosea* which acts as a collateral host for the virus (Uppal *et al.*, 1940; Capoor and Varma, 1950a).

**v) Management**

Kharif okra should be sown early between last week of May and second week of June under Konkan conditions for obtaining higher yield with less YVMV incidence. Removal of malvaceous hosts, and weeds (especially *Hibiscus tetraphyllus* and *Althea rosea*) around okra field helps in reducing the disease incidence. Rouging of diseased plants as soon as they are noticed in the field helps to reduce the disease incidence. Infested crop should not be used for seed collection. The appearance of YVMV can be delayed by mulching the soil with coloured polythene sheet.

Sowing of 4 to 5 rows of sorghum or pearl millet or maize all round the okra field at least 60 days before sowing okra has been found beneficial for the management of YVMV. Border cropping along with 3 to 4 foliar sprays of either dimethoate or monocrotophos both at 0.1% at 10 days interval has been found more effective (Singh, 1990).

Seed treatment with 0.025% imidachloprid is effective in controlling the vector. Carbofuran at 1.0 kg a.i./ha should be applied to the soil at the time of sowing seeds. The white fly population and yellow-vein mosaic virus incidence could be minimized significantly by 4 to 5 foliar sprays of either dimethoate or monocrotophos both at 0.1% at 10 days interval followed by 1 or 2 sprays 0.2% mineral oil (Sastry and Singh, 1983a; Singh *et al.*, 1977).

Use resistant okra cultivars like Arka Anamika, Arka Abhay, Pusa Sawani, Harbajan, Parbhani Kranti to manage YVMV.

**Integrated Methods**

- Using YVMV resistant varieties (Arka Abhay, Arka Anamika, Parbhani Kranti, Varsha Uphar) for planting.
- Destruction of YVMV affected plants from time to time.

### 3.7.2.2. Enation Leaf Curl Virus

The disease was first observed at Bangalore during 1986 by Singh and Dutta. This disease is most destructive virus disease which is severe during kharif season. The disease is caused by Gemini virus and transmitted by white fly.

**i) Symptoms:** The characteristic symptoms include thickening of the veins, warty and rough texture of leaves, leaf curling and twisting of the main stem and lateral branches. Twisting of leaf petiole was found to be prominent. In severe cases, newly emerged leaves also exhibit bold enations, thickening and curling. The virus infected plant is unable to bear its weight. The stem and leaf petiole are bent and zig zag in shape. All flowers do not set the fruits. If the fruit is set, seeds are not fertile and aborted. The characteristic small green outgrowths are observed on vein and vein lets on the lower side of the leaves (Singh and Dutta, 1986).

**ii) Transmission:** The virus is readily transmitted by grafting. The natural transmission occurs through white fly, *Bemisia tabaci* (Singh and Dutta, 1986).

**iii) Management**

- The infected plants should be rouged out as soon as they are noticed in the field.
- Early sowing of crop during end of June or 1st week of July.
- Removal of Malvaceous hosts around the field.
- The spread of the disease can be reduced by foliar sprays of dimethoate or monocrotophos both at 0.05% at 10 days interval.
- Okra cvs. VRO-5 and VRO-6 are tolerant to enation leaf curl virus.

## 3.8. CABBAGE AND CAULIFLOWER

Cabbage (*Brassica oleracea* var. *capitata*) is an important winter vegetable of cole group. It is a rich source of Vitamins A,B and C and also contains minerals. It is being cultivated in 3.121 million hectares producing 67.315 million tonnes of cabbages with an average yield of 21.568 tonnes per hectare (Bijay Kumar, 2007) (Table 3.15). It covers about 4% of total area under vegetables. India comes next to China in cabbage production. In India, it is being cultivated in 0.267 million hectares producing 5.922 million tonnes of cabbage with an average yield of 22.1 tonnes per hectare (Bijay Kumar, 2007) (Table 3.16).

**Table 3.15. Major cabbage producing countries of the world**

| Country | Area (in '000 ha) | Production (in '000 MT) | Productivity (in MT/ha) | % share of world production |
|---|---|---|---|---|
| China | 1719 | 33817 | 19.5 | 50.2 |
| India | 267 | 5922 | 22.1 | 8.8 |
| Russia | 166 | 4051 | 24.5 | 6.0 |

| | | | | |
|---|---|---|---|---|
| Korea | 42 | 2603 | 61.8 | 3.9 |
| Japan | 53 | 2287 | 42.9 | 3.4 |
| Ukraine | 71 | 1475 | 20.9 | 2.2 |
| Poland | 39 | 1382 | 35.6 | 2.1 |
| Indonesia | 64 | 1293 | 20.1 | 1.9 |
| USA | 30 | 1100 | 36.9 | 1.6 |
| Romania | 55 | 1012 | 18.5 | 1.5 |
| Others | 615 | 12375 | 20.1 | 18.4 |
| Total | 3121 | 67315 | 21.5 | |

Source: FAO (except Indian data) (Indian data source: Indian Horticulture Database 2006)

**Table 3.16. Area, production and productivity of leading cabbage growing states in India (2005-06)**

| State | Area (in'000 ha) | Production (in '000 MT) | Productivity (MT/ha) |
|---|---|---|---|
| West Bengal | 74.7 | 1982.7 | 26.5 |
| Bihar | 33.7 | 931.6 | 27.6 |
| Orissa | 36.6 | 578.3 | 15.8 |
| Assam | 24.3 | 480.3 | 19.8 |
| Maharashtra | 15.0 | 381.0 | 25.4 |
| Gujarat | 19.0 | 329.2 | 17.3 |
| Uttar Pradesh | 7.0 | 209.9 | 30.0 |
| Haryana | 12.3 | 172.2 | 14.0 |
| Karnataka | 6.4 | 134.2 | 20.8 |
| Chattishgarh | 6.8 | 114.0 | 16.8 |
| Madhya Pradesh | 3.8 | 75.0 | 20.0 |
| Others | 27.7 | 533.1 | 19.2 |
| Total | 267.4 | 5921.6 | 22.1 |

(Source: Indian Horticulture Database 2006)

Cauliflower (*Brassica oleracea* var. *botrytis*) is the most popular winter vegetable among cole crops. It is rich in Vitamin C. It is being cultivated in 0.292 million hectares producing 5.260 million tonnes of cauliflowers with an average yield of 18.0 tonnes per hectare (Bijay Kumar, 2007) (Tables 3.17 and 3.18). With the development of new varieties, it is now being grown in non-traditional areas like Andhra Pradesh, Tamil Nadu and Kerala.

**Table 3.17. Major cauliflower producing countries of the world**

| Country | Area (in '000 ha) | Production (in '000 MT) | Productivity (in MT/ha) | % share of world production |
|---|---|---|---|---|
| China | 363 | 7369 | 20.3 | 40.5 |
| India | 292 | 5260 | 18.0 | 28.9 |

| | | | | |
|---|---|---|---|---|
| USA | 69 | 1288 | 18.6 | 7.1 |
| Spain | 25 | 442 | 17.4 | 2.4 |
| Italy | 18 | 431 | 23.7 | 2.4 |
| France | 27 | 375 | 13.8 | 2.1 |
| Mexico | 25 | 359 | 14.2 | 2.0 |
| Poland | 14 | 239 | 17.0 | 1.3 |
| UK | 19 | 219 | 11.7 | 1.2 |
| Pakistan | 11 | 206 | 18.2 | 1.1 |
| Others | 130 | 1992 | 15.3 | 11.0 |
| Total | 995 | 18180 | | |

Source: FAO (except Indian data) (Indian data source: Indian Horticulture Database 2006)

**Table 3.18. Area, production and productivity of leading cauliflower growing states in India (2005-06)**

| State | Area (in'000 ha) | Production (in '000 MT) | Productivity (MT/ha) |
|---|---|---|---|
| West Bengal | 65.6 | 1666.2 | 25.4 |
| Bihar | 59.8 | 938.5 | 15.7 |
| Orissa | 45.1 | 638.9 | 14.2 |
| Maharashtra | 13.0 | 328.9 | 25.3 |
| Haryana | 21.0 | 302.8 | 14.4 |
| Gujarat | 16.1 | 277.8 | 17.2 |
| Assam | 16.7 | 247.8 | 14.8 |
| Madhya Pradesh | 8.6 | 137.4 | 16.0 |
| Uttar Pradesh | 6.7 | 134.4 | 20.0 |
| Punjab | 5.6 | 131.4 | 23.5 |
| Chattishgarh | 9.6 | 123.5 | 12.9 |
| Others | 24.0 | 332.7 | 13.9 |
| Total | 291.8 | 5260.1 | 18.0 |

(Source: Indian Horticulture Database 2006)

### 3.8.1. Bacterial Diseases

#### 3.8.1.1. Black Rot, *Xanthomonas campestris* pv. *campestris*

The disease was first observed in the USA on cabbage in 1891. In India, the disease was first described in 1928 on cabbage from Pune, Maharashtra. Now the disease is known to be prevalent in most of the crucifer growing areas of the country. Severe losses due to the disease in heavily manured soils near big towns is reported (Patel *et al.*, 1949a). Black rot also occurs on cauliflower, radish, turnip, knol khol and mustard.

**i) Symptoms:** The first sign of the disease appears near the leaf margins, characterized by chlorosis, which progresses towards the centre of the leaf blade in 'V' shape (Fig. 3.25). In affected portions, veins and vein lets turn brown and finally black. In case of severe infection, the leaves wither. The vascular blackening may extend to the main stem causing systemic spread of the infection. Infected plants become stunted, often one side of cotyledons turn yellow to black, bend down and drop-off prematurely. Many lower leaves are also shed off early. Young plants may be killed in the seed bed itself. In cabbage and cauliflower, if the heads are already formed, soft-rot of the head may set in.

**ii) Epidemiology:** Black rot is seed-borne. The pathogen survives for longer periods in naturally infected seed stored at room temperature or at lower temperature (10°C) (Bandopadhyay and Chattopadhyay, 1986a). The pathogen survives in infected plant debris in soil (Trivedi and Patel, 1972). The bacterium also survives in resident form on a weed, *Centenella asiaticum* (Kishun and Chand, 1988d). The pathogen spread to the neighbouring healthy plants by rain drops and enter the leaves through natural openings and wounds. Infected seeds and transplants are mainly responsible for widespread of the disease. Cabbage planted in September-October and January, and cauliflower in October showed higher disease incidence than cabbage planted in November and February, and cauliflower in February (Bandopadhyay and Chattopadhyay, 1986b).

**Fig. 3.25.** Black rot affected plants of cabbage and cauliflower.

### iii) Management

- Hot water treatment of seeds at 50°C for 30 minutes (Shekhawat *et al.*, 1982).
- Seed treatment with 0.1% mercuric chloride for 30 min, 100 ppm plantamycin (Kishun, 1981a, 1984a), 100 ppm agrimycin-100 (Rao and Kishun, 1986), 100 ppm streptocycline (Shah *et al.*, 1985) or 0.01% aureomycin (Trivedi and Patel, 1972).
- Drenching of seed beds with 5% commercial formalin or1% Bordeaux mixture.
- Spraying cauliflower with 100-200 ppm agrimycin-100 (Bera, 1986), 500 ppm carbendazim (Saini and Parashar, 1981); and cabbage with erythromycin (Mondel and Mukherjee, 1978) are found effective in reducing the disease.

- Three years crop rotation with non-host crops. Crop rotation with rice was found to be best in reducing the black rot disease.
- Strict field sanitation will help in checking the disease.
- Rouging of diseased plants helps in reducing the disease incidence.
- *Bacillus subtilis* was found effective as seed treatment, seedling root-dip treatment and soil drenching.
- Cauliflower cvs. Pant Shubra (Singh *et al.*, 1987), Pusa Ice (Sharma *et al.*, 1977) and Pusa Snowball K-1 are reported to be resistant.

### 3.8.1.2. Cauliflower Bacterial Leaf Spot, *Pseudomonas maculicola*

**i) Symptoms:** A bacterial disease of crucifers has on occasion been a severe problem in cauliflower. *Pseudomonas maculicola* especially attacks cauliflower but will also infect broccoli, Brussels sprouts, and other crucifers. Usually in July, leaves become covered with small (1 mm) irregularly shaped brown spots with yellow halos. This stage is difficult to see on the upper leaf surface, whereas the leaf underside clearly is "peppered" with water-soaked brown spots (Fig. 3.26).

**Fig. 3.26.** Advanced stages of bacterial leaf spot of cauliflower showing brown, papery areas surrounded by yellow leaf tissue. Note the holes in the leaf blade, giving the plant a ragged appearance.

Eventually the many small spots coalesce into large brown papery areas with yellow borders which tear, giving the plant a ragged appearance. Like those which cause black rot, leaf spot bacteria survive in diseased crop residue for at least two years, are carried by seed, and have similar methods of dispersal. However, *P. maculicola* is favoured by cool weather, developing optimally at 22-23°C.

**ii) Management:** Control measures for black rot are also effective for bacterial or peppery leaf spot of cauliflower.

### 3.8.1.3. Bacterial Soft Rot, *Erwinia carotovora*

**i) Symptoms:** Bacterial soft rot is caused by *Erwinia carotovora* and other secondary bacteria. Fleshy tissue is invaded through wounds caused by insects, man, machinery, cold temperature, wind-blown soil or hail and is rapidly decayed into a watery, stinking mess at high temperatures (Fig. 3.27). Cauliflower curd is sometimes invaded by both seed corn maggots and soft rot bacteria, although it is uncertain which agent is primary (Fig. 3.27). Broccoli heads having depressions in which water collects often become infected by soft rot bacteria in the field. Tip burn in cabbage heads can lead to subsequent breakdown by soft rot bacteria. Soft rot is a severe problem in storage and transit as well as in the field. Every effort should be made to minimize wounding during harvest and to avoid rinsing or chilling crucifers with unclean water.

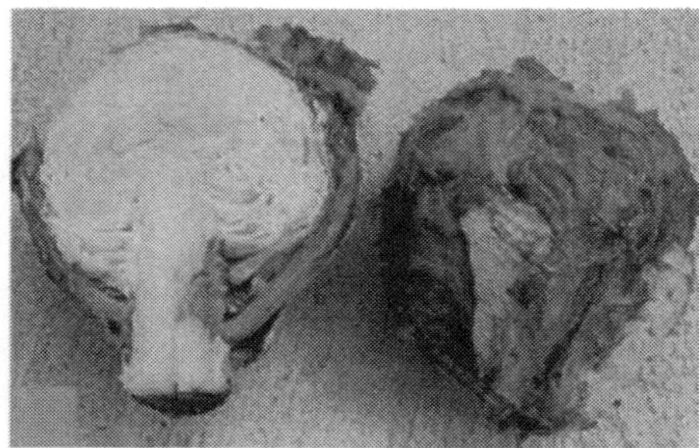

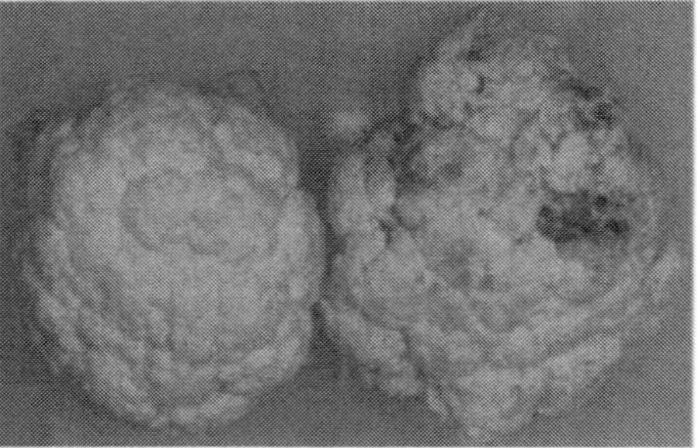

**Fig. 3.27. Left** - Brown, mushy, soft-rotted tissue on the cabbage head is caused by bacterial soft rot organisms. **Right** - Cauliflower head on right has bacterial soft rot bacteria and is also infested with seed corn maggot larvae in the brown, decayed pockets.

**ii) Integrated Management of Bacterial Diseases:** A generalized control program for bacterial diseases of cruciferous crops is as follows:

#### a) For Producing Transplants or Direct-Seeding

- Use seed which is certified, free from black rot bacteria, and has been treated with hot water or streptomycin to control seed-borne infection. Efforts to rid the seed of black rot bacteria have in the past involved mercuric chloride and hot water treatment. Streptomycin seed treatment is recommended for Brussels sprouts, broccoli, cauliflower, cabbage and rutabaga. Seeds are placed in cheese cloth bags and soaked in 500 ppm streptomycin solution for an hour, rinsed for two minutes in running tap water, re-soaked in 0.5% solution of a freshly prepared sodium hypochlorite (household bleach) solution for 30 minutes and rinsed again in running tap water for two minutes. The seeds are then dried.

- Hot water treatment, if carefully done, is still a very effective method of controlling not only seed-borne *Xanthomonas*, but also black leg and bacterial leaf spot of cauliflower. Even though hot water seed treatment may reduce germination, simply conduct your own germination test. Calculate the percentage germination after seed treatment, and sow extra seed to supply enough plants for your needs. Sow seed in soil which has been fumigated (seed beds) or sterilized (greenhouse). If fumigation is uneconomical, sow seedbeds in land which has not grown crucifers for at least two years.
- Be sure seedlings are not too densely planted, and provide optimum conditions of ventilation, watering, fertility, temperature, and light for growth.
- Locate field-grown seed beds away from existing crucifer crops and avoid run-off from the latter or land which was recently in cruciferous crops to prevent introduction of disease.
- Maintain weed-free seed beds.
- Inspect the seedlings regularly, removing and destroying localized infections ("hot spots") early.

**b) In Addition, Transplant or Direct Seed into Fields which have**

- Warm, well-drained soil, with pH greater than 7.2 if club root has been a problem.
- Adequate weed control in field headlands, especially of volunteer cruciferous crops, and cruciferous weeds, such as wild mustards, shepherd's purse, pepper-grasses, wild radish.
- Not grown crucifers for at least three crops, and which do not have evidence of former cruciferous crop residue on or in the soil.
- Continued use of pesticides as necessary for insect and disease control.
- Adequate sanitation practices such as removing infected plants if practical, removal and destruction of diseased crop refuse promptly, burying cull piles of crucifers remotely from growing areas, working in the field only when the foliage is dry and avoiding sprinkler irrigation of the diseased fields.

## 3.9. PEA

Pea (*Pisum sativum*) is an important vegetable crop widely grown throughout the world. As a cool season crop, it is extensively grown in temperate zone; but restricted to cooler altitudes in the tropics and winter season in the sub-tropics. A rich source of proteins (25%), amino acids and sugars (12%), green peas are an all time favourite vegetables. It is rich in Vitamins A,B,C and minerals. It is being cultivated in 0.314 million hectares producing 2.299 million tonnes of peas with an average yield of 7.3 tonnes per hectare (Bijay Kumar, 2007) (Tables 3.19 and 3.20).

**Table 3.19. Major Peas Producing Countries of the World**

| Country | Area (in '000 ha) | Production (in '000 MT) | Productivity (in MT/ha) | % Share of world production |
|---|---|---|---|---|
| India | 314 | 2299 | 7.3 | 29.5 |
| China | 241 | 2210 | 9.2 | 28.3 |
| USA | 86 | 859 | 10.0 | 11.0 |
| France | 31 | 360 | 11.7 | 4.6 |
| Egypt | 28 | 290 | 10.4 | 3.7 |
| UK | 32 | 133 | 4.2 | 1.7 |
| Turkey | 17 | 122 | 7.0 | 1.6 |
| Algeria | 24 | 110 | 4.5 | 1.4 |
| Hungary | 14 | 85 | 6.2 | 1.1 |
| Netherlands | 5 | 84 | 16.5 | 1.1 |
| Others | 303 | 1248 | 4.1 | 16.0 |
| Total | 1094 | 7799 | | |

Source: FAO (except Indian data) (Indian data source: Indian Horticulture Database 2006)

**Table 3.20. Area, production and productivity of leading peas growing states in India (2005-06)**

| State | Area (in'000 ha) | Production (in '000 MT) | Productivity (MT/ha) |
|---|---|---|---|
| Uttar Pradesh | 139.0 | 1146.8 | 8.3 |
| Madhya Pradesh | 19.4 | 212.9 | 11.0 |
| Himachal Pradesh | 16.3 | 177.0 | 10.9 |
| West Bengal | 21.7 | 129.4 | 6.0 |
| Jammu & Kashmir | 7.3 | 112.2 | 15.4 |
| Punjab | 17.2 | 103.3 | 6.0 |
| Haryana | 11.1 | 84.9 | 7.6 |
| Uttarakhand | 9.6 | 60.9 | 6.3 |
| Chattishgarh | 7.2 | 50.8 | 7.1 |
| Bihar | 8.1 | 50.5 | 6.2 |
| Orissa | 4.8 | 42.0 | 8.7 |
| Others | 52.4 | 128.0 | 2.4 |
| Total | 314.0 | 2298.7 | 7.3 |

(Source: Indian Horticulture Database 2006)

### 3.9.1. Bacterial Diseases

#### 3.9.1.1. Bacterial Blight, *Pseudomonas pisi* (*Pseudomonas syringae* pv. *pisi* and/or *P. syringae* pv. *syringae*)

**i) Symptoms:** Small, dark-green marks that look soaked with water appear on the leaflets (Fig. 3.28). These lesions enlarge and merge together (coal-esce), but are limited in their expansion by the leaf's veins.

The lesions on leaves turn yellowish and are later brown and distinctly papery. Lesions on pods are sunken and turn an olive-brown (Fig. 3.28). Lesions can also develop on stems near ground level, beginning as 'water-soaked' areas. When these lesions coalesce they cause the stem to shrivel and die.

**Fig. 3.28.** Bacterial blight on pea leaves and pods.

Stem infection may spread upwards into the leaves and stipules. Infected seedlings are killed very readily, but even advanced plants can be killed by the disease.

Heavily infected seed may be discoloured. However, there is no visible effect on seed with light infection.

**ii) Spread:** Spread of the bacteria is by seed or survival in pea trash. Rain-splash and wind-borne water droplets can also spread the bacteria.

Excessive rainfall and strong winds are the most favourable environmental conditions. And plants damaged by frosts and heavy rain are most susceptible to attack.

The disease is seed-borne. High humidity and rains facilitate disease development. The optimum temperature for bacterial growth is about 27.8° C.

**iii) Management**

- Blight is most severe in wet seasons, so be vigilant when conditions are most favourable to the disease.
- Purchase clean seed and do not use seed from plants that have any disease symptoms.
- Seed treatments are not effective for the control of bacterial blight.
- Do not grow peas in the same place each year, or even adjacent to an area of pea stubble. In fact, peas should not be grown in the same place more than once in three years.
- Early sown crops are more vulnerable than late sown crops. Never sow earlier than recommended for your district.
- Bacteria will survive on pea stubble, so if infection is suspected, bury the pea trash 10 cm below the soil surface, which reduces the survival time of the bacteria significantly.

- Remove all infected plant parts, bag and put in the rubbish bin.
- Peas should then only be grown on a rotation of one in four years.
- There are copper-based compounds available for the control of this disease in field.

### 3.9.2. Viral Diseases

Peas are susceptible to a large number of aphid-transmitted viruses, which can produce diseases individually or in combination. The main viruses infecting pea are pea enation mosaic, pea streak and bean yellow mosaic.

#### 3.9.2.1. Pea Mosaic

**i) Symptoms:** The diseased plants are pale, weak and dwarfed. The young leaves show a general mottling whereas the older ones display distinct irregular whitish areas. The tendrils are curled abnormally and the apical leaves remain folded. Invariably small specks of light brown discoloured areas are seen on petioles, stems and tendrils. The diseased plants flower later than the healthy ones and the flowers are smaller in size and fewer in number. Comparatively fewer pods are set on diseased plants and they are smaller in size and contain a few seeds (Srinivasan and Nariani, 1966).

**ii) Transmission:** The disease is caused by marble pea mosaic strain of pea virus 2A. The virus is mechanically sap transmissible. Six species of aphids viz., *Aphis craccivora, A. gossypii, A. evonymi, Myzus persicae, Rhophalosiphon pseudobrassicae* and *R. maidis* transmit the virus.

**iii) Host Range:** Besides peas, the virus infects *Vicia faba, V. arbonensis, V. sativa, V. binensis, Lathyrus odoratus, L. ochrus, L. sativus, Trigonella foenum-gracum, T. corriculata, Crotolaria juncea, Lupinus augustfolius, L. album, Cicer arientum* and *Trifolium subterraneum*. Local lesions are produced on *C. amaranticolor* host.

**iv) Management**

Pea cvs. Canners Perfection, Horal, Hundred Fold and Little Marvel are immune to this virus. The pea varieties Dwarf and Kelvedon Wonder are found to be highly resistant.

#### 3.9.2.2. Pea Seed-Borne Mosaic

**i) Symptoms:** The main symptoms are mild to severe mosaic, pale plants, deformed and reduced leaf lamina. In most cases, the leaves have deep indentations followed by marginal rolling and leaf curling. Flowering is delayed/reduced and consequently pods are few, small and distorted (Thakur *et al.*, 1985).

**ii) Transmission:** The virus is sap transmissible. In nature, it is transmitted by aphids such as *Aphis craccivora, Acyrthosiphum pisum* and *Myzus persicae*. The virus is also transmitted by seed to an extent of 20-40%.

**iii) Management:** Selection of virus-free seeds from healthy plants.

### 3.9.2.3. Pea Enation Mosaic Virus (PEMV)

**i) Symptoms:** Infected pea plants develop mosaic and chlorotic vein flecking (appears as translucent windows) and veinal enations (blister like outgrowths), which are very characteristic for pea enation mosaic virus. Plants are stunted, and proliferation of basal branches is common. Pods are distorted, split open, and may show prominent enations.

**ii) Host Range:** In addition to pea, PEMV also infects broad bean, sweet pea, and alfalfa. It probably over seasons in many common perennial legumes.

**iii) Transmission:** The virus is spread in nature most efficiently by the pea aphid (*Acyrthosiphon pisum*) and to a lesser extent by the green peach aphid (*Myzus persicae*). The virus is transmitted in a persistent (circulative) manner.

**iv) Management**

- Control aphids.
- Plant resistant varieties, if available.
- Remove perennial legumes bordering the planting area to reduce the primary virus reservoir.

### 3.9.2.4. Bean Yellow Mosaic Virus (BYMV)

Bean yellow mosaic virus is transmitted by at least 20 aphid species in a non-persistent manner. The symptoms depend upon the strain of virus involved. The pea strain, for example, causes a very bright yellow mosaic whereas the more typical isolates produce a dull light and dark green mosaic. Symptoms are masked or delayed at low temperatures (below 15.6°C), but develop normally at 18.3° to 23.9°C. Seed transmission is absent or very rare in pea and several other legumes.

**Management**

- Control aphids.
- Plant resistant varieties, if available.
- Remove perennial legumes bordering the planting area to reduce the primary virus reservoir.

## 3.10. FRENCH BEAN

French bean (*Phaseolus vulgaris*), also known as common, haricot, kidney, string, salad, runner or snap bean, is an important leguminous vegetable. It is consumed as tender pods, shelled green beans and dry beans. It is a nutritious vegetable which contains proteins (1.7g), calcium (50 mg), phosphorus (28 mg), iron (1.7 mg), carotene (132 mg), Thiamine (0.08 mg), Riboflavin (0.06 mg) and Vitamin C (24 mg/100 g of edible pods). It is largely grown in hilly areas of Himachal Pradesh, Jammu and Kashmir and north-eastern states during summer, winter and autumn crop in parts of Uttar Pradesh, Maharashtra, Karnataka and Andhra Pradesh.

### 3.10.1. Bacterial Diseases

#### 3.10.1.1. Common Bacterial Blight, *Xanthomonas campestris* pv. *phaseoli*

Common blight affects the foliage and pods of beans and causes significant losses in both yield and seed quality. The disease typically develops when contaminated seed is planted, when plantings are made in fields with a history of the disease, and when the climate is consistently hot and wet or humid.

**i) Symptoms:** Common and fuscous blights are commonly first seen as small, angular, light green, water-soaked or translucent spots (lesions) on the leaves. During warm, wet conditions the lesions rapidly enlarge and merge. As they develop the centers become dry, brown, and surrounded by a distinct, narrow, zone of yellow tissue. In highly susceptible varieties, the lesions continue to expand until the leaves appear scorched or sun scalded (Fig. 3.29). Such leaves soon become ragged and torn by wind and rain. Later, they wither and drop off.

Pod lesions start as round, water-soaked dots that enlarge, merge, dry, and form sunken, irregular, frequently reddish brown blotches (Fig. 3.29). When severe, entire pods may be badly shriveled and die. Seeds in such pods either fail to develop or are shriveled. Seeds in less severely affected pods develop normally and show no signs of disease. Other seeds may become slightly wrinkled. When seeds containing the bacteria are planted, many fail to germinate. Those that do germinate produce seedlings with blight lesions on the cotyledons, stems, and first true leaves.

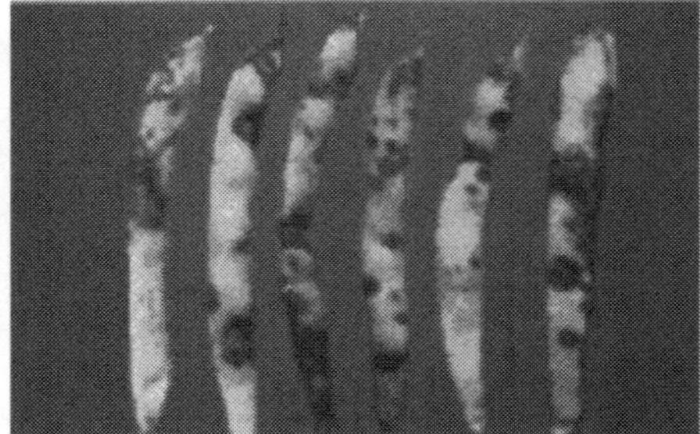

**Fig. 3.29.** Common bacterial blight of bean

The stems of seedlings may have water-soaked, sunken areas that enlarge and develop into reddish streaks. Any time during the season affected stems commonly crack and become girdled by water-soaked cankers or rot. The tops may break over during a rain or strong wind. In humid weather, a yellowish bacterial ooze, that later dries to form a crust, may be evident on the lesions on infected pods, leaves, stems, and cotyledons.

**ii) Host Range:** Common bacterial blight, caused by the bacterium *Xanthomonas campestris* pv. *phaseoli*, affects green or snap, wax, field, and lima beans as well as the scarlet runner, mung, Tepary, urd, moth, hyacinth, and civet or Sieve beans, the Washington or white-flowered lupine, and fenugreek (*Trigonella*). Fuscous blight, caused by fuscous variant of the bacterium, occurs on field beans, civet, and scarlet runner beans.

### 3.10.1.2. Halo Blight, *Pseudomonas syringae* pv. *phaseolicola*

Halo blight, caused by the bacterium *Pseudomonas syringae* pv. *phaseolicola*, affects green or snap, wax, lima, scarlet runner, and civet beans, and kudzu.

**i) Symptoms:** Leaf symptoms appear several days after infection as small (1/8 to 1/4 inch, or 3 mm), water soaked, tan brown, angular spots, each surrounded by a relatively wide, diffuse, pale green to yellow green tissue (halo) (Fig. 3.30). Halos do not develop or tend to disappear if the temperature is above 21 to 23°C. The spots then become reddish brown to brown, and dry. Infection foci generally remain small. In cases of severe leaf infection, plants develop a generalized systemic chlorosis. The systemic chlorosis is particularly pronounced at 18 to 23°C. Leaves with systemic infections lack halos. The intervinal tissues of these leaves are yellow, with the veins remaining dark green. At 7-10 days after infection, bacteria ooze from the sub stomatal cavities to give lesions a greasy, water-soaked appearance. Bacteria are thus available for secondary spread of the disease.

Infected plants may defoliate, wilt, and die. Diseased plants may be dwarfed, with the upper leaves crinkled and mottled. Dark green, water-soaked areas may also be visible on the stems.

Young seedlings infected with halo blight develop a chlorosis (yellowing), which may be general or confined to the young trifoliate leaves at the top of the plant. The chlorosis is caused by a toxin produced by the halo blight bacterium.

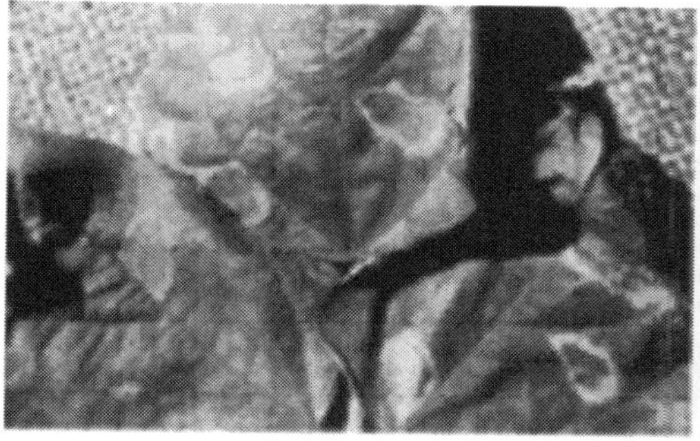

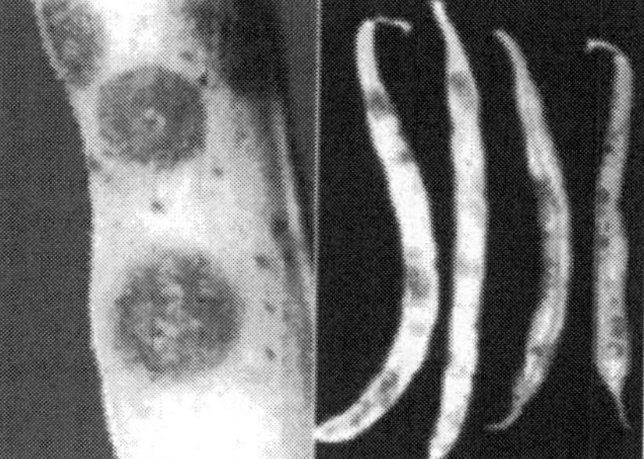

**Fig. 3.30.** Halo blight of beans. Left, Lesions and bacterial crust of halo blight on pod (Courtesy A.W. Saettler); Ceter, Halo blight on pods (Courtesy A.F. Sherf).

Pod infections develop rapidly under cool, wet conditions resulting in oval to circular, dark green, water-soaked, "greasy" lesions of 3/8 inch (9 mm) in diameter. Halos do not develop around pod lesions (Fig. 3.30). As pods mature and turn yellow, pod lesions may remain green and may exhibit crusty bacterial ooze on the surface. Developing seed may be shriveled or discoloured if lesions expand to involve the pod surface. With age the lesions become slightly sunken and reddish brown.

### 3.10.1.3. Bacterial Brown Spot, *Pseudomonas syringae* pv. *syringae*

Bacterial brown spot is caused by the bacterium *Pseudomonas syringae* pv. *syringae*.

**i) Symptoms:** On lima beans, small, light gray spots with yellowish brown borders develop on the upper leaf surface (Fig. 3.31). Water-soaking is not as pronounced as it is in common and halo blights. Later, the gray centers may crack open or drop out, leaving small holes. The veins on the under leaf surface turn red or reddish brown. The spots on the stems and pods are somewhat more elongated than those on the leaves. In severe cases much of the foliage may be killed.

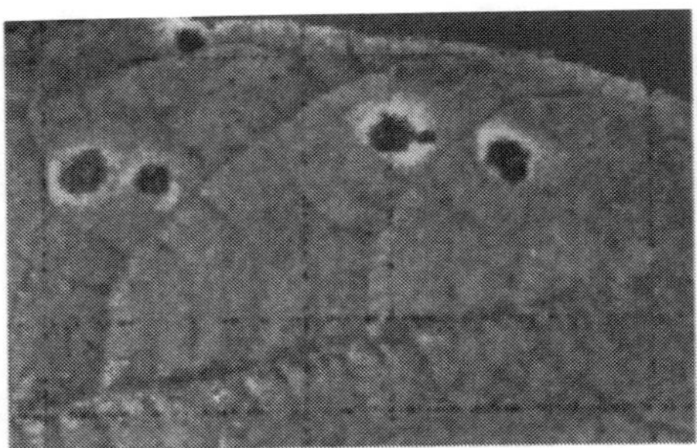

**Fig. 3.31.** Bacterial brown spot on leaf and pods, caused by *Pseudomonas syringae* pv. *syringae*. (Courtesy A.W. Saettler, H.F. Schwartz).

On green and wax bean leaves, small (1/8 to 3/8 inch or 3 to 9 mm in diameter), oval, brown lesions surrounded by a narrow, yellow-green halo develop. The lesions may merge and their centers fall out giving leaves a tattered appearance. When the bacterium becomes systemic, tan sunken lesions with a reddish brown margin form on the stems and petioles. Bacterial exudate (ooze) and water-soaking are rare prior to lesion development. Pods develop somewhat circular, dark green, water-soaked areas which gradually enlarge, suddenly become depressed and tan, then dark brown with a distinctive, reddish brown margin. Pods infected when young are bent at an acute angle or are twisted. Bacterial ooze is white to cream coloured.

**ii) Host Range:** In addition to attacking lima, green or snap, and wax beans; the bacterium infects a wide range of host plants. These hosts include civet, hyacinth, yard-long, and broad beans, soybeans, cowpea, clovers, alfalfa, corn, sorghums, Sudan grass, buck wheat, kudzu, lilac, ash, poplar, Forsythia, wild and cultivate stone fruits (*Prunus* spp.), apple, pear, *Citrus* species, rose, flowering stock, hibiscus, jasmine, Florida velvet bean, and many others. Fortunately only bean isolates are highly virulent on beans.

### iii) Disease Cycle of Bacterial Diseases

The disease cycles of all three pathogens are very similar. The bacteria are all disseminated in infected seed and from plant to plant and field to field by wind-driven and splashing rains, sprinkler irrigation, surface-drainage water, insects, birds, large animals, humans, farm machinery, tools, and other agencies. The bacteria may survive for 6 to 18 months in plant refuse (on or above the soil surface and under dry conditions), in bean cull piles within or near fields, on volunteer plants from a previous crop, and even on the surface of weeds. Long-distance spread is generally by seed, but there is evidence that minute aerosol particles from rain splash carrying the bacteria can float in the air for more than two hours. Wind-borne particles carrying bacteria could move many miles during rain storms.

There are two types of field infection. Primary infection involves the invasion of bean seedlings as they emerge from the soil. This occurs when the seed is infected or when an emerging seedling comes into contact with infected plant material. Secondary infection is due to spread from a primary source to other growing plants; often involving only one or two isolated plants, or it can cover an entire field or area. In severely diseased fields nearly all infection is secondary.

Bacteria enter a plant through natural leaf openings (stomata and hydathodes), and through wounds created by hail, blowing soil particles, insects, or cultivator injury. The brown spot organism often infects the leaves through rust pustules; thus, rust and bacterial brown spot are often found together in the same lesion.

When an infected seed sprouts and the seedling emerges, bacteria ooze to the surface of the diseased cotyledons, are splashed to a neighbouring plant, enter the stomata, and infect. Stomata or hydathode entry also occurs from bacteria that survive on surface debris and are rain splashed to the leaves and stems. Symptoms appear two to five days after penetration for common, fuscous, and halo blights. Once inside the plant the bacteria may move systemically, by way of the vascular bundles, to the leaves, stems, pods, and into the seed. The bacteria are carried by rain splash to neighbouring leaves and pod surfaces. They invade growing pods largely along the suture, and enter the seed through the vascular tissue. The bacteria remain on or just below the surface of the seed. The bacterial pathogens have been recovered from 3- to 14-year-old seed. The halo blight organism was found to be viable after passing through sheep fed on diseased bean plants.

Usually only a few seeds in a seed lot serve as a source of primary inoculum. One infected seed in 16,000 is thought to be sufficient to supply inoculum for a severe outbreak of halo blight where essentially every plant in one or more fields becomes infected. Seed from clean fields can be surface contaminated by small dust particles which adhere to the surface of the seed during threshing and milling.

Common blight and bacterial brown spot are favored by cloudy damp weather and relatively high air temperatures (28 to 32°C), while halo blight thrives under damp and cooler conditions (18 to 22°C).

**iv) Management of Bacterial Diseases**

Plant only certified, pathogen-free seed grown in semi-arid areas of Idaho and California. In these areas the humidity and rainfall is normally too low for the bacteria to infect plants. The use of pathogen-free seed cannot be over emphasized. All seed stock of reputable companies is treated with streptomycin to reduce or eliminate surface contamination. Do not enter fields or gardens and cultivate or handle plants wet with dew or rain. Use a suggested herbicide to keep fields and border areas weed-free.

Practice good sanitation. Whenever feasible, collect and burn (or cover completely by thorough disking followed by deep ploughing) all infected plant debris as soon after harvest as possible. Fields subject to wind and water erosion should be planted to a nonsusceptible cover crop before winter. Planters, harvesters and other equipment should be sanitized by spraying with a disinfectant such as chlorine dioxide, sodium hypochlorite, or a quaternary ammonium compound before moving from an infected to a blight-free field. Storage areas should also be sanitized.

Grow beans in the same area or field only once in three years or longer. In the rotation exclude all beans, soybeans, and cowpeas, or other plants affected by one or more of these diseases.

Foliar sprays with a fixed copper compound, applied weekly starting four days after seedling emergence or at the first sign of disease have given fair to good control when continued to harvest. Thorough coverage of the plants is required. Properly applied aerial sprays are superior to applications by ground equipment. Sprays provide good control of bacterial brown spot and halo blight, but only moderate control of common blight.

Efforts to develop bean varieties resistant to bacterial blights with acceptable quality and other characteristics have not been very successful partly because of the appearance of new races of the organisms. Grow bean varieties recommended for your area that are somewhat resistant to all three bacterial diseases.

### 3.10.2. Viral Diseases

#### 3.10.2.1. Common Mosaic Virus

This is one of the most important virus diseases among the legumes in India.

**i) Symptoms:** The important symptoms of the disease are severe mosaic mottling of the leaves and various types of malformations. Trifoliate leaves affected with mosaic usually have irregular shaped, light yellow and green areas of various sizes, a type of mottling characteristic of mosaic in general. Infected leaves may be narrower than the normal ones, with downward cupping caused by unequal growth of the tissue. Infected plant produce very few, small and abnormal fruits.

**ii) Transmission:** The virus is transmitted by mechanical sap inoculation, pollen and infected seeds. Several species of aphid vectors viz., *Aphis gossypii, A. medicaginis, A. rumicis, A. spiraecola, Brevicoryne brassicae, M. acrosiphum pisi, M. gei* and *Myzus persicae* transmit the virus under natural conditions.

**iii) Host Range:** The leguminous plants susceptible to infection to virus are *Phaseolus acutifolius* var. *latifolius, P. lunatus, P. calcaratus, P. aureus, Lagenaria macrocorpus, Vicia faba, V. lathyrodaes, Lespedesa striata* and *Melilotus alba*.

**iv) Management**

- Select virus-free seeds for sowing.
- Rogue infected plants from the field.
- Remove weed hosts in and around the bean field.
- Apply carbofuran at 1.0 kg a.i./ha at the time of sowing followed by 3 to 4 sprays of dimethoate or monocrotophos both at 0.1% at 10 days interval (Singh *et al*., 1981).
- Use resistant French bean cvs. GM 108, 1140, Silver, PI 217179, Sinilac, Sea Farer and Idle Refugi (Sastry *et al*., 1981).

### 3.10.2.2. Golden Mosaic Virus

This is a very widespread and serious disease of beans and other legumes. This disease was first reported by Capoor and Varma (1948) from India. As much as 70-80% losses in yield have been observed in case of early infection (Singh and Sastry, 1980a).

**i) Symptoms:** The disease is characterized by scattered, slightly discoloured patches on leaves, which gradually turn bright yellow (Fig. 3.32). At times, the whole leaf turns golden yellow. The chlorophyll of the leaves is partially or completely destroyed. Affected plants are usually dwarf and pod production is greatly reduced.

**Fig. 3.32.** Yellow vein mosaic virus infection on French bean leaves.

**ii) Transmission:** The virus is readily transmitted by bud grafting. In nature, the disease is transmitted by the white fly, *Bemisia tabaci*.

**iii) Host Range:** Besides bean, the virus infects *Phaseolus lunatus, P. limensis, P. aureus, Dolichos biflorus* and *Canavalia ensiformis*.

**iv) Management**

- Remove the infected plants from the field.
- Soil application of carbofuan, fensulfothion or disulfoton or disyston or phorate all at 1.5 kg a.i./ha at the time of sowing followed by 3-4 foliar sprays of dimethoate or monocrotophos both at 0.05% at 10 days interval significantly controlled the whitefly population and the disease spread. Spraying of 2% mineral oil has proved very effective in checking the spread of the disease (Singh *et al.*, 1981).

### 3.10.2.3. Seed-Borne Mosaic

**i) Symptoms:** Field symptoms include chlorosis of primary leaves whereas mild mosaic is produced on trifoliate leaves. Infected leaves become rough, leathery, crinkled and puckered with downward curling of tips and margins. Severely affected plants remain stunted with a lighter foliage colour. Premature flower drop occurs in case of early infections. At times pods are distorted with deformed seeds. Severe loss in pod yield occurs in case of early infection (Lakshmi, 1990; Bhattiprolu, 1991).

**ii) Etiology:** The disease is incited by a strain of cucumber mosaic virus (Bhattiprolu, 1991).

**iii) Transmission:** The disease is readily transmitted by sap to French bean. Aphids namely *Aphis craccivora* and *Myzus persicae* transmit the virus in a non-persistent manner. The disease is highly seed transmitted, at times 100%. However, seed transmission vary in different cultivars. Seeds near the pedicel show higher levels of virus transmission than the upper ones.

**iv) Host Range:** Besides legumes, the virus infects plants belonging to members of Cucurbitaceae and Solanaceae (Table 3.21).

**Table 3.21. Reaction of plant species found susceptible to seed-borne CMV of French bean**

| Family | Plant species | Symptoms |
|---|---|---|
| Chenopodiaceae | *Chenopodium amaranticolor* | Chlorotic lesions. |
| Cucurbitaceae | *Cucumis melo* | Mosaic. |
| | *C. sativus* | Mosaic, crinkling and upward curling of leaves. |
| Leguminaceae | *Phaseolus acontifolia* | Prominent veinal necrosis only on inoculated primary leaves, followed by interveinal chlorosis of uninoculated trifoliate leaves. |
| | *P. lunatus* | Mild mosaic. |
| | *Vigna angularis* | Chlorosis of leaves. |

| | | |
|---|---|---|
| | *V. umbellate* | Faint mosaic mottling. |
| | *V. unguiculata* cv. C 152 | Pin point necrotic as well as chlorotic local lesions followed by systemic mosaic. |
| Solanaceae | *Nicotiana glutinosa* | Faint chlorosis, reduced leaf lamina. |
| | *N. tabacum* cv. Xanthi | Faint chlorosis, but no reduction in leaf size. |
| | *N. tabacum* cv. Godavari Special | Faint chlorosis. |

**v) Management**

- Use of virus-free seeds.
- Control of aphid vectors.

### 3.10.2.4. Leaf Crinkle

**i) Symptoms:** The disease is characterized by expansion and crinkling of the leaves and malformation of the inflorescence. In later stages, the leaves show conspicuous crinkling. Secondary branches develop numerous short but thick internodes. Leaf stalks too become conspicuously short and thick. Inflorescence becomes malformed and sterile. Floral parts develop into leafy structures (Singh, 1982).

**ii) Transmission:** The virus is easily sap transmissible. Seed transmission up to 17% has been reported. Two species of beetles – *Epilachna vignitioctopunctata* and *E. dodcatigma* transmit the virus.

**iii) Host Range:** The virus can infect *Vigna mungo, V. radiata, V. unguiculata* and *V. aconitifolia.*

**iv) Management**

- Rouging of infected plants from the field.
- Selection of seeds from apparently healthy plants.

### 3.10.2.5. Southern Bean Mosaic

**i) Symptoms:** The first systemic symptoms are characterized by a mild mottle of the trifoliate leaves similar to that caused by bean common mosaic virus (BCMV). The mottling increases in intensity about the time the plants reach maturity. Vein banding is common, The interveinal tissues being lighter green than the tissue adjacent to the veins. The leaves may be puckered and blistered. Reduction in leaf size and malformation of the leaf are noted on very susceptible varieties. The symptoms on pods are more pronounced than those produced by BCMV. They appear as irregular-shaped, dark green, water-soaked blotches on green podded varieties and the greenish yellow areas on wax-podded varieties.

**ii) Transmission:** The virus is highly infectious. It can be transmitted by touching the infected plants. The virus is seed-borne in beans. A high percentage of immature seed embryos of bean carry the virus. The virus is also transmitted by bean leaf beetle, *Ceratoma trifurcata* 48 hr after acquisition feeding period.

**iii) Management**

- The virus in the embryo can be inactivated by very rapid dehydration of the seeds.
- Selection of seeds from healthy plants.
- Use of resistant/tolerant varieties like Blue Lake, Decature, Great Northern UI Nos. 15, 29 and 123, Kentucky Wonder, Kentucky Wonder White, Pinto UL Nos. 16, 21 and 81.

### 3.10.2.6. Phyllody

**i) Symptoms:** The disease is characterized by transformation of floral parts into green leafy structures. The infected plants are stunted. The pods proliferate too much and there is an excessive proliferation of branches' which gives witches' broom appearance to the plant. Leaves are drastically reduced in size. A few of the flowers on an infected plant look yellow in colour but do not open and at maturity give rise to abnormal seedless pods.

**ii) Etiology:** Mycoplasma-like organisms are associated with the disease. They vary in shape viz., spherical, ellipsoidal and filamentous and measure from 60 to 1000 nm in diameter. They are abundantly present in the phloem sieve elements.

**iii) Transmission:** The disease is transmitted by grafting but not by mechanical sap inoculation. No insect vector is known yet.

**iv) Host Range:** The pathogen can infect *Phaseolus vulgaris, P. mungo, P. aureus, P. calcartus, Vigna sinensis* and *V. reticulata* by grafting.

**v) Management:** The disease symptoms can be suppressed temporarily by foliar application of 500 ppm oxytetracycline hydrochloride solution at weekly intervals (Singh, 1991).

## 3.11. COWPEA

Cowpea (*Vigna unguiculata*) also known as southern pea and black-eye pea, is one of the most important vegetables. It is cultivated for its long, green or purplish pods to be cooked as vegetable or for dry seeds used as pulse. Its foliage is used as fodder or green manure. In India, cowpea is grown almost throughout the country.

### 3.11.1. Bacterial Diseases

### 3.11.1.1. Bacterial Blight, *Xanthomonas campestris* pv. *phaseoli/ vignicola*

Bacterial blight is the major disease particularly in cowpea. Presently, the disease is prevalent in most of the cowpea growing areas in India. The disease is responsible for losses in yield up to 90% (Kishun, 1989). The disease appears during the months of July to September in rainy season.

**i) Symptoms:** This disease appears as tan to brown angular leaf spots with yellow margins on leaves, pods, and stems. Infected cotyledons become red and shriveled. In case of severe infection, the cotyledons, primary leaves and growing points are killed in the initial stage. Small, translucent, water-soaked spots appear on leaves. It may cause severe defoliation during periods of high humidity. As the spots develop, the centre turns yellow to brown and dries with a narrow, yellow halo around it. Reddish streaks and cankers appear on stem, reducing plant vigour and yield.

**ii) Epidemiology:** The pathogen is seed-borne. The bacterium survives in seed up to 250 days (Shah *et al.*, 1989). The disease spreads up to 62% plants with an initial inoculum load of 1% infected seed at 20-80% RH and temperature range of 25-34°C (Shekhawat and Patel, 1977). Intermittent rain has been found favourable for the disease development (Shah *et al.*, 1989). Cowpea plants were found most susceptible up to the age of 30 days (Kishun, 1989).

**iii) Management:** Use of healthy disease-free seed for sowing and clean cultivation helps to reduce the disease incidence. Proper drainage should be maintained in the field to avoid water stagnation. Three years crop rotation with non-host crops is essential and sufficient to dispose off over wintering pathogen.

Soak seed in a solution of 0.01% streptomycin and 0.25% copper oxychloride for 4 hr before sowing followed by spraying the crop during the rainy season. Pre-soaking of seeds at 15°C for 20 hours and then immersing them in water maintained at 30-60°C for 25 minutes and clean cultivation gave effective control of the disease.

Spraying of 1% Bordeaux mixture; 0.1% carbendazim + 250 ppm agrimycin-100 followed by 250 ppm agrimycin-100 are effective (Chakravarti *et al.*, 1978). Kishun *et al.* (1980) obtained good control by seed treatment with 100 ppm plantomycin followed by 2 sprays of 100 ppm plantomycin or 0.3% copper oxychloride.

Cowpea cvs. S-288 (Sreekantaradhya *et al.*, 1977), Iron (Singh and Patel, 1977), TVU 200, 347, 401, 456, 580, 726, 1460, 2460 and 4540 (Kishun *et al.*, 1980), Iron Grey (Gour *et al.*, 1981), Pundulea 68-5939 and Bistremiki were reported resistant.

### 3.11.2. Viral Diseases

#### 3.11.2.1. Cowpea Mosaic Virus

**i) Symptoms:** The virus causes mosaic, mottling, vein banding of leaves and stunting of plants (Govindaswamy *et al.*, 1970). At times, blisters are also seen on the infected leaves.

**ii) Transmission:** The virus is easily transmitted through sap. Seed transmission is up to 17.5%. Aphids mainly *Myzus persicae, Aphis fabae, A. craccivora* and *A. gossypii* transmit the virus in stylet-borne manner.

**iii) Management**

- Since the disease is seed-borne, use seeds collected only from healthy plants for sowing.
- Apply carbofuran at 1.0 kg a.i./ha at the time of sowing followed by 3 to 4 foliar sprays of dimethoate or monocrotophos both at 0.1% at 10 days interval to control aphids.
- Spraying mineral oils help in reducing the spread of the virus.
- Use resistant cowpea varieties such as P-340, P-1093, T-2, C-2509, C-42 and EC 78.

### 3.11.2.2. Cowpea Yellow Fleck

**i) Symptoms:** This is a white fly transmitted disease caused by Gemini virus. Yellow and green mosaic symptoms are seen on the apical leaves. Later, most of the leaves turn completely yellow in colour. Infected pods turned yellow from normal green colour. Pod setting is also reduced.

**ii) Transmission:** The virus is readily transmitted by grafting. In nature, the white fly, *Bemisia tabaci* transmit the disease up to 20%.

**iii) Management:** Rouging should be practiced at the initial stage. Soil application of carbofuran at 1.5 kg a.i./ha is recommended. Foliar spray of 0.05% monocrotophos at 10 days interval up to flowering stage is effective. Application of mineral oils has been reported effective in reducing the spread of the disease.

Cowpea cvs. Arka Garima, Pusa Parvati, Contender, Kentucky Wonder are reported to be resistant (Chakraborty *et al.*, 1999).

### 3.11.2.3. Cowpea Banding Mosaic Virus

**i) Symptoms:** Mosaic mottling accompanied by crinkling and reduction in leaf size and typical green bands along the veins are the symptoms observed under field conditions. The diseased plants remain green for longer period and set fewer pods.

**ii) Transmission:** The virus is easily sap transmitted. Addition of reducing agents increases the percentage of transmission. Seven aphid species transmit the virus. *Aphis craccivora* is the most efficient vector. Seed transmission up to 3% has been observed.

**iii) Management**

- Use disease-free seeds for sowing.
- Hot water treatment of seeds at 65°C for 15 min followed by 2, 4 or 6 days incubation at 30°C.
- Cultural practices like time of sowing, mixed cropping, barrier cropping and density of plant population help in reducing the disease incidence.

- Soil application of disulfoton, phorate or aldicarb in combination with paraffin oil sprays has been found quite effective in reducing the spread of the disease in field.

### 3.11.2.4. Cowpea Necrosis Virus

**i) Symptoms:** The field symptoms include mild mosaic, severe crinkling and curling of leaves. The leaf lamina is severely reduced giving a shiny appearance with thickened leaves and veins.

**ii) Transmission:** The virus is easily sap transmitted. Among several species of aphids transmitting the disease, *Aphis craccivora* is the most efficient vector followed by *A. fabae* and *A. gossypii*.

**iii) Management:** Cowpea cvs. Pusa Dophasli, Pusa Barsati and Pusa Phalguni have been reported to be resistant.

### 3.11.2.5. Cowpea Chlorotic Spot Virus

**i) Symptoms:** Irregular chlorotic spots, which later coalesce on primary and trifoliate leaves, vein clearing in young leaves, reduction in leaf lamina, dwarfing, early defoliation and poor pod formation are the common symptoms observed in the field.

**ii) Transmission:** This disease is caused by a strain of TMV. The virus is easily sap transmitted. Seed transmission has been reported up to 30-43%. No vectors are known.

**iii) Management:** Use virus-free seeds for sowing.

### 3.11.2.6. Cowpea Witches' Broom

**i) Symptoms:** The characteristic symptoms include smaller leaves, excessive branching, shortening of internodes and typical witches' broom appearance of the plant. At later stages, plants trail on the ground and produce phylloid flowers. The disease incidence varies from 1 to 6% in different seasons (Varma *et al.*, 1978).

**ii) Transmission:** The disease is transmitted by grafting and by *Cuscuta* sp.

**iii) Management**

- Soil Application of granular insecticides like disulfoton is effective for the management of the disease.
- Temporary remission of the disease can be obtained by spraying oxytetracycline hydrochloride solution (Varma *et al.*, 1978).

### 3.11.2.7. Cowpea Tobacco Ring Spot Virus

**i) Symptoms:** The symptoms include concentric chlorotic and necrotic ring spots, veinal necrosis of leaves, necrotic streaking of stem and petiole, top necrosis, defoliation, stunting and concomitant death of the plants. The disease incidence has been reported up to 60% under field conditions (Patil and Mali, 1982).

**ii) Transmission:** The virus is easily transmitted by mechanical sap inoculation. No insect vector is known. The virus is not seed-borne.

**iii) Host Range:** *Tetragonia expansa, Amaranthus caudatus, Ricinus communis, Vigna unguiculata, Datura fistulosas, D. metel* and *Nicandra physaloides* produce chlorotic and necrotic local lesions on artificial inoculation.

*Zinnia elegans, Citrullus lunatus, C. vulgaris* var. *fistulosus, Cucumis sativum, Luffa acutangula, L. cylindrica, Memordica charantia, Vigna cylindrica, Hibiscus cannabis, Nicotiana tabacum* cvs. W.B. and Xanthi produce systemic ring spots, ring mottle, top necrosis, epinasty, streaking and death of plants.

**iv) Management:** Cowpea cvs. C-152, L-1552, V-16, V-37, U-88-9-1-1 are reported to be resistant to ring spot virus.

### 3.11.2.8. Bean Yellow Mosaic in Cowpea

**i) Symptoms:** The disease was first reported from India by Mali and Kulthe (1980). The infected leaves exhibit chlorotic bands along the major veins and line pattern. Severe mosaic mottling, puckering and distortion of leaf are also common symptoms.

**ii) Host Range:** The virus infects *Chenopodium amaranticolor, C. murale, C. quinoa, Dolichos biflorus, Glycine max, Gompherena globosa, Phaseolus lunatus, P. vulgaris* and *Vigna mungo.*

**iii) Transmission:** The virus is mechanically sap transmitted. In nature, the virus is transmitted by the aphid, *Aphis gossypii*. The seed transmission has been reported up to 14-41%.

#### iv) Management

- Selection of seeds from apparently healthy plants.
- Soil application of carbofuran at 1.5 kg a.i./ha at the time of sowing followed by 2-3 sprays of 0.05% monocrotophos or 0.02% phosphamidon at 10 days interval.

## 3.12. CLUSTER BEAN

Cluster bean (*Cyamopsis tetragonoloba*) is also popularly called as 'guar' in India. Its tender pods being edible and are used in preparation of curry. The green pods are rich in Vitamins A and C. Since it is draught tolerant, it can be grown in summer. It is also grown as a forage and green manure crop. Its seeds are reported to be good cattle feed.

### 3.12.1. Bacterial Diseases

### 3.12.1.1. Bacterial Blight, *Xanthomonas campestris* pv. *cyamopsidis*

At present, the disease is prevalent at Delhi, Haryana, Punjab, Rajasthan and Uttar Pradesh. Loss in yield due to this disease is reported between 19 to 85% (Gupta, 1978; Gandhi and Chand, 1985b; Lodha and Gupta, 1981, 1986).

**i) Symptoms:** Two types of symptoms viz., leaf spot and blight are produced by the single pathogen (Srivastav and Rao, 1963). Several small, circular, water-soaked spots are formed on leaflets. They are oily and olive coloured on the lower surface. Spots, when fully developed become jet black in colour and several of them coalesce to form necrotic patches, surrounded by chlorotic halo. The infection from blighted leaves spread to stem through petiole and in advance cases, stem gets cracked. The pods also show heavy spotting.

**ii) Epidemiology:** Pathogen is seed-borne (Srivastav and Rao, 1963). The disease spreads to new areas by infected seeds and is destructive to rainy season crop (Srivastav and Rao, 1970). The intensity of the disease is increased with increase in amount of rain fall and number of rainy days (Karwasra *et al.*, 1981, 1984) and with increase in the nitrogen level from 20-40 kg/ha (Gandhi and Chand, 1985a). Younger plants are reported to be more susceptible (Singh and Swarup, 1987; Gandhi and Chand, 1988).

**iii) Management**

- Use of healthy seed from summer grown irrigated crop (Srivastav and Rao, 1970).
- Seed treatment with hot water at 56°C for 10 min (Singh and Swarup, 1986).
- Cluster bean cvs. G 85, 102, 126, and 155 are reported to be highly resistant; whereas HG 75, 182, HFG 119, 340 (Gandhi and Chand, 1987), G 28, 89, 98, 100, 119, 127, 139, 158, 166, 172, 223, 246, 253 and 254 (Chahal *et al.*, 1979) are moderately resistant.
- Gupta (1978) reported good control of the disease by spraying with 100-250 ppm streptocycline and 100-500 ppm agrimycin-100. Spraying of 100 ppm streptocycline followed by 100 ppm agrimycin-100 or 2000 ppm ceresan wet was also found effective (Singh and Swarup, 1986).

### 3.12.2. Viral Diseases

#### 3.12.2.1. Cluster Bean Mosaic Virus

The disease was first reported during 1985 from Bangalore, Karnataka.

**i) Symptoms:** The disease is characterized by severe mosaic mottling, leaf distortion, blisters and shortening of leaves, stunted growth and dark green appearance of the plants.

**ii) Host Range:** The virus infects sword bean, *Nicotiana rustica, Phaseolus angularis* and *P. calcaratus*.

**iii) Transmission:** The virus is easily sap transmissible. Several aphid species transmit the virus and *Aphis craccivora* is the most efficient vector. The virus is also transmitted by *Aphis gossypii* and *Myzus persicae*. The virus is seed-borne up to 50-80%.

**iv) Management**

- Use disease-free seeds for sowing.
- Seed treatment in hot water at 55°C for 15 min followed by incubation at 25°C for 4 days.
- Soil application of carbofuran or disulfotan or phorate each at 1.5 kg a.i./ha at the time of sowing followed by 2-3 sprays of 0.05% dimethoate or 0.02% phosphamidon at 10 days interval.

#### 3.12.2.2. Cluster Bean Yellow Mosaic

**i) Symptoms:** The important symptom of this disease includes yellow spots on young leaves in the initial stages of infection. As the disease progresses, the leaves exhibit irregular yellow patches alternating with normal green colour of the leaves. Sometimes either the apical or middle or basal part only shows yellowing, while other parts remain normal green. Often it is also observed in some leaflets, veins remain normal while inter-veinal areas show yellowing. The intensity of yellowing varies from faint to bright yellow. The leaf size is reduced considerably and the plants are stunted. Infected leaves become thick and brittle (Prasad Rao *et al.*, 1982).

**ii) Transmission:** The virus is transmitted in nature by the white fly, *Bemisia tabaci*.

**iii) Management:** Soil application of granular insecticides like carbofuran or disulfotan at 1.5 kg a.i./ha followed by 2-3 sprays of monocrotophos or dimethoate each at 0.05% would reduce vector population and also the disease incidence.

## 3.13. FIELD BEAN

### 3.13.1. Viral Diseases

#### 3.13.1.1. Yellow Vein Mosaic

The disease was first reported from India by Capoor and Varma (1950b).

**i) Symptoms:** The leaves of the diseased plants develop bright yellow patches interspersed with green areas. Initially, only a few yellow patches are seen on the leaf lamina but later on the entire leaf becomes yellow in colour. However, sometimes few patches of green colour are seen on the leaf. In severe cases, the entire leaf looks golden in colour. The plants which are infected in early stages remain dwarf with small leaves and less number of branches. Infected plants produce less number of fruits.

**ii) Transmission:** In nature, the virus is transmitted by the white fly, *Bemisia tabaci*.

**iii) Host Range:** The virus infects almost all leguminous vegetables except *Phaseolus lunatus* (Capoor and Varma, 1950b).

**iv) Management**

- Removal of infected plants from field helps in reducing the disease incidence.
- Soil application of carbofuran at 1.5 kg a.i./ha at the time of sowing seeds, followed by 2-3 foliar sprays of 0.05% dimethoate to kill the white fly vector is effective in controlling the disease.
- Use resistant varieties.

### 3.13.1.2. Little Leaf (Mycoplasma)

This disease was reported for the first time from Bangalore by Singh (1989).

**i) Symptoms:** The characteristic symptoms consists of small leaves, proliferation of side shoots and phyllody. Later on, the leaves become smaller and smaller measuring less than 1 sq. cm. There is a tendency of proliferation of axillary branches. The leaf tip is most obtuse and rarely acute. The infected leaves are thin, glabrous and light green in colour. The internodes of the axillary branches are short, giving a witches' broom appearance of the plant. The infected plants exhibit phyllody very often. The petals, stamens and pistil are reduced completely or partially to green bud. Ovary is partially or completely suppressed. The infected plants normally do not bear fruits. In case there is any fruit formation, it is small and deformed.

**ii) Transmission:** The disease is caused by mycoplasma-like organism. The pathogen is readily transmitted by grafting. No insect vector is known yet.

**iii) Management:** The disease symptoms can be temporarily suppressed by spraying the infected plants with 500 ppm of oxytetracycline hydrochloride at 10 days interval.

## 3.14. LIMA BEAN

### 3.14.1. Bacterial Diseases

### 3.14.1.1. Common Bacterial Blight, *Xanthomonas phaseoli*

**i) Symptoms:** The initial symptom of bacterial blight infection appears as water-soaked spots on the lower leaf surface. Lesions usually turn brown with time and often are surrounded by a narrow yellow halo. As the lesions enlarge they often run together resulting in extensive brown, dead areas. When severe, leaves often appear burned. In time, leaves become ragged and may drop prematurely.

Symptoms on pods appear as circular, sunken, reddish-brown lesions that vary in size. Pod lesions are usually surrounded by a dark reddish-brown margin. Infected seeds are usually shriveled, discoloured, and exhibit poor germination and vigour.

**ii) Persistence and Transmission:** Spores of bacterial blight can survive inside lima bean seed and in plant debris in the soil. Seed contamination is an effective means of both local and widespread dissemination of the pathogen. Seedlings arising from infected seed harbour large numbers of bacteria. The pathogen can survive on alternate weed hosts and volunteer bean plants in the area.

The bacteria survive for about one year in plant debris. Survival is generally longer in residue left at or near the soil surface. High humidity, rain, or both favour rapid development of the disease in the field. Bacterial spores are spread by wind blown rain and splashing irrigation water or through contact between adjacent leaves wet from rain or dew.

Development of bacterial blight is favoured by temperatures in the range of 85° to 95°F.

**iii) Management :** Bacterial blight of lima beans is best controlled by using the following strategies:

- Use certified disease-free seed. Do not save seed from the previous season if bacterial blight was present.
- Use crop rotation. Do not plant lima beans in the same field for at least 2 years. Control volunteer beans during this rotation period.
- Eliminate weeds that may act as reservoir hosts for the bacteria.
- A copper fungicide/bactericide may be needed if the disease is present and weather conditions favour its development.

### 3.14.1.2. Halo blight, *Pseudomonas syringae* pv. *phaseolicola*

**i) Symptoms:** The initial symptom of halo blight appears as water-soaked spots on the lower leaf surface. A distinctive greenish-yellow halo forms around the initial infection site that remains relatively small. When severe, the leaves take on a generalized yellow or chlorotic appearance that may be misidentified as a nutritional or pH disorder.

Pod symptoms generally consist of a reddish-brown discolouration that may also appear water-soaked. As pods mature and turn yellow, pod lesions remain green and may exhibit crusty bacterial ooze on their surface. Infected seed are often shriveled and discoloured.

**ii) Persistence and Transmission:** Spores of halo blight can survive inside lima bean seed and in plant debris in the soil. Seed contamination is an effective means of both local and widespread dissemination of the pathogen. Seedlings arising from infected seed harbour large numbers of bacteria. The pathogen can survive on alternate weed hosts and volunteer bean plants in the area.

The bacteria survives for about one year in plant debris. Survival is generally longer in residue left at or near the soil surface. High humidity, rain, or both favour rapid development of the disease in the field. Bacterial spores are spread by wind blown rain and splashing irrigation water or through contact between adjacent leaves wet from rain or dew.

Halo blight is considered a cool-season disease and is favoured by temperatures in 65° to 75°F range.

**iii) Management:** Halo blight of lima beans is best controlled by using the following strategies:

- Use certified disease-free seed. Do not save seed from the previous season if halo blight was present.
- Use crop rotation. Do not plant lima beans in the same field for at least 2 years. Control volunteer beans during this rotation period.
- Eliminate weeds that may act as reservoir hosts for the bacteria.
- A copper fungicide/bactericide may be needed if the disease is present and weather conditions favour its development.

### 3.14.2. Viral Diseases

#### 3.14.2.1. Yellow Vein Mosaic

The disease was first reported by Capoor and Varma (1948a).

**i) Symptoms:** The disease is characterized by conspicuous patches of bright yellow interspersed with deep green areas on lamina of leaflets on infected vines. As the leaves mature, the chlorotic areas are often scorched and turn brown because of sun burning. Sometimes the entire leaf becomes fully bright yellow in colour. The infected plants produce very poor pod yield.

**ii) Transmission:** The virus is sap transmitted. In nature, it is transmitted by the white fly, *Bemisia tabaci*.

**iii) Host Range:** Besides lima bean, the virus also infects *Phaseolus lemensis*, French bean, horse gram and jack bean or sword bean (Capoor and Varma, 1948a).

**iv) Management**

- Old infected perennial creepers should be cut down and removed before planting the new crop.
- Rouging of infected plants from the field as soon as they are noticed.
- Soil application of carbofuran at 1.5 kg a.i./ha at the time of sowing seeds followed by 2-3 sprays of 0.2% methyl parathion or 0.05% dimethoate at 10 days interval.

## 3.15. WINGED BEAN

### 3.15.1. Viral Diseases

#### 3.15.1.1. Witches' Broom (Mycoplasma)

The disease was first observed by Singh (1989a) from Bangalore. Loss in yield has been estimated to be cent per cent in case of early infection.

**i) Symptoms:** The characteristic symptoms of this disease are small leaves, stunted plant growth, witches' broom and phyllody. The leaves become very small measuring less than 1 sq. cm. The infected leaves are thin, glabrous and light green in colour. The internodes of the axillary branches become very short resulting in witches' broom appearance of the plant. There is proliferation of axillary branches which gives bushy appearance to the plant.

The flowers from the infected plants are invariably phylloid. The calyx, corolla, stamens and pistil are reduced partially or completely to a smaller bud and turn green. The ovary is transformed into a vegetative branch. The entire flower becomes leafy in structure. Usually there is no fruit formation in the infected plants. At times, one or two fruits are formed in plants which are infected at later stages. The fruits from such plants are small and deformed without any market value.

**ii) Transmission:** The disease is caused by mycoplasma-like organism (Singh, 1989a). The pathogen is transmitted by grafting. No seed transmission has been noticed. It is not mechanically transmitted. No known insect vectors yet.

**iii) Management:** The symptoms of the disease can be suppressed by spraying 500 ppm of oxytetracycline hydrochloride at 10 days interval.

## 3.16. SWORD BEAN

### 3.16.1. Viral Diseases

#### 3.16.1.1. Sword Bean Mosaic

The disease was first recorded by Singh (1983) from Bangalore, Karnataka.

**i) Symptoms:** The characteristic symptom of the disease is severe mosaic mottling of the leaves. Initially the infected leaves exhibit mild mottling followed by severe mosaic. Later on conspicuous blisters are seen on the infected leaf surface. The leaves also show veinal necrosis and yellowing. At times, the leaves become discoloured and turn brown. The texture of the leaf becomes rough. Many a times, the leaves are distorted and the size of the leaf is drastically reduced. Very few fruits are produced on infected plants and they are very small, thin, distorted, and exhibit longitudinal mottled streak. The infected fruits usually do not set seeds.

**ii) Host Range:** The virus also infects *Canavalia ensiformis, C. gladiata, Glycine max,* French bean and mung bean.

**iii) Transmission:** The virus readily sap transmitted. Seed transmission up to 70-80% has been reported. The virus is also transmitted by aphids such as *Aphis gossypii* and *Myzus persicae* in a non-persistent manner.

**iv) Management**

- The disease can be controlled by the selection of seeds from apparently healthy plants.

- Four to five foliar sprays of phosphamidon or oxydemeton methyl each at 0.02% at 10 days interval are effective in reducing the aphid vector population and the disease spread.

## 3.17. AMARANTHUS

### 3.17.1. Viral Diseases

#### 3.17.1.1. Amaranthus Mosaic Virus

**i) Symptoms:** The diseased plants exhibit slightly smaller leaves, showing crinkling and mosaic mottling of dark green areas interspersed with light green patches. The leaves are also reduced in size (Ramakrishnan *et al.*, 1971).

**ii) Transmission:** The virus is sap transmissible. *Aphis gossypii* acts as the vector of this disease.

**iii) Host Range:** The virus attacks *Amaranthus gangeticus, Achyranthes aspera, Celosia cristata, Gompherina globosa, Nicotiana glutinosa* and *Petunia hybrida.*

**iv) Management**

The disease incidence can be minimized by the application of 3 to 4 sprays of dimethoate or monocrotophos both at 0.5% at weekly intervals. The spraying should be stopped at least 20 days before first harvest.

## 3.18. SPINACH

### 3.18.1. Bacterial Diseases

#### 3.18.1.1. Spinach Bacterial Soft Rot, *Erwinia carotovora*

**i) Symptoms:** It is one of the most important diseases. Its symptoms include water soaked tissue and muddy-green or greasy appearance of leaves. Rapid decay occurs and the tissue becomes wet and mushy. This bacterium is found in the soil and in plant debris. It can enter into the plant through mechanical injury, insect injury, disease lesions and other skin punctures.

**ii) Management**

- Practice rotation with maize, beans, small grains and grasses.
- Care at harvesting and handling to avoid bruising
- A storage temperature just above freezing (0°C) and a relative humidity below 90% does much to reduce soft rot losses
- Storage rooms, dump tanks and boxes should be disinfected each season with copper sulfate

### 3.18.2. Viral Diseases

#### 3.18.2.1. Spinach Mosaic

This disease is caused by a strain of cucumber mosaic virus

**i) Symptoms:** The disease is characterized by light and dark green patches and mosaic mottling of infected leaves. The leaves are reduced in size, distorted and invariably show chlorotic sectors which are almost white. The infected plants show stunted growth (Naqvi and Mahmood, 1975).

**ii) Transmission:** The virus is easily transmitted by sap inoculation. In nature, the disease is transmitted by aphid vector, *Myzus persicae* in a non-persistent manner.

**iii) Host Range:** The symptoms on plants artificially inoculated with virus are presented in Table 3.22.

**Table 3.22. Host range of spinach mosaic**

| Host | Symptoms |
|---|---|
| *Achranthes aspera* | Chlorotic patches, vein clearing, mosaic, leaf deformation, and stunting. |
| *Amaranthus caudus* | Chlorotic patches, mosaic, and leaf deformation. |
| *A. leucocarpus* | Mosaic, leaf deformation, and stunting. |
| *Beta vulgaris* | Chlorotic spots. |
| *Benincasa hispida* | Mosaic. |
| *Lagenaria leucantha* | Chlorotic patches, and mosaic. |
| *Datura stramonium* | Vein clearing, mosaic, leaf deformation, and stunting. |
| *Lycopersicon glandulosum* | Apical chlorosis, mosaic, and leaf deformation. |
| *Nicotiana glutinosa* | Vein yellowing, chlorosis, leaf deformation, and stunting. |
| *N. rustica* | Mosaic, and leaf deformation. |
| *N. tabacum* cv. Harrison's Special | Mosaic, and leaf deformation. |
| *N. tabacum* cv. Samsun | Vein yellowing, and chlorosis. |
| *Solanum melongena* | Chlorotic concentric rings. |

**iv) Management:** The disease incidence can be checked by 1 or 2 sprays of dimethoate or monocrotophos both at 0.05%. The spraying should be stopped at least 20 days before the first harvest.

#### 3.18.2.2. Cucumber Mosaic Virus (CMV)

It is transmitted by the green peach aphids. Symptoms begin as a mottling of the younger inner leaves, which later change to a yellow colour. The symptoms gradually appear on outer leaves, which also change to yellow. The affected leaves curl and wrinkles. Severely affected leaves die. If a plant is affected at the seedling stage, its growth is stunted. The dwarfing, yellowing, corrugation and leaf death are conspicuous symptoms of the disease different from other diseases attacking spinach. Intensity of the virus increases under long days and intense light. This virus affects, in addition to spinach, a wider group of vegetables, flowers, weeds and ornamentals than any other virus. At least 34 plant families are included as hosts.

**Management**

- Use resistant varieties (e.g. "Early Hybrid 7")
- Control aphid vectors throughout the growing season.
- Control weeds

## 3.19. PUMPKIN

Pumpkin (*Cucurbita moschata*) occupies a prominent place among vegetables owing to its high productivity, nutritive value, good storability, long period of availability, better transport qualities and extensive cultivation in sub-tropical and tropical parts of the world. Fruits are rich in Vitamin A. It is also consumed as processed and stock feed. The fruits can be used in preparing sweets, candy, fermented into beverages. Yellow or orange-fleshed pumpkins are rich in carotene. In India, it is grown mainly in Assam, West Bengal, Tamil Nadu, Karnataka, Madhya Pradesh, Uttar Pradesh, Orissa and Bihar.

### 3.19.1. Viral Diseases

#### 3.19.1.1. Pumpkin Mosaic Virus

This is a strain of watermelon mosaic virus reported on pumpkin from Bangalore, Karnataka (Singh, 1979a; 1981b). If the plants get infection in the early stage of their growth (10-15 days old), the plant growth is arrested, the size and number of leaves are reduced considerably. Fruits from infected plants weigh on an average 0.5 kg in comparison to 2.5 to 3.0 kg in healthy plants. The virus infection also reduces vitamin A content of the fruits (Singh, 1979a).

**i) Symptoms:** The infected leaves initially show vein-clearing symptoms. Later, the subsequent leaves show a conspicuous vein-banding consisting of a narrow border of dark green parallel with each of the main veins of the leaf, the remainder of the leaf remaining usual lighter green. A peculiar serration is noticed in some foliage in which tips of the veins protrude beyond the edges of the leaf, sometimes giving toothed appearance. The diseased leaves may also have deformations which vary from an elongation to a rather showing outlined cordate pattern in mild cases. The leaves become smaller than the normal size. The fruits from infected plant exhibit various deformations and dark green lines. Fruits are blistered, small and misshapen. Carotene content of the infected fruit is reduced.

**ii) Transmission:** The virus is easily sap transmitted and to a limited extent seed-borne (2-6%). Several species of aphids viz., *Aphis gossypii, A. craccivora, Macrosiphum sonchii, Myzus persicae* and *Rhopalosiphum maidis* acts as vectors of the virus. Among these, *Aphis gossypii, A. craccivora,* and *Myzus persicae* are very efficient vectors. The aphids can acquire and transmit the virus within 5-10 seconds. Even a single aphid can transmit the virus (Singh, 1981d, 1982a, 1982b).

**iii) Host Range:** The virus infects plants belonging to Cucurbitaceae viz., ash gourd (*Benincasia hispida*), watermelon (*Citrullus lunatus*), round melon (*Citrullus vulgaris* var. *fistulosus*), musk melon (*Cucumis melo*), long melon (*Cucumis melo* var. *utilissima*), cucumber (*Cucumis sativus*), *Cucurbita maxima*, pumpkin (*C. moschata*), squash (*C. pepo*), bottle gourd (*Lagenaria siceraria*), ridge gourd (*Luffa acutangula*), sponge gourd (*L. cylindrica*), bitter gourd (*Memordica charantia*) and snake gourd (*Trichosanthes anguina*) (Singh, 1979a, 1981b).

**iv) Management**

- Spray nuvacron or dimecron each at 0.05% at 10 days interval.
- Collect seeds from healthy plants for planting.
- Spray Krishi oil to check the spread of the disease in the field.

### 3.19.1.2. Yellow Vein Mosaic

The disease was first reported by Vasudeva and Lal (1943), and Varma (1955) from India.

**i) Symptoms:** The characteristic symptoms of the disease are yellowing of the veins and mosaic. Initially, the virus causes vein clearing on the young leaves. Lateran, mosaic symptoms are produced on the leaves. Leaves become smaller as compared to healthy ones. Fruits from infected plants are small and deformed.

**ii) Transmission:** The virus is transmitted in nature through the white fly, *Bemisia tabaci*. The virus is also sap transmitted.

**iii) Host Range:** Besides pumpkin, the virus also infects cucumber, musk melon, water melon, *Citrullus colocynthis*, squash and *Lagenaria vulgaris*.

**iv) Management**

- Spray insecticides to keep the disease under check.
- Straw mulching or colour mulching delays the onset of infection.
- The cucumber cv. Kyoto-3-Feet is resistant to this virus.

## 3.20. CUCUMBER

India is considered to be the home of cucumber (*Cucumis sativus*). Cucumber is cultivated for fresh consumption or as pickling cucumber for preservation, marinated with vinegar, salt, dill or other spices. It is an important salad crop cultivated both in north and south and lower as well as higher hills in India.

### 3.20.1. Bacterial Diseases

#### 3.20.1.1. Angular Leaf Spot, *Pseudomonas syringae* pv. *lachrymans*

**i) Symptoms:** This bacterial disease can occur on most cucurbits, including cucumber, muskmelon, pumpkin, and winter squash. The disease is less common on cucumber because resistant varieties are widely grown. Leaf spots are variable in size and may be angular in shape because leaf veins limit enlargement of spots. Initial symptoms appear as brown water-soaked spots, as illustrated on butternut squash (Fig. 3.33). Under moist conditions, especially in early morning, small droplets of bacterial ooze can come from the water-soaked areas; which, when dry, appear as a white residue. The initial brown spots may be surrounded by a yellow halo, become white, and, as they dry, tear away from the healthy tissue, producing irregular holes in the affected leaves. Fruit spots are smaller than leaf spots and are usually circular. These spots, as shown on the processing squash Delicious, are usually superficial but can develop into deeper rots if secondary soft-rot bacteria invade.

**ii) Host Range:** The disease attacks gherkin, muskmelon, pumpkin, squash, vegetable marrow and watermelon.

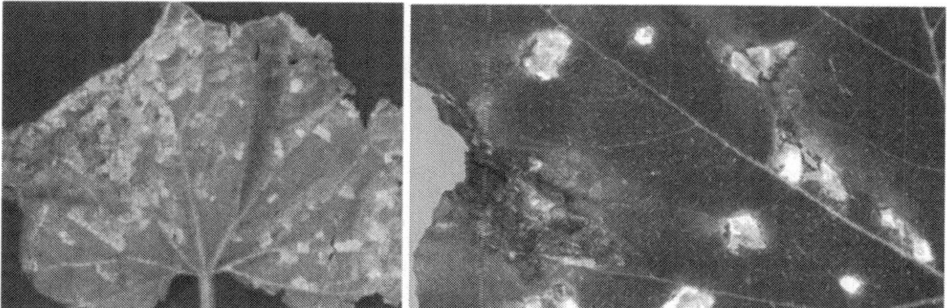

**Fig. 3.33.** Angular leaf spot on cucumber.

**iii) Epidemiology:** The bacteria can survive for possibly 2 years in soil or debris from diseased plants. It is also known to be associated with seed. Bacteria can enter through wounds or stomata. Moisture in the form of rainfall or relative humidity greater than 95 per cent for several days is required for infection and later disease development. Drainage water can spread the bacteria in the soil. Angular leaf spot disease is favoured by wet conditions, frequently associated with rainfall and overhead irrigation. Optimum temperature for disease development is 23.9 to 27.8°C. Dry weather for 2 weeks can arrest the disease.

**iv) Management :** Controls include the selection of resistant cucumber varieties, use of disease-free seed, and a 2 year rotation out of all cucurbits. Practice crop rotation. A crop rotation of 1 to 2 years is recommended. Check with your seed supplier for the current list of resistant cucumbers. Copper fungicides can be applied to slow disease spread during particularly wet periods but can be dropped if dry weather continues for 2 weeks. The best way to avoid this disease is to keep the foliage dry. Drip irrigation and soaker hoses work great to accomplish this. Avoid run-off water from nearby cucurbit fields and overhead irrigation.

### 3.20.1.2. Bacterial Wilt, *Erwinia tracheiphila*

Bacterial wilt, caused by the bacterium *Erwinia tracheiphila,* is a common and destructive disease of cucurbits. Losses from bacterial wilt vary from the premature death of occasional plants to as high as 75 percent of a crop. It is a serious threat to commercial melon and cucumber production in some parts of the world, including parts of North America.

**i) Symptoms:** The expression of bacterial wilt symptoms varies with different crop species. Cucumber and melon are severely affected by the disease. Individual runners or whole plants wilt and die rapidly (Fig. 3.34). Affected runners appear dark green at first and then become necrotic as the wilt becomes irreversible. Symptoms appear at all stages of plant development, but wilting is most severe early in the season, when the plants are growing rapidly.

**Fig. 3.34.** Cucumber plant infected with bacterial wilt.

Summer squash and pumpkin also develop bacterial wilt. Plants may wilt dramatically during the heat of the day but partially recover by morning. Foliage of affected plants often appears chlorotic. A distinct necrosis and chlorosis develops around leaf margins before the entire plant collapses and dies. Young pumpkin plants normally lose vigor and die within 2 weeks of the initial signs of wilt. Summer squash may remain vigorous and continue to produce for weeks after wilt symptoms appear.

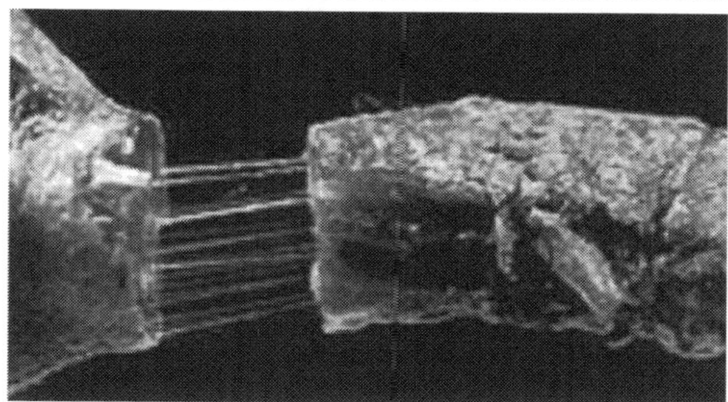

**Fig. 3.35.** Sticky strand test on cut stems, with bacterial slime streaming from xylem tissues.

A common diagnostic test for bacterial wilt involves cutting a wilted runner close to the crown of the plant, rejoining the cut surfaces for a moment, and then slowly drawing apart the cut ends (Fig. 3.35). The presence of bacterial slime (masses of bacteria streaming from xylem tissues) extending from one cut surface to the other is a positive indication of bacterial wilt. A variation of this test involves placing the cut runner or crown in water. Bacteria will ooze from the exposed vascular elements of infected plants in 5-10 min, forming milky strands flowing from the cut surface into the water.

**ii) Causal Organism:** The causal agent of bacterial wilt is *Erwinia tracheiphila*, a gram-negative, rod-shaped, motile bacterium with peritrichous flagella. It is transmitted by insect vectors, primarily the striped cucumber beetle (*Acalymma vittatum*) and the spotted cucumber beetle (*Diabrotica undecimpunctata howardi*). The western spotted cucumber beetle (*D. undecimpunctata undecimpunctata*), the banded cucumber beetle (*D. balteata*), and other insects that cause wounds, such as grass hoppers, may also transmit the bacteria.

**iii) Disease Cycle:** The bacteria over winter in the digestive tracts of hibernating, adult striped and spotted cucumber beetles. In the spring the bacteria are introduced into the leaves of cucurbits by the beetles as they feed. The bacteria enter cucurbit tissue only through wounds, such as those produced by beetle feeding. When the beetles are absent, bacterial wilt does not occur. The pathogen is not seed-transmitted and does not survive in soil.

Weather conditions have an indirect effect on the disease, in that environmental conditions which favour the over wintering, feeding, and reproduction of the cucumber beetles subsequently affect the prevalence of bacterial wilt. A good winter snow cover followed by warm temperatures in March and April can be expected to increase the number of over wintering beetles and therefore increase the early incidence of bacterial wilt.

When a beetle feeds on a diseased leaf, its mouth parts become contaminated. The wilt bacteria are then introduced into the next several plants on which the beetle feeds. Beetles remain infectious for at least three weeks. Infection of cucurbit plants occur only when sufficient water from dew, rain, or irrigation is present to allow the bacteria to penetrate the inner leaf tissue via feeding wounds. Wilting may be visible in several days, and the entire plant is invaded in 12 to 15 days. Field symptoms are usually apparent approximately one month after the beetles appear.

Bacterial wilt develops when daily temperatures average between 7 to 22°C with an optimum of 11 to 15°C. The bacterium grows best at temperatures of 25 to 30°C, and under conditions of high relative humidity. Bacterial isolates differ in their virulence on each species of cucurbit. Some isolates that are highly virulent on cucumbers are also quite virulent on squashes while others that are less virulent on cucumbers usually are not virulent on squashes. However, distinct cucumber and squash strain of the bacterium may exist.

**iv) Epidemiology:** The incubation period in the field from infection to symptom expression ranges from several days to several weeks and is influenced primarily by the age of the plant and the point of inoculation. The rate of bacterial wilt in the field is most rapid while plants are young and succulent. Several factors contribute to a dramatic decrease in the rate of spread as plants mature. Beetle populations naturally decline in midsummer; the older, more brittle leaves and stems are not preferred food sources; and bacteria introduced into new growth at vine tips must travel greater distances in the plant to induce wilting. Weather has little effect on wilt incidence, but it may influence the rate of symptom expression. Ideal growing conditions for crops, including warm weather, adequate soil moisture, plentiful sunlight, and balanced nutrient concentrations, also appear to favour the development of the disease.

**v) Management :** Provide season-long control of spotted and striped cucumber beetles and other insects, such as grass hoppers and squash bugs, with insecticides. Start applications as the plants start to crack the soil, before the leaves appear, even if no beetles are evident. Frequent applications are necessary, especially in the seedling stage, to keep the foliage free of beetle-feeding wounds. Applications may be needed at 4 to 5 day intervals. Repeat after rains, especially if beetles are present. Early-season sprays or dusts are the most important step in controlling bacterial wilt. Make treatment in late afternoon or evening to avoid damage to bees. The use of a systemic, soil-applied insecticide will provide moderate control for 5 to 6 weeks. Supplemental foliar insecticide treatments, however, will be necessary.

Roguing wilted plants and the use of trap crops have been suggested to slow the progress of the disease, they are of no practical value unless appropriate insecticide programs are implemented. Bait formulations, which consist of volatile attractants, a feeding stimulant, and a small amount of insecticide, are being developed and show promise.

## Integrated Management

- Since this bacterium is transmitted systemically, copper sprays are of no value. Cucumber beetles must be controlled by appropriate insecticides (Bifenthrin, Carbaryl, Endosulfan, Imidacloprid, Kaolin, Methomyl, Permethrin, Pyrethrin).

- Scout twice weekly at seedling stage for cucumber beetles. Treat when beetle numbers reach the threshold of 1 beetle per 100 feet of row.
- Use crop rotation to reduce beetle numbers.
- Rogue infected plants.
- Spun-bonded row covers will exclude beetles.
- Plant a sprayed perimeter trap crop of Blue Hubbard squash to protect more susceptible crops.
- Resistant cultivars of cucumber are being developed. No resistance has been identified in melon.
- Bait formulations consisting of attractants, feeding stimulants, and insecticide are being developed and tested.

### 3.20.2. Viral Diseases

#### 3.20.2.1. Cucumber Mosaic Virus (CMV)

**i) Symptoms:** This disease infects almost all the cucurbits. The young leaves develop greenish-yellow areas and they are more translucent than those in the remaining parts of the leaf. Yellow mottle is seen on the leaves (Fig. 3.36). This is followed by leaf distortion and stunting of plants. Infected fruits are mottled with yellowish-green colour and blistered. Mosaic virus causes deformation of the cucumber plants, with small leaves and underdeveloped plants. Discolouration and spotting on the leaves are observed. This disease is spread by aphids.

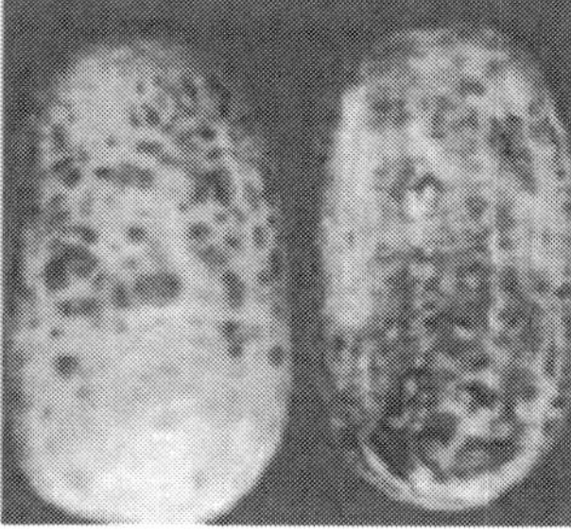

**Fig. 3.36.** Cucumber mosaic virus on leaves and fruits.

**ii) Transmission:** It is not seed transmitted except through seed of perennial wild cucumber (*Echinocystis lobata*) and chickweed (*Stellaria media*). CMV is transmitted by sap, seed and several species of aphids, especially *Aphis gossypii* and *Myzus persicae* in a non-persistent manner. So make sure to control these pests (Kennedy *et al.*, 1982).

**iii) Host Range:** It has a very extensive host range including such varied species as bananas, carrots, cowpeas, lupine, lilies, onions, passion fruit, potatoes and tomatoes.

**iv) Management**

- Select seeds from absolutely virus-free plants.
- Remove weed hosts from the field.
- Rouging of infected plants from time to time.
- The fruits from infected plants should also be destroyed.
- Do not compost plants infected with the virus.
- Crop rotation with non-host plants should be followed.
- Use barrier crops like sunflower, sorghum and pearl millet all round the cucumber crop to prevent the vectors entering the field.
- Mulch the soil with saw dust or wood chips or with aluminium foil which repels the aphid vectors and reduce the CMV infection up to 87.9% (Loebenstain *et al.*, 1975).
- Spray with either dimethoate or monocrotophos each at 0.1% at weekly interval to control aphid vectors. Spraying of mineral oils at 0.75 to 1.0% at weekly intervals has been found promising (Loebenstain *et al.*, 1966; Zitter, 1978).

### 3.20.2.2. Cucumber Green Mottle Mosaic

**i) Symptoms:** The disease is caused by a strain of tobacco mosaic virus (TMV). The characteristic symptoms are slight clearing of veins and crumpling of the young leaves, followed by a light or dark green mottle, together with blistering and distortion of the leaves and stunting of the plant. The symptoms first appear on the younger leaves. The mottle is independent of the season, but leaf distortion is more severe when the plant is growing slowly in winter. Occasional yellow flecks occur on leaves showing the green mottle and may, on fully developed leaves, be a prominent symptom. The fruit is usually unmarked, though it may be slightly mottled (Vasudeva and Nariani, 1952).

**ii) Transmission:** The virus is easily sap transmitted and to a limited extent through seeds. No insect vector is known.

**iii) Host Range:** The virus is confined to the members of the cucurbitaceae only viz., cucumber, bottle gourd, snake gourd, squash, watermelon, pumpkin and ridge gourd.

**iv) Management**

- Use disease-free seeds for planting. Seeds collected from apparently healthy plants may also be used.
- The hands of workers as well as the implements should be washed thoroughly during field operations.
- Treat the seeds in dry heat at 70°C for 3 days.
- Fumigate soil with methyl bromide.

### 3.20.2.3. Cucurbit Phyllody (Mycoplasma)

This disease caused by mycoplasma-like organism was recorded by Singh (1979b) from Bangalore.

**i) Symptoms:** The characteristic symptoms of this disease are shortening of internodes and phyllody of normal flowers. The floral parts like corolla, androecium and gynoecium are transformed into green leaf-like structures. Infected plants do not attain full size. Some infected plants bear clusters of phylloid flowers. The infected plant usually do not set fruits. Nevertheless, if fruits are formed, they are small, abnormal and distorted. The infected plant often show dull pale colour.

**ii) Transmission:** The pathogen is transmitted through grafting. However, in nature, the transmission occurs through leaf hopper vector.

**iii) Host Range:** The disease was recorded on cucumber, bottle gourd, snake gourd, ridge gourd and bitter gourd. Besides, the disease was also noticed on a weed host, *Melothria maderaspatana*. The pathogen survives on this perennial weed from one season to another.

**iv) Management**

- All the infected plants should be removed from the field as soon as they are noticed.
- Spray the plants with systemic insecticides like 0.05% dimethoate or 0.02% metasystox or 0.05% nuvacron at 10 days interval to control the insect vector population.
- Spray with 500 ppm oxytetracycline hydrochloride at 8 days interval.

## 3.21. BITTER GOURD

### 3.21.1. Viral Diseases

### 3.21.1.1. Bitter Gourd Mosaic Virus

**i) Symptoms:** The symptoms on the naturally infected bitter gourd plants are confined mostly to leaves. The leaves in the secondary branches produced at the apical end of the plant show the symptoms of the disease. Scattered, small, irregular yellowish patches are seen on the leaves. Some leaves show vein clearing in one or two lobes of the leaf. Severely infected plants show reduction in their size and elongation and/or suppression of one or two lobes. Young developing leaves are completely distorted, filiform and malformed with considerable reduction in their size. Some of the leaves show marked reduction in the development of lamina resulting in a shoe string effect (Nagarajan and Ramakrishnan, 1971).

**ii) Transmission:** The virus is transmitted by 5 species of aphids, viz., *Aphis gossypii, A. malvae, A. nerii, Brevicoryne brassicae* and *Myzus persicae.*

**iii) Host Range:** The virus can infect *Cucurbita moschata, C. maxima, C. pepo, C. melo, Citrullus vulgaris, Lagenaria vulgaris* and *Trichosanthes anguina*.

**iv) Management**

- Spray the crop just after germination with nuvacron or dimecron both at 0.05% at 10 day intervals to prevent aphid vectors.
- Spraying of 1.5% mineral oil minimizes the virus spread.

### 3.21.1.2. Bitter Gourd Witches' Broom

The disease was first reported by Singh (1985) from Bangalore, Karnataka. When the plants get infection in the early stages, they do not bear any fruits and the loss is 100%. However, in late infections, the yield loss is about 30-50%.

**i) Symptoms:** The characteristic symptoms are malformation and proliferation of auxilliary buds. Diseased plants show many abnormally small leaves which fail to attain full size. The diseased plant bears many flowers and blossom earlier than healthy plants. Internode bud sprouts give rise to many bud-like chlorotic leaves. Flowers on infected plants show characteristic green phyllody symptoms. A variety of abnormalities ranging from near abnormal flowers showing green patches on the yellow petals or greenish yellow flower to a completely vegetative green leafy structure can be observed. Fruits from these flowers are very small, thin, cylindrical, deformed and completely seedless. Plants showing severe witches' broom symptoms do not bear fruits.

**ii) Transmission:** The disease is easily transmitted by grafting. Leaf hoppers have been suspected as vectors of this disease.

**iii) Host Range:** The virus also attacks other cucurbits like *Lagenaria vulgaris, Trichosanthes anguina* and *Luffa acutangula*.

**iv) Management**

- Application of carbofuran at 1.5 kg a.i./ha at the time of sowing seeds followed by 5-6 sprays of dimecron or nuvacron or metasystox all at 0.05% at 10 days interval keeps the vector population under check.
- Spraying of oxytetracycline hydrochloride at 500 ppm at weekly intervals is effective in suppressing the symptoms of the disease.

## 3.22. WATERMELON

Watermelon (*Citrullus lunatus*) is an important cucurbitaceous summer vegetable grown for ripe fruits. Though it can be grown in garden land, it is a major river-bed crop of Uttar Pradesh, Rajasthan, Gujarat, Maharashtra and Andhra Pradesh. An excellent desert fruit, it is relished by rich as well as poor. The fruit juice makes an excellent refreshing and cooling beverage after adding a pinch of salt and black pepper. The fruits contain 92% water, 0.2% protein, 0.3% minerals and 7.0% carbohydrates in a 100g edible flesh.

### 3.22.1. Bacterial Diseases

#### 3.22.1.1. Bacterial Fruit Blotch

**i) Symptoms:** The fruit blotch bacterium can cause seedling blight, leaf spot, and a fruit disorder. Initial symptoms on seedlings consist of dark water-soaking of the lower surface of cotyledons (seed leaves) and leaves. Lesions develop and are often accompanied by a yellow halo. Lesions may form on the hypocotyl (the part of the seedling below the seed leaves), causing death of the plant. Foliar symptoms can occur throughout the growing season but are difficult to detect. Leaf lesions, which act as a reservoir of bacteria for subsequent fruit infection, are tan to reddish brown and usually appear along the midrib. Defoliation is not common. Fruit symptoms begin as small, greasy-looking, water-soaked areas a few millimeters in diameter. Spots enlarge rapidly to irregularly shaped, dark green, water-soaked lesions of several centimeters. In a few days these lesions may expand to cover the entire upper surface of the fruit. Fruit lesions do not extend into the flesh of the melon initially. In time, the centers of the lesions turn brown and crack. White bacterial ooze, or an effervescent exudate, can often be seen on the fruit. Fruit decay usually follows.

**ii) Epidemiology:** The bacteria can be introduced into a field through infested seed, infected transplants, volunteer watermelons, or from natural spreading of alternate hosts. Seed harvested from infected fruit often harbor the pathogen. Undetected infestations can spread rapidly in transplant houses, resulting in high numbers of infected plants going into fields. Volunteer watermelons act as hosts in fields infested with blotch the previous year. Bacterial fruit blotch is favoured by warm, moist environmental conditions. Hot, sunny days accompanied by thunder showers appear to favour its development and spread. Under favourable conditions, the disease spreads rapidly; by harvest an entire field may be infected. Symptoms do not appear to develop under cool, rainy weather. The bacteria, once present in a field, can be spread by wind, rain, and equipment. Young fruit (2 to 3 weeks old) are most susceptible with bacteria entering through stomata (small openings in the epidermis). Once the wax layer forms, infection of mature fruit can occur only through wounds.

**iii) Management:** Preventing the introduction of the pathogen into a field is the best method of control for fruit blotch. Other control strategies include the following:

- Obtain seed and transplants only from a reliable source and inspect transplants for fruit blotch symptoms.
- Plow under crop debris, including watermelon culls, from a field previously infected with fruit blotch.
- Plant the current field as far away as possible from last year's site.

- Destroy volunteer watermelons and wild cucurbit weeds in and around the current season's crop.
- Plant less susceptible fruit. Those with light green coloured rinds, such as 'Charleston Gray' are more susceptible; those with a solid dark green rind, such as 'Sugar Baby' are more tolerant. However, the level of tolerance currently available will not ensure freedom from the disease when environmental conditions favour its development.
- Avoid working infested fields when they are wet.
- Apply copper fungicides prior to fruit set to reduce disease incidence. Two to three applications are necessary. Be aware, however, that copper can result in stunted watermelons.

### 3.22.2. Viral Diseases

#### 3.22.2.1. Watermelon Mosaic virus

The disease was first reported from the USA in 1932. In India, the virus was first reported by Bhargava and Joshi (1960).

**i) Symptoms:** Diseased vines are conspicuous because the tips of the runners and a proliferation of shoots from around the crown protrude stiffly above the general level of the vines, giving a 'petunia-like' appearance. The tips and shoots have shorter than normal internodes which produces crowding of the young leaves stunted and rolled. The characteristic symptoms are mosaic, mottling (Fig. 3.37) and blistering of leaves and stunted growth of the infected plants. Flowers on severely infected plants show abnormalities of the petals which are of unequal size and greenish in colour. Stamens are distorted and unevenly developed and the flowers may fall of prematurely. Melons produced on diseased vines are mottled and sometimes deformed. Diseased vines bear few fruits of small size.

**ii) Transmission:** The virus is easily transmitted by a mechanical means. Insect vectors include several species of aphids viz., *Aphis gossypii, A. craccivora, A. cirthosiphose, A. spiraecola, Myzus persicae, Brachycaudatus relichtysi* and *Cavariella aegopodii*

**iii) Host Range:** *Cucurbita moschata, C. maxima, C. pepo, Cucumis sativus, C. melo, Citrullus vulgaris, Luffa acutangula, L. cylindrical, Lagenaria vulgaris, Memordica charantia, Benincaspa hispida, Trichosanthes anguina* and *Zinnia elegans* act as symptomless carrier of this virus.

**iv) Management**

- Rogue out infected plants from the field as soon as they are noticed.
- Control weed hosts mainly *Melothria pendula,* which act as a source of inoculum.
- Spray the crop 4 to 5 times with either dimethoate or monocrotophos both at 0.1% at 10 days interval to kill aphid vectors.

- Spread of the virus is significantly reduced by spray of mineral oils (Witcher and Smith, 1980; Kibata, 1979).
- Grow 5 to 6 rows of barrier crops such as sorghum, maize or pearl millet around the watermelon crop 50 to 60 days before sowing to prevent the entry of vectors.

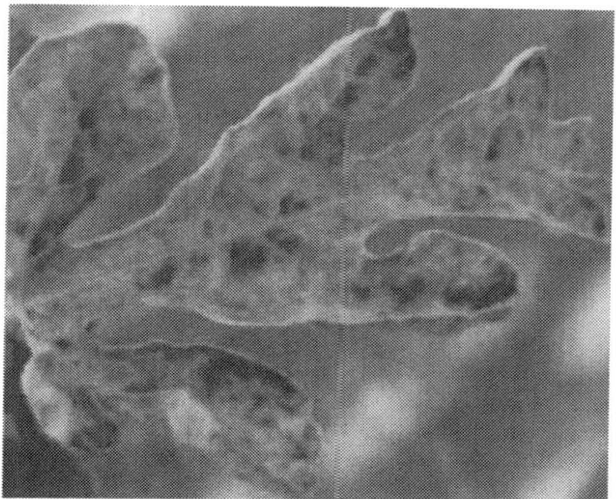

**Fig. 3.37.** Watermelon mosaic virus.

### 3.22.2.2. Watermelon Bud Necrosis

This disease was first observed at Hessaraghatta, Bangalore, Karnataka, during 1990 on watermelon.

**i) Symptoms:** The initial symptom of the disease is necrosis of the buds. The buds turn brown to black and die. This is followed by longitudinal streaks on the vine commencing from the tip downwards. These streaks may vary in size from 10 to 50 mm in length. The streaks first appear brown and later turn to black colour. The leaf symptoms include brown or black spots covering the entire leaf lamina and leaf petiole. Streaks are also noticed on leaf petioles, fruit stalk and tendrils. Conspicuous broken or complete concentric rings are produced on fruits. The rings vary in size from 5 to 30 mm in diameter. The margins of these rings are corky in texture and raised. The disease causes severe losses in yield of fruits.

**ii) Transmission:** The virus is transmitted by mechanical sap inoculation. Thrips transmit this virus under natural conditions.

**iii) Management:** The control measures used are same as described under tomato spotted wilt.

## 3.23. SQUASH

### 3.23.1. Viral Diseases

### 3.23.1.1. Squash Mosaic Virus

**i) Symptoms:** It is mechanically transmitted. It is transmitted through seeds of melons and squash. It is also transmitted by spotted, striped and banded cucumber beetles, which attack cucumbers in the Americas. The Epilachna beetle (*Epilachna chrysomelina*) a pest of cucumber in Africa, is also vector of squash mosaic virus. Its host range includes cucurbits, peas, coriander, and salad chervil.

**ii) Management**

- Use tolerant / resistant varieties if available.
- Remove infected plants (disinfect hands and tools with 70% alcohol after contact with infected plants).
- Do proper weeding.
- Control insect vectors. A sustainable approach of controlling aphids is important to prevent aphids reaching the crops and transmitting virus.
- In case of squash mosaic virus use disease-free seeds.

## 3.24. CHOW-CHOW

### 3.24.1. Viral Diseases

#### 3.24.1.1. Chow-Chow Mosaic

**i) Symptoms:** The disease was first reported by Singh (1981a) from Bangalore, Karnataka. The disease symptoms include severe mosaic mottling, puckering and distortion of leaves. The lateral veins of the leaves coalesce with mid vein making the leaves smaller in size. The infected plants does not grow well and produce only one or two fruits which are small and deformed.

**ii) Transmission:** The virus is readily transmitted by sap inoculation. Three species of aphids – *Aphis craccivora, A. gossypii* and *Myzus persicae* transmit the virus in a non-persistent manner.

**iii) Host Range:** The host range of this virus is restricted to the family Cucurbitaceae – *Cucurbita maxima, C. moschata, Cucumis sativus, Citrullus vulgaris, C. lanatus, Lagenaria siceraria, Memordica charantia* and *Trichosanthes anguina.*

**iv) Management:** The spread of the disease can be minimized by 4 to 5 foliar sprays of either dimethoate or monocrotophos each at 0.05% or methyl parathion at 0.02% at 15 days interval.

## 3.25. RADISH

Radish (*Raphanus sativus*) is a root vegetable which can be grown as an intercrop with other crops. Its leaves are rich in Vitamins A and C. It can be eaten raw and also cooked. It has some medicinal value. Radish is cultivated throughout India. West Bengal, Bihar, Uttar Pradesh, Punjab, Assam, Haryana, Gujarat and Himachal Pradesh are major radish-growing states.

### 3.25.1. Bacterial Diseases

#### 3.25.1.1. Bacterial Root Rot, *Erwinia rhapontici*

The disease was first reported from Bangalore, Karnataka, during 1983 and its incidence ranged from 15 to 50% (Kishun, 1983b).

**i) Symptoms:** The characteristic symptoms of the disease are rotting of pith tissues resulting in cavity formation and wilting of plants. The disease is more severe in seed crop grown from cut roots.

**ii) Epidemiology:** The pathogen is soil-borne and the disease spreads through implements, irrigation water, insects, etc.

**iii) Management:** Root dip treatment with 100 ppm of agrimycin-100 is found most effective in reducing the disease incidence (Kishun, 1983b).

### 3.25.2. Viral Diseases

#### 3.25.2.1. Radish Mosaic Virus

The disease is caused by a strain of turnip mosaic virus. The disease was first reported by Thakur and Dube (1981) and later by Ahlawat and Chenulu (1984). The disease incidence ranges from 7 to 63% (Ahlawat and Chenulu, 1984). The virus causes losses ranging from 3.3 to 100.0% in root weight and up to 95.8% in seed weight depending on the time of infection.

**i) Symptoms:** The affected plants are conspicuous with stunting, reduced leaf lamina and root size. Diseased leaves show interveinal chlorosis, vein banding and dark green patches on light green leaves. Curling and puckering of leaves are not uncommon Mosaic pattern is also seen on the stem, petioles and pods. Seeds from diseased plants show poor germination. The virus has a wide host range.

**ii) Transmission:** The virus is readily sap transmitted. The virus is not transmitted through seeds in radish. Several species of aphids, *Aphis gossypii, A. spiraecola, Brevicoryne brassicae, Lipaphis erysimi, Myzus persicae, Rhapalosiphum maidis, R. padi Schizaphus graminum, Toxoptera citricidus* and *T. odinae* have been found to transmit the virus in a non-persistent manner (Ahlawat and Chenulu, 1982). *Lipaphis erysimi* and *Myzus persicae* are reported to be efficient vectors (Thakur and Dube, 1981).

**iii) Host Range:** The virus has a wide host range including *Alyssum martimum; Brassica campestris* var. *sarson; B. campestris* var. *dichotoma; B. campestris* var. *toria; B. hirta; B. juncea* cv. Kalimpong Local and Varuna; *B. niger, B. oleracea* var. *italica* cv. Wong Bok; *B. rapa* cvs. Pusa Sweta, Snow Ball, Golden Ball; *B. rugosa; Capsella bursa pastroris, Cardamine impatiens, Cheiranthus cheiri; Crambe abyssinica; Eruca sativa; Iberis umbellata; Nastrutium montarum; Raphanus sativus; Chenopodium album; C. amaranticolor; C. morale; C. quinoa; Nicotiana debneyi; N. tabacum* cv. KY 57; *Physalis floridana; Petunia hybrida; Escholtzia californiica; Papaver somniferum; Phaseolus aureus; Chrysanthemum sagetum; Hekipterum roseum* and *Zinnia elegans.*

**iv) Management:** The disease spread can be minimized by soil application of carbofuran at 1.5 kg a.i./ha at the time of sowing seeds, followed by 2-3 foliar sprays of dimethoate or monocrotophos both at 0.05% at 10 days interval.

### 3.25.2.2. Radish Phyllody

**i) Symptoms:** The diseased plants assume a dull grey to light violet colouration and can be easily distinguished from a distance. The floral parts become green-violet and leafy. The sepals and petals become green, thick knob headed leafy structures. The ovules are transformed into leafy structures which push open the ovary. If the infection occurs in the nursery, then the whole plant is affected but delayed infection results in partial phyllody. The incidence of this disease varies from 5 to 20% in radish seed crop.

**ii) Etiology:** The radish phyllody is caused by mycoplasma-like organism.

**iii) Transmission:** The disease is transmitted by sesamum jassid, *Orosius albicinctus*.

**iv) Host Range:** The pathogen has a very wide host range. The most important hosts are sesamum, radish, turnip, toria, sarson, raya, green gram and Bengal gram. Some weeds like *Fagonia cretica* and *Spermacoce hispida* are the reservoirs of the pathogen.

**v) Vector-Pathogen Relationship:** The sesamum jassid requires minimum of 4 hr of feeding on a diseased plant to acquire the pathogen in sufficient quantity. The jassids become capable of transmitting the pathogen in 2 to 3 weeks after acquiring the mycoplasma. If an infective jassid feeds on a healthy plant fore more than 30 min, it injects sufficient quantity of mycoplasma to make the plant diseased. About two months after injection, the radish plants show symptoms. Both male and female jassids can transmit the pathogen. The nymphs can not inherit the disease from their infective parents but they acquire the pathogen by feeding on the diseased plants.

**vi) Management**

- Since the majority of the plants get infection in the nursery, one or two sprays of monocrotophos or phosphamidon each at 0.05% or oxydemeton methyl at 0.02% kill the jassids.
- Soil application of carbofuran or phorate at 1.5 kg a.i./ha followed by irrigation is effective against the disease.

## 3.26. BEET ROOT

### 3.26.1. Viral Diseases

### 3.26.1.1. Purple Leaf virus

The virus was first recorded at Pune during 1958 (Capoor and Sharma, 1965).

**i) Symptoms:** The infected plants are stunted and leaves have a tendency to stand erect and come closer, unlike in healthy plants where the leaves are broad, long and profuse. The leaves of infected plants show an unusual intense purple colour, while the young emerging leaves show it prominently. Few leaves develop minute necrotic lesions all over the lamina.

**ii) Transmission:** The virus, a strain of TMV, is readily transmitted by mechanical sap inoculation.

**iii) Host Range:** The virus also infects *Cassia tora, Nicotiana glutinosa, Chenopodium amaranticolor, C. album* and *Gompharena globosa. Spinacea oleracea* shows chlorotic rings on inoculated leaves followed by yellow vein-banding in leaves developing subsequently.

**iv) Management:** Rouging of virus-infected plants and removal of weed hosts may help in minimizing the disease incidence.

## 3.27. CARROT

Carrot (*Daucus carota*) is another root vegetable rich in Vitamin A and some minerals. It is grown all over India. Main carrot growing states are Uttar Pradesh, Assam, Karnataka, Andhra Pradesh, Punjab and Haryana. It is taken raw as well as in cooked form. It is made into pickles and sweet meat. Carrot juice is a rich source of carotene (a precursor of Vitamin A) and contains appreciable quantities of thiamine and riboflavin. It is sometimes used for colouring butter and other foods. Green carrot leaves are highly nutritive, rich in protein, minerals and vitamins and used as fodder and also for preparation of poultry feed.

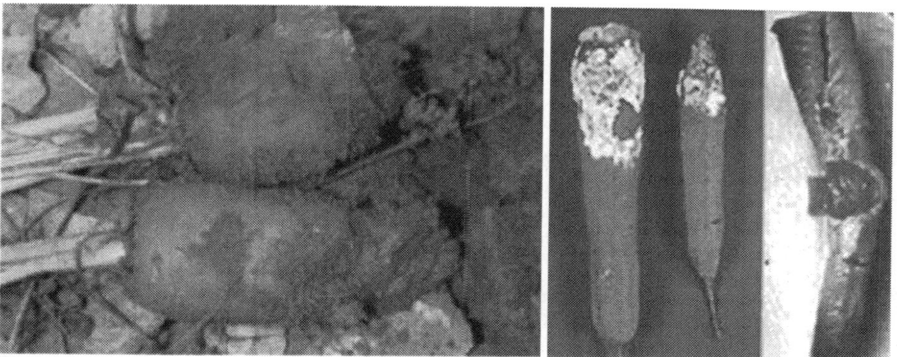

**Fig. 3.38.** Soft rot on carrot.

### 3.27.1. Bacterial Diseases

#### 3.27.1.1. Soft Rot, *Erwinia carotovora* var. *carotovora*

**i) Symptoms:** The disease causes a soft, watery, slimy rot. The rotted tissues are grey to brown and may have a foul odour. It decays the core of the root. Prolonged wet weather favours disease development. It is a serious transit and storage problem if affected carrots are not discarded. In the field, tops of rotted carrots turn yellow and wilt as roots break down. Water soaked irregular lesions appear on tubers (Fig. 3.38). Mostly the disease occurs after the harvest.

**ii) Epidemiology:** Bacteria survive in decaying refuse and enter the root principally through cultivation wounds, harvest bruises, freezing injury, and insect openings. After infection, high humidity is essential for progress of the disease. When soft rot occurs in the field, it usually follows a period of water logging in low areas following excessive rain or irrigation. Carrots, potatoes, onions, crucifers, and celery are only a few of the many plants attacked.

**iii) Management**

- Follow crop rotation of cereals and fodder grasses.
- Spraying carbendazim at 2g/litre of water at an interval of 8-10 days is effective for the management of the disease.
- Destroy by burning of infected plants.
- Carefully handle carrots at harvesting to minimize bruising.
- Discard affected carrots before transport and storage.
- Store carrots in well ventilated places.

### 3.27.1.2. Bacterial Leaf Blight, *Xanthomonas campestris* pv. *carotae*

Bacterial leaf blight of carrot is caused by *Xanthomonas campestris* pv. *carotae* and is a common disease wherever carrot is grown. It is particularly important in areas of frequent rainfall or extensive sprinkler irrigation.

**i) Symptoms:** Symptoms appear primarily on leaf margins as small, yellow, angular leaf spots which expand, turn brown to black with a yellow halo, and become dry and brittle. Leaflets may become distorted and curled (Fig. 3.39). Symptoms can extend into petioles, produce a yellow-brown, gummy exudate, and occur on flower stalks. Infected umbels can be completely blighted and seed infection can occur.

*X. campestris* pv. *carotae* is a common contaminant of carrot seed and can persist in association with infected plant debris in the soil for up to one year. The bacterium is spread by splashing rain, irrigation water, insects, animals, and machinery.

**ii) Management**

- Start with certified, disease-free seed or treat seed with hot water (52°C for 25 minutes).
- Avoid overhead irrigation.
- Rotate out of carrots for 2-3 years.
- Plow under infected crop debris after harvest to hasten decomposition.

Application of copper bactericides can slow disease development, especially if applications begin when plants are young.

**Fig. 3.39.** Bacterial leaf blight on carrot.

## 3.28. TURNIP

### 3.28.1. Bacterial Diseases

#### 3.28.1.1. Bacterial Soft Rot, *Erwinia carotovora* pv. *carotovora*

The disease was noticed in Pusa, Bihar, during 1931 (Prasad, 1931b).

**i) Symptoms:** In early infection, foliage is normal in appearance but in advanced cases, they turn yellow and drop down. On roots, soft, water-soaked patches, paler than healthy portion are visible. Inside the soft patches, tissues are reduced to white pulp from which a putrid smell emnates.

**ii) Management:** Prasad (1931) reported reduction in the disease incidence by rotation of turnip with non-host crops.

### 3.28.2. Viral Diseases

#### 3.28.2.1. Turnip Mosaic Virus

**i) Distribution:** The virus is cosmopolitan in nature. It is known to occur throughout the United States, Europe, South Africa, Asia, Japan, Australia and Taiwan.

**ii) Symptoms:** The first indication of infection is often a premature yellowing and loss of older leaves of individual plants or patches of plants (Fig. 3.40). In severe cases, the entire field will become infected and yellow all at once. Younger leaves produced after a plant becomes infected are stunted, wrinkled, and display a distinct mosaic pattern of yellow areas surrounded by normal green. Distinct symptoms on leaves of turnip develop approximately three weeks after plants become infected. The accelerated loss and replacement of leaves results in an elongated "goose necked" appearance of the roots. Infected turnip become stunted, and loss of leaves makes mechanical harvesting difficult. Roots of plants infected early in the season do not reach normal size.

**Fig. 3.40. Left** - Turnip plant infected with TuMV. Note the older yellow leaves and younger deformed leaves with mosaic pattern of yellow and green areas. **Right** - Virus-infected plant on the right has fewer and smaller leaves and a smaller root compared to the healthy plant on the left.

**iii) Source of Infection:** The virus over winters only in living plant tissue. Important sources of infection early in the season include infected volunteer turnip and infected turnips dumped from storage in the spring, winter rape seed fields and volunteer plants, and possibly a few species of wild plants. Even small amounts of over wintering virus can be important, and severe outbreaks of the disease often occur where volunteer turnips are growing in fields of wheat, barley, or other crops in close proximity to turnip fields. Strains of TuMV have been isolated in the spring from many wild plants, such as dame's violet, yellow rocket, garlic mustard, and water cress, but these strains generally do not infect turnip. During the summer, the virus infects several cruciferous crops and some annual weeds. In addition to turnip and winter rape seed, TuMV can be very devastating on chinese cabbage and other chinese vegetables, horse radish, and mustard crops.

**iv) Infection of Winter Rape Seed:** In the fall, aphids move from turnip fields to winter rape seed, and fields of winter rape seed planted within two or three kilometers of turnip often become highly infected (80 to 100%) with TuMV. Leaves of rape seed infected with the virus display a distinct yellow and green mottling (mosaic); infected plants become stunted and yellow prematurely; and seed pods become shrunken and twisted. Infection with TuMV decreases yield of winter rape seed and may reduce winter hardiness.

Winter rape seed is an important over wintering host for TuMV and turnips should be isolated from this crop as much as possible. Early plantings of turnip within about two kilometers of winter rape seed are often infected earlier and more severely, while late-planted turnips within 5 to 10 kilometers of winter rape seed are at risk because of the spread of TuMV directly from winter rape seed or from previously infected early turnips and other summer hosts. Spring canola is rarely infected with TuMV, and it is not important in the spread of the disease to turnip.

**v) Transmission and Spread:** TuMV is transmitted by aphids notably green peach aphid (*Myzus persicae*) (Fig. 3.41) and cabbage aphid (*Brevicoryne brassicae*) and is readily transmitted mechanically. The virus is transmitted by aphids in a non-persistent manner (short virus transmission period of 10 to 30 seconds) with no latent period (the time from start of an acquisition feeding until the vector can infect healthy plants with the virus). It can be acquired in less than one minute and can be inoculated in less than a minute. The transfer of viruses by aphids usually occurs over short distances (up to a few hundred metres), particularly down-wind and involves migrating alates (winged aphids). Weather conditions and temperature influence aphid activity and migration patterns which in turn affect dissemination of TuMV. Dry and warm conditions favour aphid reproduction and dissemination and hence early and increased spread of the virus. Cool, wet and windy conditions reduce the reproduction and movement of aphids and hence spread of the virus.

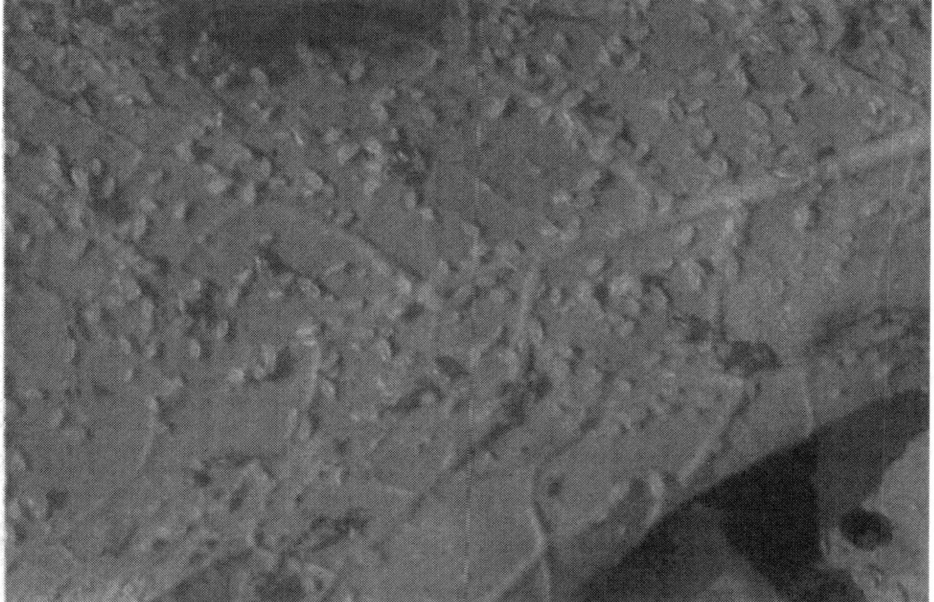

**Fig. 3.41.** Green peach aphids on the under surface of a turnip leaf. This common aphid is one of many species able to spread the disease.

Once primary infections are established in fields, TuMV may spread relatively rapidly from plant to plant if aphids are not controlled. Seed transmission of the virus has not been observed.

### vi) Management

#### a) Cultural Methods

**Effect of Planting Date on Crop Loss:** A rapid increase in infection usually begins in early July when large numbers of winged aphids become active. Aphid flight continues from this time until the end of the growing season. Ideally, the main crop of turnips is planted during the final two weeks of June. Turnips planted at this time size during the cool fall weather, which improves the quality of the roots and prevents growth cracks and oversized roots. Losses from TuMV are greatly reduced, however, if the crop is sown no later than the third week of June and preferably prior to mid-June. Turnips planted after these dates are infected at a young age, and crop loss is often severe.

Applying additional fertilizer, particularly nitrogen, will not help turnip withstand damage from the disease, and may cause roots to break down rapidly in storage.

Listed below is a summary of recommended cultural control measures:

- Turnips should not be ploughed under in the fall, as this helps the roots survive the winter. Turnips left in the field after harvest should be disked and left on the surface to freeze. Volunteer plants that survive the winter should be destroyed early in the spring with herbicide sprays, removal by hand, or cultivation.
- Culls from storage should be dumped early and exposed to freezing weather. Culls surviving in piles can be killed with kerosene or diesel fuel. To prevent the spread of other turnip diseases, culls spread out on fields to freeze should not be placed on land that will be used to grow turnips for at least three years.
- Late-season fields located near earlier plantings often become highly infected, and they should be isolated as much as possible from earlier turnips. Whenever possible, isolate turnips from other cruciferous crops, particularly winter rape seed.
- Whenever possible, plant turnips prior to mid-June, or at the latest by the end of the third week of June. Planting earlier prevents young highly susceptible plants from being exposed at the beginning of the infection period in early-July.
- Market virus-infected turnips as soon as possible. Storing infected roots may be unreliable, and it is best to use virus-free roots for long-term storage.
- Turnips under stress or grown improperly will be more severely affected by the disease.

#### b) Control of TuMV with Weekly Oil Sprays

Weekly applications of oil (70 Superior Oil) are effective for controlling infection of turnips if used properly. Oil acts as a protectant, and prevents aphids from transmitting the virus by interfering with virus particles on aphid mouth parts. Proper application of oil requires the use of high spray volumes up to 1,100 litres per hectare of a 1 to 2% oil solution. Apply oil using T-jet nozzles, preferably more than one per row. The use of drop nozzles will also improve coverage.

Weekly applications of oil should begin around the last week of June and continue at least until the end of July, preferably until the end of August. When plants are growing rapidly, a 5-day interval between sprays provides better coverage of new growth. Continuous coverage is very important, and plants must be protected as soon as they emerge. When plants are small, spray volumes can be reduced as long as leaves are thoroughly covered, but it is important that the amount of oil is reduced to maintain the proper oil concentration. Phytotoxic problems may occur if oil concentration exceeds recommended rates. Turnips may be damaged if oil is applied in full sunlight when temperatures exceed 25°C, or when plants are visibly stressed. Do not apply oil in combination with other spray materials. If insecticides are required, apply them 24 hours before or after the oil.

## 3.29. MUSHROOMS

Mushrooms have achieved significant importance due to their nutritive and medicinal values. Mushrooms, the edible fleshy fungi, represent one of the few valuable sources of nutritive and palatable food. They are rich in proteins (35-48%), vitamins and minerals. They are weight reducers and useful for diabetic patients. Though, these fungi have been naturally occurring and being utilized in India since time immemorial, the commercial cultivation is only of recent times. It was only in 1964 that the first successful crop of mushroom was grown at Solan in Himachal Pradesh. Since then, research on various aspects of mushroom cultivation has been strengthened and today it is one of the important cottage industries, not only in the state of Himachal Pradesh but its commercial cultivation has also been initiated in most of the other states and union territories of India.

At present, it is estimated that 40,000 tonnes of mushrooms are produced annually. The major share (85%) is contributed by button mushroom (*Agaricus bisporus* and *A. bitorquis*). Oyster mushroom (*Pleurotus* spp.) is popular in southern states and Maharashtra and milky mushroom (*Calocybe indica*) in Tamil Nadu, Andhra Pradesh and Karnataka (Table 3.23).

**Table 3.23. State-Wise Production of Cultivated Mushrooms**

| State | Production in MT | Type of mushroom |
|---|---|---|
| Tamil Nadu | 10000 | *Agaricus/Volvariella* |
| Maharashtra | 7000 | *Agaricus/Pleurotus* |
| Haryana | 4000 | *Agaricus* |
| Punjab | 4000 | *Agaricus* |
| Utter Pradesh | 4000 | *Agaricus/Pleurotus* |

| | | |
|---|---|---|
| Andhra Pradesh | 2500 | *Agaricus/Pleurotus* |
| Madhya Pradesh | 2000 | *Agaricus/Pleurotus* |
| Goa | 1000 | *Agaricus* |
| North Eastern States | 1000 | *Agaricus/Lentinus/* |
| Orissa | 700 | *Pleurotus/Volvariella* |
| Karnataka | 600 | *Agaricus/Volvariella* |
| Himachal Pradesh | 500 | *Agaricus* |
| Jammu & Kashmir | 400 | *Agaricus* |
| West Bengal | 250 | *Agaricus/Pleurotus* |
| Kerala | 100 | *Pleurotus/Volvariella* |
| Rajasthan | 50 | *Pleurotus* |
| Others | 1900 | *Agaricus/Pleurotus* |
| Total | 40000 | |

Source: National Centre for Mushroom Research & Training, Solan (1996-97)

### 3.29.1. Bacterial Diseases

#### 3.29.1.1. Bacterial Blotch or Brown Blotch, *Pseudomonas tolaasi*

This is most common and serious bacterial disease of white button mushrooms which is widespread in nature. This disease causes heavy crop losses in marketable mushrooms, especially during the months when temperature and relative humidity are high.

**i) Symptoms:** The characteristic symptom of this disease is the occurrence of dark brown, irregularly shaped spots on surface of the pileus. These spots may initially be light brown in colour but eventually become dark brown. The spots are typically slightly sunken and with a silken sheen. Severely affected mushrooms may become distorted and the entire crop may turn dark brown in colour with splitting of affected areas. In very severe cases, stalk may also be affected. The intensity of symptom development is higher due to more wetting of cap during watering.

**ii) Epidemiology:** In favourable conditions for disease development like high temperature, high humidity and frequent wetting of caps, epidemic levels of blotch disease can occur. In between the crops, the pathogen survives on surfaces, in debris, on tools and on various structures. The chief means of spread of bacteria is through flies, mites and moving equipments in cropping rooms. The bacteria can also be dispersed on debris, on mushroom spores, in water droplets, by flies and on hands of pickers. Watering will disperse bacteria very readily when infection is present in the beds.

**iii) Management:** The treatment of casing material with formalin or steam before use is essential. Frequent and prolonged wetting of caps should be avoided during watering. Use of bleaching powder (sodium hypochlorite) as a routine spray to the beds will help to keep the bacterial population below critical threshold levels of infection. The recommended effective concentration for the use of chlorine at each watering is 150 ppm.

### 3.29.2. Viral Diseases

#### 3.29.2.1. Mushroom viruses (Virus-1, Virus-2, Virus-3, Virus-4, Virus-5)

The viruses are widely distributed all over the world wherever mushrooms are grown. Severe outbreaks of virus diseases on many mushroom farms have been reported with severe crop losses.

**i) Symptoms:** Varied symptoms such as distortion of sporophore, elongation of the stalk, tilting of caps and reduction in size of caps on normal sized stalks are caused by virus infection. Affected beds may show patchy appearance, associated with death of mushroom mycelium. Premature opening of sporophores and reduction in cropping are other symptoms of viral diseases.

**ii) Epidemiology:** Most of the mushroom viruses are spread by mushroom spores and mycelium. The viruses can remain viable for 6 to 9 years in mushroom spores, depending upon the prevailing temperature of the growing areas. Mushroom spores become air-borne, but spread can also be by workers and flies. Infection of the mycelium at spawning or during spawn run and before casing results in maximum yield reduction. The compost may also become contaminated by infected mushroom spores during cool down after pasteurization. Mushroom viruses can also survive in mycelium which is left attached to wooden trays at the end of the crop. Mushroom spawn is one of the potential sources of virus infection.

**iii) Management:** Maintenance of strict hygiene on a mushroom farm is very essential. The entry points to points of pasteurization chamber and growing rooms should be protected by fungal filters effectively. Cook out the beds after completion of cropping with steam, especially when wooden shelves/trays are used. Use mushroom spawn free from viruses.

*Agaricus bitorquis* is resistant to mushroom virus infection and can hence be used to break the disease chain in monoculture system on a mushroom farm.

**Table 3.24. Vegetable crop varieties resistant to bacterial and viral diseases**

| Vegetable crop | Disease/s | Resistant varieties |
|---|---|---|
| Tomato | Bacterial wilt | Arka Abha, Arka Alok, Arka Shreshta, Arka Abhijit, Megha, Shakthi, Sun 7610, Sun 7611, BT-1, BT-10, LE 79-5. |
| | Leaf curl virus | Avinash-2, Hisar Anmol, H-44, H-36, H-86, H-88. |
| Brinjal | Bacterial wilt | Arka Nidhi, Arka Keshav, Arka Neelkant, Swarna Shree, Swarna Shyamali, Surya, SM |

| | | |
|---|---|---|
| | | 6-6, SM 6-7, BB-7, BB- 44. |
| | Little leaf | Pusa Purple Cluster (Field resistant). |
| Chilli | TMV, CMV, Leaf curl | Pusa Sada Bahar, Punjab Lal, Pusa Jwala. |
| | Mosaic, leaf curl | Pant C-1. |
| | Viruses | Arka Harita, Arka Meghana. |
| French bean | Angular leaf spot, Mosaic | Pant Anupama. |
| | Bacterial blight | Arka Anoop. |
| Cowpea | Bacterial blight | Pusa Komal. |
| Field bean | Viral diseases | Pusa Sem-2, Pusa Sem-3. |
| Okra | YVMV | Pusa Sawani, Arka Abhay, Arka Anamika, Hisar Unnat, DVR-1, DVR-2, IIVR-10, Varsha Upkar, P-7, Pusa A-4, Parbhani Kranti (T), Punjab Kesari. |
| Cucumber | Cucumber Green Mottle Virus | DURM-1 |
| Cabbage | Black rot | Pusa Mukta. |
| | Black leg | Pusa Drum Head. |
| Cauliflower | Black rot | Pusa Snowball K-1. |
| | Black rot & Curd blight | Pusa Shubhra. |
| | Curd blight | Pusa Synthetic. |
| Onion | Basal rot | Arka Pitamber, Arka Kirtiman, Arka Lalima. |

MR – Moderately Resistant T - Tolerant

## Table 3.25. Bacterial disease Management in Vegetable Crops

| Crop | Pest | Symptoms | Pesticide Dose/litre | Remarks |
|---|---|---|---|---|
| Solanaceous crops (Tomato, brinjal, chilli, capsicum) | Bacterial wilt | Wilting of entire plant, oozes white milky streak in water | Bleaching powder (Soil application before transplanting) at 15 kg/ha | 15 days before transplanting |
| Cabbage/ Cauliflower | Black rot | Blackening of stem leading to rotting and foul smell | Streptocycline (seed soaking for 25 min)-0.1 g | Seed treatment |
| | Bacterial black spot | Light brown to black spores on leaves | Streptocycline-0.1 g (100 ppm ) | Rainy season |
| French Bean | Angular leaf spot | Light brown angular spots on leaves | Mancozeb-2 g<br>Carbendazim-1 g<br>Thiophanate Methyl-3 g | September-October |
| Cowpea | Bacterial blight | V- shaped yellow water soaked spots on leaves | Streptocyline-0.1 g (100 ppm ) | Rainy season |

**Table 3.26. Bacterial biocontrol agents for the management of diseases of Vegetable crops**

| Horticultural crops | Disease | Biocontrol agent | Mode of application |
|---|---|---|---|
| Tomato | Bacterial wilt, *Ralstonia solanacearum* | *Pseudomonas fluorescens, Glomus mosseae, G. fasciculatum* | Soil treatment |
| Potato | Bacterial wilt, *Ralstonia solanacearum* | *Bacillus cereus, B. subtilis* | Soil treatment |
| | Tuber rot, *Erwinia carotovora* | *P. fluorescens* | Seed/tuber treatment |
| | Scab, *Streptomyces scabies* | *T. viride, T. harzianum, B. subtilis* | Soil treatment |
| Onion | Wilt, *Erwinia carotovora* | *T. viride* | Seed treatment |

## 3.27. Management of Viral Diseases in Vegetables

| Crop | Pest | Symptoms | Pesticide Dose/litre | Remarks |
|---|---|---|---|---|
| Tomato | Leaf curl | Curling of leaves | Carbofuran-1 kg a.i./ha (Soil application in nursery) Carbofuran-1.5 kg a.i./ha (Soil appln. in field) Dimethoate-1.5 ml Monocrotophos-1.5 ml | Raise nursery under nylon net (40 gauze) during summer months, spray once in 10-15 days in main field. |
| | PotatoVirus Y/ Cucumber mosaic virus | Mosaic mottling | Carbofuran-1 kg a.i./ha (Soil application in nursery) Carbofuran-1.5 kg a.i./ha (Soil appln. in field) Dimethoate-1.5 ml Monocrotophos-1.5 ml | |
| | Spotted wilt | Necrotic spots on the leaves & stem and concentric rings on fruits | Carbofuran-1 kg a.i./ha (Soil application in nursery) Carbofuran-1.5 kg a.i./ha (Soil appln. in field) Dimethoate-1.5 ml Monocrotophos-1.5 ml | |
| Chilli/ Capsicum/ Brinjal | Leaf curl/ mosaics (Transmitted by aphids) | Mosaic mottling as in tomato | Carbofuran-1 kg a.i./ha (Soil application in nursery) Carbofuran-1.5 kg a.i./ha (Soil appln. in field) | Soil application followed by insecticide sprays every fortnight |

| | | | | |
|---|---|---|---|---|
| | | | Monocrotophos-1.5 ml<br>Methyl demeton-2 ml | |
| | Spotted wilt | Necrotic spots on the leaves & stem and concentric rings on fruits | Carbofuran-1 kg a.i./ha (Soil application in nursery)<br>Carbofuran-1.5 kg a.i./ha (Soil appln. in field)<br>Monocrotophos-1.5 ml<br>Methyl demeton-2 ml | Soil application followed by insecticide sprays every fortnight |
| Okra | Yellow vein mosaic | Young leaves start curling and whiteflies on the lower surface of leaves | Carbofuran-1 kg a.i./ha<br>Monocrotophos-1.5 ml<br>Methyl demeton-2 ml | Soil application followed by insecticide sprays every fortnight |
| | Enation leaf curl | Enations on leaves, bending of petiole | Carbofuran-1 kg a.i./ha<br>Monocrotophos-1.5 ml<br>Methyl demeton-2 ml | Soil application followed by insecticide sprays every fortnight |
| Cucurbits | Mosaics(Aphid transmitted) | Aphids on young leaves and curling of leaves | Carbofuran-1 kg a.i./ha<br>Monocrotophos-1.5 ml<br>Methyl demeton-2 ml | Insecticide sprays every fortnight |
| | Bud necrosis in water melon | Necrotic spots on leaf, streaks on vines | NSKE-4%<br>Dimethoate-2 ml<br>Monocrotophos-1.5 ml<br>Methyl demeton-2 ml | Sprays are to be given every fortnight |

Chapter 4

# ORNAMENTAL CROPS

India is known for growing of traditional flowers such as jasmine, marigold, chrysanthemum, tuberose, crossandra and aster. Commercial cultivation of cut flowers-rose, orchids, gladiolus, carnation, gerbera, anthurium and lilies-has also become popular.

With the liberalization of Indian economy, the domestic floricultural trade, which hitherto remained an unorganized sector, became an organized sector with the active participation of corporate giants. The establishment of state-of-art hi-tech floricultural units in early 90's for intensive cultivation of cut flowers contributes about Rs. 300 crores of valuable foreign exchange today. Various government agencies are actively involved in the promotion of this sector and are highly committed to resolve the teething problems. The world demand for floricultural products is approximately US $ 40 billions and cut flowers contribute nearly 60% (US$ 25 billions) to the global trade. India's share in the world flower market is about 0.7% which is contributed mainly from open cultivation.

Floriculture is estimated to cover an area of 126,000 ha with production of 694,000 MT of loose flowers during 2006. The area under cut flowers has increased significantly in recent years and the production has reached 2762 million numbers (Table 4.1). Another 500 ha of climatic control greenhouse are available for growing quality flowers for export. Many corporate houses (more than 100) have set up 100% export oriented units with an investment of more than Rs. 1000 crores. India is bestowed with natural advantages like favourable weather conditions, soil, cheap labour which are likely to make a real mark in international trade. With this, India is emerging as one of the powerful forces in world flower trade.

**Table 4.1. State-wise area and production of flowers (1999-2000)**

| State / UT's | Area (in ha ) | Production (in MT ) | |
|---|---|---|---|
| | | Loose | Cut |
| Andhra Pradesh | 18087 | 126609 | --- |
| Delhi | 3450 | 21007 | --- |

| | | | |
|---|---|---|---|
| Haryana | 2250 | 40500 | 760 |
| Himachal Pradesh | 133.3 | 976 | 181 |
| Jammu & Kashmir | 42 | 7 | 17.4 |
| Karnataka | 20801 | 124440 | --- |
| Madhya Pradesh | 1387 | 832 | --- |
| Maharashtra | 6600 | 28336 | --- |
| Orissa | 175 | 868.5 | 9.3 |
| Tamil Nadu | 18120 | 142250 | --- |
| U.P. Plains | 160 | --- | 160 |
| West Bengal | 13227 | 14644 | 5678 |
| Others | 4174.7 | 8723.5 | 1.0 |
| Total | 88607 | 509193 | 6806 |

Floriculture is an intensive type of agriculture wherein the income per unit area is much higher than any other agricultural product if it is done in a scientific way. The area used for production of ornamental plants is low, but the cash returns are high because of the intensive cropping. The flowers grown in large areas include jasmine, rose, marigold, chrysanthemum, crossandra, champak, tuberose, gladiolus and aster. New crops like lilies, tulips, anthurium, carnation, gerbera, etc., are also finding potential place in the foreign market. With the rapid development in urban planning and hotel industry, there has been an increasing demand for cut flowers and ornamentals. The cut flowers followed by live plants, cut foliage, seeds and bulbs have got an immense potential for export to Western countries where the cost of production is comparatively high. India exports roses, gladioli, carnations, chrysanthemums, lilies and jasmines to Gulf, Germany, France, England, Netherlands and Singapore. Our share in the export market has registered a sharp increase from Rs.14.45 crores in 1991-92 to Rs.299.41 crores in 2005-06.

## 4.1. ROSE

Rose (*Rosa* spp.) is the most ancient and popular flower grown the world over. In India, it is commercially grown for cut flowers and rose oil. Rose flowers without stems and loose flower petals are used in traditional markets for making garlands, for offering in temples, while the florist shops sell cut roses with stems mainly for bouquets and floral arrangements. In recent times, about 60 units have been established under joint ventures around Bangalore, Pune, Nasik, Mumbai, Hyderabad, Gurgaon (Haryana), Chandigarh and Saharanpur (Uttar Pradesh) for growing roses in greenhouses for export of cut flowers to Japan, Major protected rose growing areas of include Ambala (Punjab), Bangalore (Karnataka), Hyderabad (Andhra Pradesh), Kodaikanal (Tamil Nadu), Nasik (Maharashtra), New Delhi and Pune (Maharashtra). The Netherlands, Germany and other European countries. Besides, the Damask rose (*Rosa damascena*) and Edward rose (*R. bourboniana*) are cultivated for rose *attar* and other products – *gulkand, gulabjal,* and *pankkurj*. It is cultivated in about 6,000 ha area in the states of West Bengal, Uttar Pradesh, Gujarat, Haryana, Jammu and Kashmir, Tamil Nadu, Karnataka, Rajasthan, Madhya Pradesh, Andhra Pradesh and Punjab.

As is the case with other plants, maintaining the health of roses makes them less susceptible to diseases and better able to withstand disease attack. Cultivars that are more resistant to certain diseases should be used. Humidity encourages fungal disease, as does poor air circulation. Avoid planting roses too closely; avoid overhead watering; and prune to maintain a fairly open structure.

### 4.1.1. Bacterial Diseases

#### 4.1.1.1. Crown Gall, *Agrobacterium tumefaciens*

**i) Symptoms:** The bacterium *Agrobacterium tumefaciens* that live in the soil causes crown gall. The bacteria enter a plant through the roots or through wounds at the root area. The bacteria cause abnormal cell growth, which produces the galls. The galls that develop on stems or just under the soil surface are round and rough textured. Round growths about 5 cm in diameter appear at the base of the plant (Fig. 4.1). Young galls are soft and white or pale green; older ones are dark and woody. Affected plants lose vigour and may become stunted, with abnormally small leaves and fewer flowers. If the disease affects the plant whilst it is young the plant may be affected to the degree where it will not produce blooms. All affected plants wilt readily and grow poorly. The bacterium is spread on garden tools.

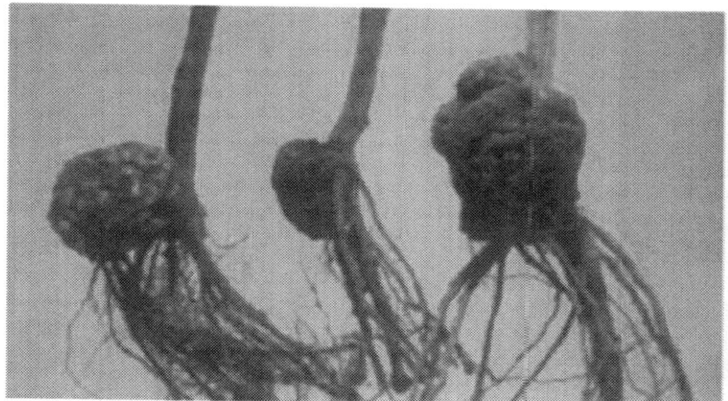
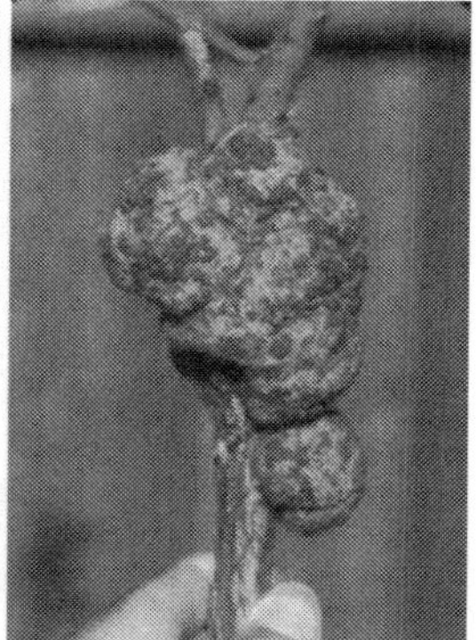

**Fig. 4.1.** Crown galls on stems of rose.

**ii) Management:** Commercial formulation of non-tumerogenic strain of *Agrobacterium rhizogenes* (K-84 strain) is being used successfully to control crown gall.

### 4.1.2. Viral Diseases

#### 4.1.2.1. Rose Mosaic Virus

**i) Symptoms:** This disease is caused by a complex or viruses and is characterized by yellow patterns on the leaves. The patterns vary considerably, ranging between all-over fine blotches to patterns of lines in waves (Fig. 4.2). The patterns may appear on a few or many leaves. Plants are infected by this virus at propagation using infected plant material.

**ii) Management:** Clean nursery practices, production of disease-free planting material and removing and destroying all infected plants helps in checking the disease.

**Fig. 4.2.** Mosaic virus on Rose leaves.

## 4.2. CARNATION

Carnation (*Dianthus caryophyllus*) flower is valued for its excellent keeping quality, wide array of colours and forms, and ability to rehydrate after continuous transportation. Moderate climatic control measures that are economical can deliver quality carnations at the internationally competitive prices year round. In India, Carnation and gerbera are commercially grown mostly under cover around big cites like Solan, Shimla, Mandi, Kullu, Chandigarh, Ludhiana, Guragaon, Kalimpong, Pune, Bangalore and Delhi. Carnation has tremendous potential for export market.

### 4.2.1. Bacterial Diseases

#### 4.2.1.1. Bacterial Wilt/ Blight, *Pseudomonas caryophylli*

**i) Symptoms:** Primary symptoms of infection include wilting of plants or shoots as well as splitting of the stems and frayed yellow streaks in the vascular system. Tissue just under the epidermis is sticky and cutting across the stem shows discolouration and bacterial ooze.

**ii) Survival and Spread:** The pathogen survives in infected plants and debris. Bacteria spread in water. The disease appears to spread rapidly at high temperatures.

**iii) Management:** Control involves using clean cuttings, rouging and removing symptomatic plants, and avoiding overhead irrigation and splashing water. Infected plants should not be composted. Sanitation is also very important since this disease is highly contagious. All equipment should be disinfested between use with 10% household bleach, 70% alcohol, or one of the commercially available compounds.

Carnation cvs. Cal Red and Cal Improved White showed high disease of resistance.

### 4.2.2. Viral Diseases

#### 4.2.2.1. Carnation Mottle Virus (CarMV)

**i) Symptoms:** CarMV is the most common and widespread virus disease of carnations. Most carnation cultivars infected with this virus do not express symptoms. In other cases, when the symptom is expressed, the infected plant may have a yellow mottle on the leaves. Flower quality and yield are reduced. "Virus free" plants have broader leaves and more vigorous growth. CarMV is highly infectious and is rapidly spread from plant to plant by foliage contact or during handling. Research has shown that up to 75% of carnations initially free of CarMV became infected within six months when grown in the same glasshouse as carnations infected with CarMV.

A flexuous rod-shaped virus measuring 718 nm x 12 nm was isolated from carnation. It causes veinal chlorosis, leaf mottling, chlorotic spots, distortion of flowers and stunting of plants. The virus belongs to the corlavirus group.

**ii) Management :** The only current method for controlling virus diseases is through the use of virus tested material in conjunction with strict hygienic practices. To prevent infection of this virus-tested material, the carnation grower should enforce hygiene regulations (for example, washing hands with hot soapy water before handling plants, aphid-proof glasshouses, sterilization of cutting and propagating instruments, etc). It is essential to educate workers about the spread of virus diseases.

#### 4.2.2.2. Etched Ring

**i) Symptoms:** Rings are usually oval or elongated; rarely concentric on older leaves and stems. Symptoms may be slight in young cuttings. No obvious effect on plant vigour.

**ii) Transmission:** The virus is transmitted by aphids. It does not spread by handling.

**iii) Management**

- Obtain virus-free plants.
- Control aphids.

#### 4.2.2.3. Necrotic Fleck

**i) Symptoms:** Reddish purple necrotic flecks, streaks, or spots appear on leaves. Symptoms are masked at low temperatures.

**ii) Transmission:** The virus is transmitted by aphids. It does not spread by handling.

**iii) Management**

- Obtain virus-free plants.
- Control aphids.

### 4.2.2.4. Ring Spot

**i) Symptoms:** The symptoms include small 1- to 2-cm rings (sometimes concentric), chlorosis, mottling, and distortion of young leaves. Plants are obviously stunted in growth.

**ii) Transmission:** The virus is transmitted through cutting knife and handling. Insect vectors are not known.

**iii) Management**

- Obtain virus-free plants.
- Sterilize cutting knife frequently.

### 4.2.2.5. Vein Mottle

**i) Symptoms:** Young leaves exhibit a vein clearing, which develops into chlorotic spots and patterns that mostly follow veins. Symptoms tend to disappear on old leaves.

**ii) Transmission:** The virus is transmitted by aphids. It does not spread by handling or cutting knife.

**iii) Management**

- Obtain virus-free plants.
- Control aphids.

## 4.3 CHRYSANTHEMUM

Garden chrysanthemum (*Dandrathema grandiflora*), also known as 'Queen of the East', is a highly versatile and accommodating ornamental with a wide range of type, size and colour in flowers. In India, large flowered varieties are grown for cut flower, making garland, wreaths and 'veni', religious offerings and for bedding and potting purposes. Its commercial cultivation is being done in Maharashtra, Rajasthan, Madhya Pradesh and Bihar. Small flowered varieties are commercially grown for loose flowers in Tamil Nadu, Karnataka and Maharashtra. In India, chrysanthemum is grown in about 4,000 ha.

Although the list of diseases affecting Chrysanthemums is long, they are relatively trouble-free given full sun, fertile well-drained soil, and adequate watering. Maximizing plant vigour by appropriate cultural practices is an important disease management strategy.

### 4.3.1. Bacterial Diseases

#### 4.3.1.1. Bacterial Blight, *Erwinia chrysanthemi*

**i) Symptoms:** Symptoms of bacterial blight extend beyond plant leaves to include water-soaked lesions on stems, darkening and death of buds and stems, blackening of terminals, wilt and collapse of upper portions of the plants (Fig. 4.3). Infected cuttings may have brown to black decay at their base.

**Fig. 4.3.** Bacterial blight on chrysanthemum.

**ii) Survival and Spread:** Bacterial blight pathogen survives in crop debris and is favoured by surface moisture, high temperature, and high humidity. It is easily spread through infested tools, hands, or plants.

**iii) Management:** The disease can be managed by starting with pasteurized growing media, use of pathogen-free cuttings, reducing humidity and increasing air circulation, avoiding wetting of the foliage, practicing good sanitation, and regularly inspecting the crop and disposing of infected plants.

#### 4.3.1.2. Bacterial Leaf Spot, *Pseudomonas cichorii*

**i) Symptoms:** Symptoms of bacterial leaf spot are tan to dark brown spots or blotches that are often bordered or ringed by yellowing tissue. Discolouration may be prominent along leaf veins or lesions may become angular as bacteria growth is limited by major veins. Leaf wilting and death often follow.

**ii) Survival:** Bacteria persist in or on infected plants, crop debris, infected seed, contaminated soil, and infested pots and tools.

**iii) Management:** Key management practices include planting pathogen-free seed and cultivars, resistant varieties, good sanitation, and avoiding overhead irrigation or handling plants when they are wet. Once plants become infected with bacteria, it is best to rogue infected plants and those near them before the disease spreads. Bactericides such as copper and antibiotics are of limited effectiveness and plants cannot be cured.

### 4.3.2. Viral Diseases

#### 4.3.2.1. Chrysanthemum Stunt Viroid (CSV)

**i) Symptoms:** This disease is highly infectious and can cause serious losses in chrysanthemum crops. Infected plants are reduced in size (Fig. 4.4) and they tend to flower earlier. Often the flowers are small, distorted and flower colour is bleached. Susceptible cultivars often have numerous conspicuous white leaf spots, termed "measles".

**Fig. 4.4.** Severe stunting of chrysanthemum plant (left) and flower distortion in chrysanthemum caused by stunt viroid infection.

**ii) Transmission:** CSV is transmitted by sap from infected plants coming into contact with healthy plants. For this reason, cultural practices such as the use of contaminated tools, knives and even the movement of staff from areas where plants are infected to healthy stock, are the main avenues for transmission of CSV in the chrysanthemum crop.

**iii) Management:** No spray or chemical treatment will eliminate a virus from an infected plant. Disease control can only be achieved by use of virus-tested planting stock in conjunction with strict hygienic practices. The movement of insects, such as thrips and aphids, should be controlled. Propagating tools, knives, etc. must be sterilized after use. Staff should be taught about the effects of virus diseases and their mode of transmission.

Washing of hands with hot soapy water before handling chrysanthemum plants helps safeguard against virus diseases that are transmitted by contact.

#### 4.3.2.2. Chrysanthemum Virus B

**i) Symptoms:** Chrysanthemums virus B can cause stunting, chlorosis (yellowing of parts of leaves), and lack of vigour. Chrysanthemums are easily infected with virus in the field.

**ii) Transmission:** Insects such as aphids and thrips transmit the disease as they feed and move from plant to plant. Cutting instruments that have not been sanitized with bleach or alcohol can also spread the virus. Once a plant has contracted a virus, it is spread systemically to nearly every plant part. The virus invades every cells of the plant. Cuttings taken from a virus infected stock plant will also contain the virus.

**iii) Management :** To remedy the virus, a virus elimination process can be implemented. Virus elimination is accomplished by harvesting the "apical meristem," the tissue occurring at the vegetative shoot tip. A meristem is where new cells are made. Meristem cells are often virus-free, possibly because new cells are made faster than the virus can invade them. To begin the process, a vegetative apical shoot of a virus-infected plant is dissected under a microscope. Leaves are removed from the shoot one-by-one with a scalpel. Each leaf is smaller than the previous one. Eventually, the apical dome containing the meristem, is revealed. It resembles a shiny drop of green jelly. The dome and surrounding area are carefully removed with a scalpel. The resulting speck on the tip of the scalpel is barely visible to the naked eye. The tissue is placed in a test tube on a specialized medium containing mineral nutrients, sugar, plant hormones, and a gelling agent. The tissue grows on this meristem-isolation-medium for 6 weeks (Fig. 4.5).

**Fig. 4.5.** Meristem tissue culture to produce virus-free plants.

During this time the cells differentiate into "micro-shoots" with leaves and stems looking the same as a normal plant except on a miniature scale. Once a micro-shoot is large enough, it is removed from the isolation media and placed on multiplication medium (similar to isolation medium, but with higher mineral content). After a few weeks the micro-shoot can be harvested and rooted like a conventional cutting. Once the plant is established and actively growing, leaf samples from the plant are tested for presence of virus. Virus-free plants from over 20 chrysanthemum cultivars including 'Kokka Senkin', 'Icy Isle', and 'Western Brushfire' have been successfully recovered. These new virus-free plants provide the stock plants to produce all the new display plants for the future. Virus-free cultures are also maintained in case of re-infection of stock plants.

#### 4.3.2.3. Spotted Wilt Virus

**i) Symptoms:** The infection is frequently one-sided in plant. Ring patterns develop on leaves of some cultivars. Other symptoms include leaf distortion and necrosis, dark necrotic streaks on stems, and flowers may be distorted and with some necrosis.

**ii) Transmission:** The virus is transmitted by thrips. It does not spread by cutting knife, but can be spread by vegetative propagation. Many weeds and perennial ornamental plants act as reservoirs of virus (dahlia, calla, nasturtium, mallow, knotweed, plantain, and others).

**iii) Management**

- Eliminate nearby susceptible ornamental plants and weeds.
- Control thrips inside and outside growing areas.
- Use virus -free cuttings.

#### 4.3.2.4. Chrysanthemum Mosaic Virus

**i) Symptoms:** The symptoms include mosaic mottling of foliage, dwarfing and deforming of leaves, stems, and flowers. Symptoms are variable, depending upon strain of virus and cultivar.

**ii) Transmission:** The virus is transmitted by vegetative propagation and by aphids.

**iii) Management**

- Obtain disease-free plants.
- Control aphids.

#### 4.3.2.5. Tomato Aspermy Virus

**i) Symptoms:** The symptoms include flower distortion and reduction in flower size, colour break in florets of red, bronze, and pink cultivars. Foliar symptoms are not apparent. Some cultivars are symptomless.

**ii) Transmission:** Transmitted by handling, cutting tools, and vegetative propagation. In nature, the virus is transmitted by aphids.

**iii) Management**

- Use disease-free cuttings.
- Control aphids.
- Remove infected plants.

## 4.4. GLADIOLUS

Gladiolus (*Gladiolus* spp.) is very much liked for its majestic spikes containing attractive, elegant and delicate florets. These florets open in sequence over a longer duration and hence have a good keeping quality of cut spikes. It is cultivated for cut flowers, garden and interior decoration, making bouquets, bedding, rockeries and pots. West Bengal, Maharashtra, Uttar Pradesh, Punjab, Haryana and Andhra Pradesh are the major gladiolus-growing states.

### 4.4.1. Bacterial Diseases

#### 4.4.1.1. Scab, *Pseudomonas marginata*

**i) Symptoms:** Scab infection is caused by the bacterium *Pseudomonas marginata*. The infection on corm husks appears as elongated lesions with black rough margins. Early corm infections start as pale yellow, water-soaked spots. As the disease progresses, circular lesions turn brown and sunken with distinct raised margins. Early symptoms on the leaves are raised reddish specks near the leaf bases. The neck of the plant rots and the plant falls over. In some cases, a gummy exudate produced by the bacteria may glue husks to the corms.

**ii) Spread:** The bacteria may spread from infected tissue in splashing water or by bulb mite feeding.

**iii) Management:** To minimize scab infection, remove and destroy infected corms or plants. Follow cultural management recommendations and chemical recommendations.

#### 4.4.1.2. Bacterial Blight

**i) Symptoms:** Bacterial blight is serious in wet weather or on plants grown in poorly drained soil. Infected areas are first irregular, water-soaked spots that later dry out and turn brown. The whole leaf becomes involved and dies. A slimy exudate may be seen.

**ii) Management:** Discard infected corms.

### 4.4.2. Viral Diseases

#### 4.4.2.1. Bean Yellow Mosaic Virus (BYMV), Cucumber Mosaic Virus (CMV)

More than seven virus diseases have been recorded in gladiolus, and of these, bean yellow mosaic virus (BYMV) and cucumber mosaic virus (CMV) are the most economically important and widespread.

**i) Symptoms:** Gladiolus plants infected with BYMV and/or CMV often have pale-yellow mottling of the leaves (Fig. 4.6). Symptoms also include "flower break" and "colour break" of flowers, but not in white, cream or yellow cultivars (Fig. 4.6). BYMV can reduce the production of gladiolus corms by about 33%. Infected plants may have a shortened life span and corms are also more susceptible to fungal attack.

**Fig. 4.6. Left** - Mosaic on leaves which is typical of BYMV in gladiolus. **Right** - Severe colour break in red variety which is typical of a dual infection with CMV and BYMV.

**ii) Host range:** Both viruses are able to infect a wide range of different plant species, including many common weeds, pasture legumes, other bulb species, and vegetables.

**iii) Spread and Source of Infection:** BYMV and CMV can infect a number of plant species apart from gladiolus, including pasture legumes and weeds which provide a field source of these viruses. The most common source of infection in gladiolus is from existing gladiolus crops that are infected with the viruses. Both viruses are spread from plant to plant by aphids which transmit these viruses in only 30 seconds. Aphids will also feed on corms and hence gladiolus corms are susceptible to virus spread during storage.

**iv) Management:** The best control strategy is to plant corms from virus tested stock. To minimize the risk of infecting healthy stock:

- Use suitable insecticides for the control of aphids on both the cultivated crop and the stored gladiolus corms.
- Rogue any plants which show symptoms and destroy this material.
- Plant virus-free gladiolus material at least 1 km (if possible) from virus-infected gladiolus crops.

### 4.4.2.2. Gladiolus Mosaic Virus

**i) Symptoms:** Mottling of leaves, whitish streaks, and color abnormality in flowers are characteristics of mosaic.

**ii) Management:** Destroy diseased plants. Control insects, especially aphids, which transmit the virus to healthy plants.

## 4.5. MARIGOLD, *Tagetes* spp.

### 4. 5.1. Viral Diseases

#### 4.5.1.1. Little Leaf (Phytoplasma)

**i) Management:** The disease can be controlled by spraying streptomycin at 5 g/litre of water.

#### 4. 5.1.2. Mosaic Virus

**i) Management:** Collection and destruction of infected plants reduces the incidence of the disease.

## 4.6. CHINA ASTER

China aster (*Callistephus chinensis*) is one of the important annual flower crops. It ranks next to chrysanthemum and marigold among traditional flowers. In recent years, it has gained more popularity due to its multifarious uses including as cut flower. The flowers fetches very good price and returns to the farmer when their production coincides with their demand during particular season/festival. It is more popular among small and marginal farmers due to its easy cultivation. It is grown for cut flowers, herbaceous borders, bedding, bouquet and pots and cultivated in most parts of the country. It is commercially grown in Karnataka, Tamil Nadu, Andhra Pradesh, Maharashtra (Pune) and West Bengal. In Bangalore and Pune alone it is being grown in an area of 500 and 400 ha, respectively.

### 4.6.1. Viral Diseases

#### 4.6.1.1. Aster Yellows (Mycoplasma)

**i) Symptoms:** Aster Yellows is caused by a mycoplasma-like organism, with symptoms similar to viral diseases. Plants are stunted, and the top leaves turn yellow; blooms are also stunted and may rot, taking on a greenish, leaf-like form (Fig. 4.7). Upper leaves are erect. Flower heads open unevenly or only partly and have a yellowish-green cast.

**ii) Host Range:** It affects China aster, calendula, coreopsis, cosmos, delphinium, daisies, marigolds, petunias, phlox, zinnia, and many other flowers and weeds. Carrots, celery and lettuce are also attacked.

**iii) Transmission:** The disease is spread by the six-spotted leaf hopper.

**iv) Management**

**a) Cultural control:** Prompt removal and destruction of affected plants when noticed. Control weeds, many of which are also hosts.

**b) Chemical control:** Chemical control aims to check the spread of the leaf hopper insect. Use insecticides to control leaf hoppers that spread the mycoplasma to neighbouring asters.

Fig. 4.7 Aster yellows.

## 4.7. NARCISSUS

### 4.7.1. Viral Diseases

#### 4.7.1.1. White Streak Virus

**i) Symptoms:** Narrow, dark green to purple streaks, which later become white to yellow-white, appear in leaves and flower stalk after flowering. Bulb size and yields are reduced by premature senescence.

**ii) Transmission:** Narcissus is the only host of the virus, which is transmitted by aphids. Symptoms do not occur until air temperatures exceed 64°F.

**iii) Management**

- Eliminate infected plants.
- Control aphids.

#### 4.7.1.2. Yellow Stripe Virus

**i) Symptoms:** Conspicuous light green to yellow streaks and mottling of leaves and flower stalk, which occurs shortly after emergence. Sometimes leaves are distorted and a colour-break of flowers occurs. Bulb yields are reduced.

**ii) Transmission:** Narcissus is the only host of the virus, which is spread by aphids. Symptoms appear early in growing season.

### iii) Management

- Eliminate infected plants.
- Control aphids.

## 4.8. ZINNIA

### 4.8.1. Bacterial Diseases

#### 4.8.1.1. Bacterial Leaf Spot

**i) Symptoms:** Reddish-brown to dark brown, angular spots up to 1/8 inch in diameter, with a prominent yellow halo are produced on leaves. Spots may coalesce (Fig. 4.8).

**Fig. 4.8.** Leaf spot on zinnia.

### ii) Management

- Avoid overhead irrigation.
- The disease is seed-borne; do not plant diseased seedlings.
- Destroy diseased plants, especially if infected when young.

## 4.9. OLEANDER (NARIUM)

### 4.9.1. Bacterial Diseases

#### 4.9.1.1. Oleander (Narium) Gall, *Pseudomonas syringae* pv. *savastanoi*

Oleander (*Nerium oleander*) is one of the most popular evergreen shrubs. The gall disease is widespread.

**i) Symptoms and Biology:** Oleander gall is caused by the bacterium *Pseudomonas syringae* pv. *savastanoi*. Galls occur on twigs, branches, leaves, flowers, and seed pods. Initially galls appear as small protuberances that subsequently develop into wart-like growths with roughened, fissured surfaces. Galls vary in size but average about 1/2 to 1 inch in diameter. Large galls are usually made up of several small galls that have grown together.

Galls are the result of the growth and multiplication of the bacterium. The bacteria enter and infect oleanders through leaf and blossom scars, wounds produced by pruning, frost injury, and natural openings.

**ii) Spread:** Rain, sprinkler water, and pruning tools can spread bacteria from diseased to healthy plants.

**iii) Management :** When procuring oleanders, examine them carefully to be sure they are free of galls. The vast majority of nursery stock is free from disease but prevention is always the most effective method of disease control. For diseased oleanders, prune out infected plant parts and apply disinfection solution (a 10 percent solution of household bleach) to each cut surface. Always dip pruning tools in the disinfectant solution between cuts to reduce the possibility of spreading the bacteria. Pruning operations should be conducted during the dry seasons to avoid infection of wounds. Avoid sprinkler irrigation.

Severe infection of large shrubs is difficult to control by selective pruning. If the entire shrub is cut down, the new succulent growth is extremely susceptible to infection. In certain situations, removal of the diseased plant and replanting may be the best method of control.

## 4.10. DAHLIA

### 4.10.1. Viral Diseases

#### 4.10.1.1. Mosaic Stunt or Dwarfing Virus

**i) Symptoms:** A common characteristic is stunt or dwarfing with a tendency to produce many shortened lateral shoots. A bushy plant with mottled leaf colour is produced. Scattered areas of yellow or light green tissue occur in the leaves at first. Conspicuous ring spots develop on leaves in some cases.

**ii) Management**

- Dig up and remove or destroy diseased plants.
- Plant only tubers from healthy plants.
- Control aphids or other vectors with approved insecticides.

#### 4.10.1.2. Ring Spot Virus

**i) Symptoms:** Scattered areas of yellow or light green tissue occur in the leaves at first. A conspicuous ring pattern develops later on the leaves.

**ii) Transmission:** Thrips transmit the virus to neighbouring plants.

**iii) Management**

- Control thrips with approved insecticides.
- Plant only tubers from healthy plants.

## 4.11. AMARYLLIS AND HIPPEASTRUM

### 4.11.1. Viral Diseases

#### 4.11.1.1. Cucumber Mosaic Virus (Amaryllis Mosaic)

**i) Symptoms:** Large yellowish green patches appear at random over the leaf surface.

**ii) Transmission:** Aphids transmit the virus to neighbouring plants.

**iii) Management :** The aphid vector can be controlled by spraying 0.2% malathion regularly at 15-days interval. Sometimes, this disease also spreads through weeds and these should be eradicated from the amaryllis plots. CMV can also be controlled by treating bulbs at temperature of 25°C and light intensity of 1000 ft candles. Aseptic shoot-apex culture was employed in an attempt to recover virus-free plant material from the infected bulbs. Bulb-cottage was a source of shoot-apices which subsequently were cultured on liquid nutrient medium at a temperature of 25°C and light intensity of 1000 ft-candles.

#### 4.11.1.2. Tomato Spotted Wilt

**i) Symptoms:** The leaves of the affected plants show numerous yellow spots and the plants ultimately die. The symptoms are similar to amaryllis mosaic virus but are coarser.

**ii) Transmission:** Thrips transmit the virus to neighbouring plants.

**iii) Management:** This disease is spread by thrips may affect many other plants of Amaryllidacae. The thrips can be controlled by spraying 0.2% Rogor or Malathion. The antibiotics Streptomycin and Terramycin suppress two of the virus diseases, tobacco ring spot and tobacco mosaic. All mosaic-infected plants should be burnt as soon as identified. The control of possible insect or mite carriers is particularly important under outdoor amaryllis culture; nearly weeds that might harbour virus diseases should also be eradicated.

## 4.12. CALADIUM

### 4.12.1. Bacterial Diseases

#### 4.12.1.1. *Xanthomonas* leaf spot, *Xanthomonas campestris* pv. *dieffenbachiae*

**i) Symptoms:** Leaf spot on caladium starts as tiny water-soaked areas which can rapidly enlarge to 1/4 inch or more (Fig. 4.9). They tend to form on leaf margins but when the bacterium infects a major leaf vein the entire leaf can collapse. Spots are frequently very black and surrounded by a bright yellow halo. Most other plants in the Aroid family such as Aglaonema, Anthurium, Dieffenbachia and Syngonium are also hosts of this pathogen.

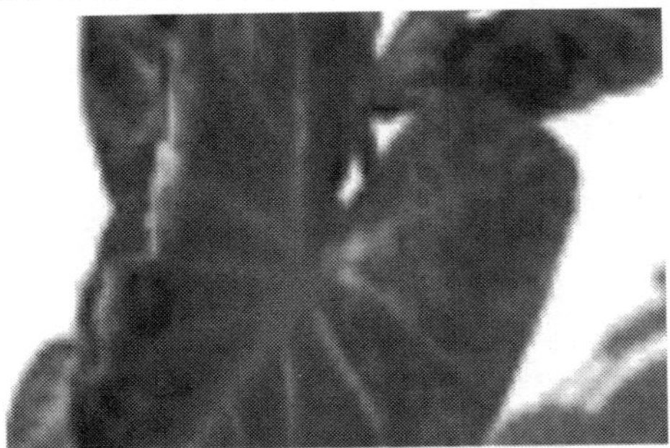

**Fig. 4.9.** Leaf spot on caladium.

**ii) Management:** Eliminate all stock plants which have *Xanthomonas* leaf spot. All bacterial diseases are very difficult to control unless plants are grown without overhead watering or exposure to rainfall. Bactericides such as copper containing compounds may be somewhat effective if used on a preventive and regular basis.

### 4.12.2. Viral Diseases

#### 4.12.2.1. Dasheen Mosaic Virus (DMV)

**i) Symptoms:** Dasheen mosaic virus can be severe when infected plants are used to grow tubers. The resulting crops will each be infected with the virus and tuber size and numbers will be reduced compared to virus - free crops. Symptoms include mosaic (Fig. 4.10), leaf distortion and stunting, and appear periodically during the year depending upon the environmental conditions.

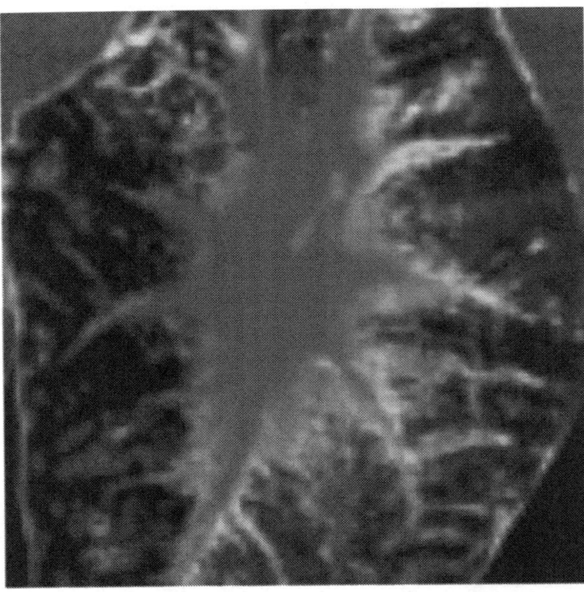

**Fig. 4.10.** Dasheen mosaic virus on caladium leaf.

**ii) Management :** DMV is sometimes spread by aphids and they should be strictly controlled during tuber production. It is very important to use pathogen-free stock whenever possible since the symptoms of DMV are not always noticeable. No chemicals have any known effects on this virus disease. Other hosts of this virus include dieffenbachia, philodendron, taro and anthurium.

## 4.13. DIEFFENBACHIA

### 4.13.1. Bacterial Diseases

#### 4.13.1.1. Bacterial Leaf Spots or Blights

Bacterial diseases caused by *Xanthomonas campestris* pv. *dieffenbachiae* (formerly called *X. dieffenbachiae*), *X. campestris* pv. *syngonii, Pseudomonas cichorii, Erwinia carotovora* sub sp. *carotovora* and *E. chrysanthemi* are all important pathogens of this foliage plant. These bacterial diseases may destroy leaves, petioles and stems rendering infected plants unsightly and unsalable.

**i) Symptoms:** *Xanthomonas* leaf spot or blight on dieffenbachias initially appears as yellowish, translucent or water-soaked specks usually first at the margins, later anywhere on the leaf blade except the midrib. The lesions enlarge to about half an inch (1.25 centimeters) and turn tan, yellow, orange- yellow, or reddish. The center is a dull watery-green, surrounded by an orange-brown border (Fig. 4.11).

**Fig. 4.11.** Bacterial blight of dieffenbachia.

Sometimes the roundish-to-elongate lesions are restricted to the leaf margins when infections occur through hydathodes. Older lesions are irregular and distinctly yellow with a reddish center. The lesions may coalesce to destroy the entire leaf. When the air is very dry, the lesions remain as small, dry, reddish brown specks. Under moist air conditions (high relative humidity) the spots enlarge and merge to cover rather large irregular areas that soon turn yellow, wilt, and dry. If the lesions are numerous, affected leaves finally turn yellow, wilt, and die. Dead leaves are a dull tan to light brown and tough, but not brittle. All parts of the leaf blade except the midrib are susceptible. Bacterial ooze usually appears on the lesions on the lower leaf surface. When the lesions are wet, the mounds of exudate are slimy; when dry, the exudate is thin, waxy, silvery-white, and easily peeled off the leaf.

**ii) Disease Cycle:** The *Xanthomonas* bacteria enter a leaf through hydathodes on the leaf margins, stomatoes on the lower surface, or wounds. Symptoms appear 7 to 18 days after infection. The newly developing leaves on an infected plant do not necessarily become infected, since the bacteria are rather slow-growing and are favoured by warm (21° to 32°C), moist conditions. The organisms are easily spread from leaf to leaf and plant to plant by splashing water, contaminated tools, insects, handling infected plants, and by propagating infected plants. The bacteria can even be in tissue-cultured plants. Other ornamental plants infected by *Xanthomonas campestris* pv. *dieffenbachiae* include Chinese evergreen, Silver Queen and other *Aglaonema* species, fancy-leaved caladiums (*Caladium* spp.), and coco yam (*Xanthosoma caracu*).

### 4.13.1.2. Pseudomonas Leaf Spots and Blights, *Pseudomonas cichorii*

**i) Symptoms:** *Pseudomonas cichorii* infects a very wide range of foliage and flowering ornamental plants as well as vegetables. Symptoms of *Pseudomonas* leaf spot and blight differ somewhat on different hosts. On most foliage plants the lesions are first minute and water-soaked but rapidly enlarge (up to over 3 cm in diameter) and turn dark brown or black, often with concentric light and dark rings and sometimes surrounded by a bright yellow halo. Most lesions are roughly circular to irregular and are rarely confined to interveinal areas. The centers of lesions are tan and brittle if allowed to dry. Lesions form anywhere on the leaf blade but are more common at the leaf margins. Under very damp conditions infected leaves may drop off.

**ii) Disease Cycle:** The *Pseudomonas* bacterium enters a leaf primarily through hydathodes and wounds. Severe symptoms can appear in as little as 3 days after infection, particularly if plants are misted before infection occurs. The number of lesions increases linearly as the misting period increases. Wounding is unnecessary for infection to occur on most hosts when moisture levels are high (such as misting). Disease severity is enhanced at 28° to 29°C with little disease developing above 32°C. The organism is easily spread from leaf to leaf and plant to plant by splashing water, contaminated tools, insects, and by handling infected plants.

**iii) Host Range:** Other foliage plants infected by *Pseudomonas cichorii* include Chinese evergreen, Silver Queen and other *Aglaonema* species, umbrella tree or schefflera (*Brassaia actinophylla*), *Caladium* spp., false aralia (*Dizygotheca elegantissima*), fiddle-leaf fig (*Ficus lyrata*), pothos (*Epipremnum aureum*), tree ivy (*Fathshedera lizei*), Japanese fatsia or aralia (*Fatsia japonica*), English ivy (*Hedera helix*), *Monstera* spp., staghorn fern (*Spathiphyllum* spp.). Vegetables that are hosts of the pathogen include cabbage, celery, chicory, endive, escarole, lettuce, and tomato.

### 4.13.1.3. *Erwinia* Leaf Spot, Blight, Stem Rot and Soft Rot

**i) Symptoms:** The most common bacterial pathogens of this foliage plant are *Erwinia carotovora* sub sp. *carotovora* and sub species of *E. chrysanthemi*. Both organisms cause soft rots or leaf spots and blight. The first symptom is usually soft, mushy, water-soaked areas at the base of the stem at or below the soil line. These lesions are gray to tan or pale brown, irregular in shape, with a distinct line separating diseased and healthy tissue. Stem and cane lesions below the soil surface usually go unnoticed until infection causes severe decay. Diseased plants commonly produce terminal leaves that are pale yellow and small. Under warm, moist conditions the stem lesions develop rapidly, causing the lower leaves to turn yellow, become soft and mushy, collapse, and die prematurely. Stem rot of cuttings is common; the lower leaves wilt and turn yellow and an entire cutting may turn into a rotted mass. The stem rot phase may develop so rapidly that entire plants or cuttings die before the bacterium has advanced into the leaves. Diseased tissue commonly has a foul, dead fish odour due to secondary microorganisms colonizing dead tissue. Leaf infections start as pinpoint spots that are water-soaked and yellow to pale brown. The lesions are sometimes surrounded by a diffused yellow halo. When the humidity is high and temperatures are warm to hot, the spots expand rapidly, becoming slimy, irregular, and sunken with light tan centers, darker brown borders, diffused yellow margins, and may involve the entire leaf in a few days. An invasion of the midrib and larger veins by the casual bacterium often results in advancement into the petiole and stem. The result is yellowing and total collapse of the leaf. Large, watery blisters may form on the lower leaf surface when conditions are humid. When the air is dry and temperatures are low, the leaf lesions continue to enlarge, become papery, brittle, yellow to tan, and commonly tear away. Some strains of both species of bacteria can produce a severe root rot of dieffenbachias causing collapse and wilting of the entire plant. Laboratory culturing of the casual organism is required to separate this from root rots caused by various pathogenic fungi. The detection of the disease is difficult since infected plants may show no visible symptoms, especially under cool and dry conditions.

**ii) Disease Cycle:** The two *Erwinia* species, like the *Xanthomonas* and *Pseudomonas* leaf spot and blight organism, enter primarily through hydathodes and wounds. Disease development is favoured by the presence of moisture and temperatures of 22° to 34°C. The bacteria are transmitted from an infected to a healthy plant by splashing water, insects, and by contaminated knives, tools, gloves, and carrying trays. The bacteria may invade the vascular (xylem) system and become systemic, causing the leaves and stems to turn soft and rot. Plants of all ages are susceptible to infection, with young leaves being more susceptible than older ones. *Erwinia chrysanthemi* can survive in greenhouse potting media with or without a host plant for a year or more and in the leaves of host or non-host plants in a greenhouse for 5 to 6 months. The same bacteria have been associated with bacterial soft rot diseases of a majority of foliage plants and numerous flowering plants. Many fleshy vegetables are also susceptible.

### iii) Management of Bacterial Diseases

- When possible, start with culture-indexed, pathogen-free plants from a reputable commercial propagator. Plant immediately into a sterilized growing medium.
- Water only the soil surface. Avoid overhead irrigation and splashing water on the leaves. Water early in the day to promote rapid drying of the foliage.
- Avoid crowding plants, heavy shade, poor air circulation, over-watering, mechanical injuries to plants, and high humidity. These factors all favour disease development.
- Separate infected plants from apparently healthy plants when disease is first noticed.
- Carefully remove and destroy seriously infected plants and plant parts. Handle these plants with throw-away plastic gloves to avoid contaminating other plants.
- Sterilize contaminated pruning knives, tools, carrying trays, and the like between plants by dipping or swabbing in a solution of 70 percent rubbing alcohol, liquid household bleach (1 part bleach in 4 parts water), or 38 to 40 percent formaldehyde solution.
- Where feasible, lower the temperature at which plants are grown to 21°C. This will slow down multiplication and spread of the causal bacteria.
- For bacterial leaf spots and blights, apply several sprays of a streptomycin formulation (such as Agrimycin) at 200 parts per million of active ingredient, 4 to 7 days apart during damp weather, starting when disease first appears. There is always the potential of getting resistance to streptomycin compounds if they are used repeatedly. Copper materials provide moderate control of *Xanthomonas* and *Pseudomonas* leaf spots and blights.
- For bacterial soft rot or stem rot, dip cuttings in a streptomycin formulation (200 parts per million) for 20 minutes. Immerse dieffenbachia cane pieces in a hot-water bath held at exactly 49°C for 40 to 60 minutes, depending on the cane diameter (40 minutes for canes 1

inch in diameter and 60 minutes for canes of 1½ inches). Extreme care is required in handling the cuttings and cane pieces after treatment to avoid recontamination. Treated plant materials should always be planted into a sterilized potting medium. Use only disinfected trays, tools, gloves, and soil mix. Wash hands with soap and hot running water before and after contacting suspected diseased plants.

- Control insects and mites using recommended chemicals.

## 4.14. ORCHIDS

### 4.14.1. Bacterial Diseases

#### 4.14.1.1. Brown Wet Spot, *Pseudomonas cattleyae*

*Pseudomonas cattleyae*, also known as 'brown wet spot' occurs mostly in phalaenopsis. However, a similar wet bacteria infection can occur on other orchids as well.

**i) Symptoms:** Plants develop brownish black, soft areas usually starting on the underside of the leaves and quickly penetrating to the leaf surface (Fig. 4.12). This affliction is a destructive rot which can spread with amazing speed, taking in the entire plant within only a few days. The cause is a combination of cool temperatures, high humidity and inadequate air movement. Prevention is easier than the cure. Keep minimum night temperatures around phalaenopsis at least 60 to 65° F, along with brisk air circulation. Brown wet spot can not develop under such conditions.

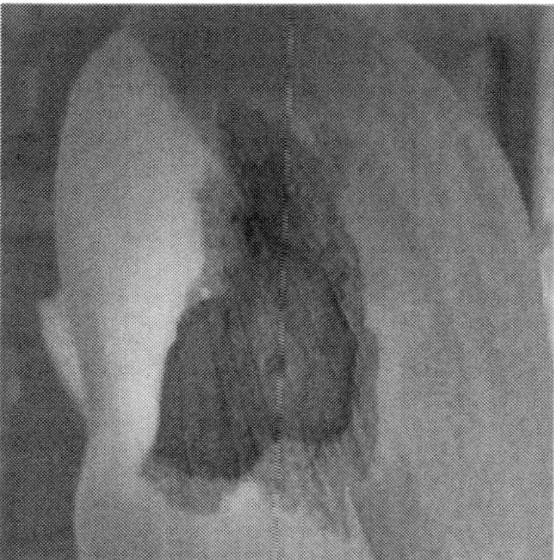

**Fig. 4.12.** Wet blight on orchid leaf.

**ii) Management :** If the disease has reached the crown portion of the plant, there is no cure even if some of the leaves look healthy. On phalaenopsis, if caught early, the affected portion of the leaf should be cut off (well below the afflicted area) with a new razorblade or some other sterilized cutting tool. Wet rot on other orchids must be treated by lancing and then cutting and/or scraping out the damaged tissue.

Dust the cut surface with sulphur. Try to rapidly dry off the injury by placing the plant before a fan. Isolate from your regular growing area to prevent any spread of the disease.

### 4.14.1.2. Bacterial Soft Rot

**i) Symptoms:** Soft rots are difficult to diagnose but whatever the causative agent, they can be devastating to an orchid collection. Orchids will be more vulnerable to infection if they are over fertilized, given insufficient light and ventilation, and if they are permitted to remain wet especially in the crown. Rots are a problem during the wet season, also after storms when plants are bruised and torn by high winds.

**ii) Management:** Shelter susceptible plants from rain. Be vigilant for rot during the wet season.

### 4.14.1.3. Bacterial Spot/Rot, *Pseudomonas* sp, *Erwinia* sp.

**i) Symptoms:** Soft, brown, smelly, fluid-filled blisters are produced on leaves and in the crown of Phalaenopsis, Paphiopedilum, and Catasetum. It is highly contagious. Crown rot will quickly kill a plant. Leaf spots can be excised and the plant saved.

**ii) Management:** Withhold water. Improve ventilation. Remove and burn the infected tissue. Be careful not to break the blisters. Consider the fertilizer being used: reduce the nitrogen and increase the potassium component. Pot Phalaenopsis so that the crown drains freely. Use Physan-20 at 7.5 ml in 5 liters as a preventive spray.

### 4.14.2. Viral Diseases

### 4.14.2.1. Cymbidium Mosaic Virus (CyMV), Tobacco Mosaic Virus (TMV), Tomato Spotted Wilt Virus (TSWV)

The three most important viruses for the orchid grower are undoubtedly Cymbidium Mosaic Virus (CyMV), Tobacco Mosaic Virus (TMV) and Tomato Spotted Wilt Virus (TSWV).

**i) Foliar symptoms:** Chlorotic streaks in Cymbidiums, to black necrotic spots and necrotic line patterns with sunken patches in Phalaenopsis, Cattleya, Dendrobium and many other genera (CyMV, TSWV) are observed on leaves (Fig. 4.13). In lavender-flowered Cattleyas, the foliage patterns are associated with pronounced red or purple pigmentation. The leaf patterns appear as spots, chevron patterns or irregular line patterns of chlorotic or red-pigmented tissues (TMV) (Fig. 4.13).

**ii) Flower symptoms:** The flowers may show necrotic streaks within a few days of the flower being open. However, flower symptoms usually show up only after about two weeks and are especially noticeable in white Cattleya flowers (CyMV).

**Fig. 4.13.** Tomato Spotted Wilt Virus (*Cattleya* hybrid) (Left); Tobacco Mosaic Virus (*Dendrobium kingianum*) (Center and Right).

Flower colour breaking is consistent with a TMV infection. Streaks of dark pigmentation are often observed in predominantly pink to lavender-flowered orchids. Colour breaking should, however, not be confused with white streaks on lavender-flowered cattleyas that are caused by a genetic abnormality.

**iii) Management**

- Virus diseases are managed by prevention. Once a plant is infected with a virus it is virtually impossible to get rid of it and the plant will have to be destroyed with the pot and potting medium.
- Do not propagate from infected plants.
- Pest control is essential to reduce the risk of virus infections.
- Weeds can be an important source of infection both as reservoirs of viruses and as hosts for vectors.
- Disinfect all cutting tools and pots between each plant that is handled as viruses can be transmitted in sap from infected plants. Flaming them or dipping them into a 10% sodium hypochlorite solution can disinfect the cutting tools. Hands should be washed with soap.
- Avoid dusting the greenhouse floor, or the garden, with tobacco dust to get rid of snails, slugs or ants. The tobacco dust may contain virus particles, which can be a source of infection. Smokers should not smoke and handle orchids at the same time.

## 4.15. ANTHURIUM

### 4.15.1. Bacterial Diseases

#### 4.15.1.1. Bacterial Wilt, *Ralstonia solanacearum*

**i) Symptoms:** The bacterium causes both systemic and localized infections. The symptoms constitute yellowing starting from the margins of outer leaves. The yellowing gradually extends to entire leaf area and finally to inner leaves. The infected plants show systemic yellowing and wilting leading to death of plants. Root and collar rot are also quite common. Often the infection results in necrotic lesions from petiole spreading along the veins. Local infections on leaves cause black necrotic lesions with greasy margins resulting in leaf distortion. On the spathe these spots are angular and irregular. The bacterium is carried through plants without expressing external symptoms. Under humid conditions, the bacterial ooze is produced on infected plants and contributes to spread of the disease through water droplets, garden tools and workers activity. The disease is more severe during summer and under excessive soil moisture conditions.

**ii) Management:** The disease can be managed by using disease-free planting stock besides adopting strict phytosanitation.

### 4.15.1.2. Bacterial Blight, *Xanthomonas campestris* pv. *dieffenbachiae*

Bacterial blight disease caused by *Xanthomonas campestris* pv. *dieffenbachiae* (= *Xanthomonas axonopodis* pv. *dieffenbachiae*) affects a broad range of ornamental and edible aroids including anthurium, colocasia (taro), Aglaonema, Syngonium, Xanthosoma, Dieffenbachia, Epipremnum, Dracaena, Alocasia, Spathiphyllum, Rhaphidophora, Caladium and Philodendron. Bacterial blight disease of anthurium was first reported in Brazil in 1960 and Hawaii in 1971. In the humid tropics, high rainfall coupled with year-round high temperatures increases the severity and spread of bacterial blight disease making it a very important impediment to the cultivation of anthurium in the Caribbean.

**i) Distribution:** The disease became a major problem in Hawaii in the 1980s and caused major decline in the production of anthurium there in the 1980s. In 1982, bacterial blight disease was accidentally introduced from Venezuela to Guadeloupe, causing severe decline of anthurium grown under traditional and greenhouse methods. This decline caused significant losses to commercial anthurium growers and was identified as an important limiting factor for the development of the crop in Martinique, Guadeloupe and indeed the rest of the Caribbean.

The disease was first reported in Jamaica in 1985 and along with the burrowing nematode, *Radopholus similis*, wiped out the anthurium industry in Jamaica. It was first identified in Trinidad in 1986 and has been responsible for the decline of the anthurium industry from 16 large anthurium farms in the early 1990s to only five large farms at present.

Bacterial blight disease has now been reported in most anthurium producing countries including the mainland United States (Florida, New Jersey and California), the Netherlands and all other anthurium producing Caribbean islands as well as South America, Tahiti and Taiwan.

**ii) Symptoms:** Foliar symptoms usually start at, or close to the leaf margins and occasionally in the centre of leaves. The first sign of infection is the appearance of irregular water-soaked spots on areas on the underside of leaves and spathes and faint chlorosis when viewed from the upper side. As infection proceeds, the spots become brown or black at the centre and bright yellow at the edges (Fig. 4.14). Bacterial blight damage spreads quickly and will eventually cover large sections of the leaf or spathe. Infections often become systemic as the bacteria enter the vascular system of the plant. A characteristic symptom of advanced systemic infection is discolouration of the vascular system. In most cases, systemic infection leads to plant death.

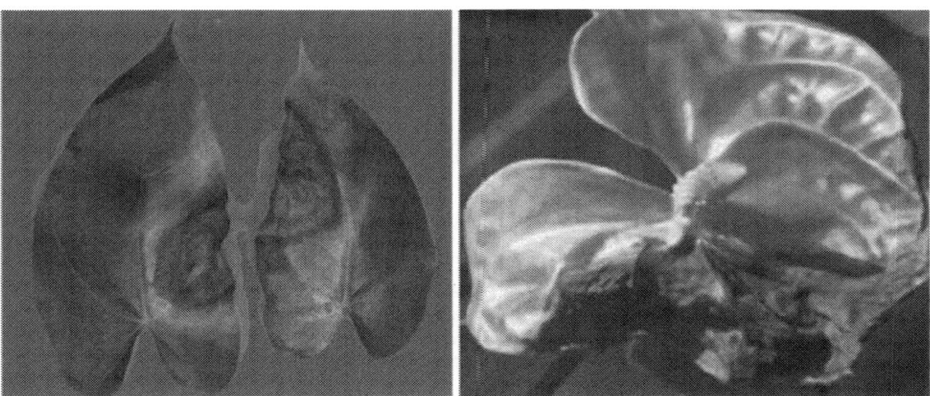

**Fig. 4.14.** Bacterial blight on anthurium.

Foliar bacterial blight disease can severely decrease farm productivity, and hence the profitability of anthurium cultivation. Infected leaves are removed, sometimes to a bare stem to reduce further disease spread. The reduced photosynthetic capacity as a result of leaf removal causes plant stunting. In cases where most of the leaves have been removed, flowers are not produced at all; plants take longer to mature or are unmarketable, being either deformed or small.

Systemic infection can occur independently of foliar infections. In such cases, the pathogen enters the plant via wounds or through the root system. The disease advances without apparent symptoms until the infected plant wilts and collapses. Symptoms of systemic infection start with yellowing of older leaves and petioles. Systemically infected leaves and flowers break off easily and may show dark-brown streaks at their base, which gradually enlarge. Cut-petioles show yellow-brown vascular bundles. Systemic infection may also show water-soaked leaf spots when the bacteria enter the leaf parenchyma from the infected vascular bundles. Many farmers and researchers have observed that some cultivars resistant to foliar bacterial blight disease succumb rapidly to systemic infection and vice versa, indicating a two-phase differential basis for the disease.

**iii) Survival and Spread:** Bacterial blight is a contagious disease that is difficult to control. The pathogen spreads by excessive water splashing (rain or irrigation), movement of infected soil, and contact with infected plant material, tools (from leaf pruning, harvesting) and persons walking through infected plants. The bacterial blight pathogen could survive for many months in tissue-cultured plantlets without showing any symptoms. Furthermore, the pathogen can survive in leaf lamina, petiole, and root residues for as long as four months when tissues are left on the ground or buried 15 cm deep; the pathogen retains its pathogenicity during that time.

### iv) Management

Preventative measures to manage bacterial blight disease at economically feasible levels include:

- Use of clean planting material.
- Installing foot baths with disinfectant at entrances to greenhouses.
- Adjusting plant density on beds to facilitate good aeration.
- Working from clean to diseased fields to minimise pathogen movement.
- Discarding and burning of infected plants.
- Disinfecting tools and clothing regularly, changing tools on different plots and avoiding exchange of material between greenhouses.
- Ensuring good drainage of the growing substrate.
- Using drip irrigation or micro-sprinklers.
- Avoiding the presence of visitors not attached to the farm.
- Removing all debris and old media (or fumigate old media with methyl bromide) and letting beds fallow for two months (or fumigate with metham-sodium) and tarped in prior to replanting.
- Producing disease-free planting material from bud- or tissue-cultured plants away from production fields and ensuring that the plants' nutritional needs are met to improve their vigour.

Where the disease has been already established, sanitation practices should be performed. In addition to adhering to all the preventative measures, include:

- Keeping plants as dry as possible.
- Removing diseased leaves early and burning them.
- Avoiding extreme climatic conditions particularly high temperatures.
- Adapting a fertilization regime to keep the production of glutamine in plants at a low level, as glutamine is a major food source for the bacteria. For instance, ammonium nitrate can be left out and desirable levels of potassium maintained.
- Chemical control of bacterial blight is ineffective and to date no chemicals have been described that can control or reduce the disease. Some chemicals used in the past have been shown to actually increase the severity of the disease.

### 4.15.1.3. Soft Rot, *Erwinia carotovora* sub sp. *carotovora*

**i) Symptoms:** Erwinia soft-rot symptoms appear at stem bases as a wet, mushy lesion. The tissue rapidly collapses, turns dark and can become rank smelling. Erwinia is most severe under hot wet conditions when controls are rarely effective.

**ii) Management:** Do not reuse pots or potting media from plants which have died from this disease since they can be contaminated with the pathogen.

**Table 4.2. Bacterial and viral diseases of ornamental crops and their management**

| Crop/Disease | Symptoms | Host range and natural spread | Management |
|---|---|---|---|
| Carnation<br><br>Mottle | Faint leaf mottle or no symptoms. Common in virtually all carnation cultivars. | Handling and cutting knife. Not transmitted by aphids. May be spread in drainage water. | Reduce spread of mottle by disinfecting tools between blocks of plants and several times a day. Obtain virus-free plants. |
| Etched ring | Rings usually oval or elongated; rarely concentric on older leaves and stems. Symptoms may be slight in young cuttings. No obvious effect on plant vigour. | Aphids. Not spread by handling. | Obtain virus-free plants. Control aphids. |
| Necrotic fleck | Reddish purple necrotic flecks, streaks, or spots appear in leaves. Symptoms are masked at low temperatures. | Aphids. | Obtain virus-free plants. Control aphids. |
| Ring spot | Small 0.5- to 1-inch (1- to 2-cm) rings; sometimes concentric. Chlorosis, mottling, and distortion of young leaves. Plants obviously stunted. | Cutting knife and handling. Insect vectors unknown. | Same as mottle. |
| Vein mottle | Young leaves exhibit a vein clearing, which develops into chlorotic spots and patterns that mostly follow veins. Symptoms tend to disappear on old leaves. | Aphids. Not spread by handling or cutting knife. | Obtain virus-free plants. Control aphids |
| Bacterial wilt *(Pseudomonas caryophylli)* | Sudden wilting of tops or individual branches. Basal stem cracks. Roots may be rotted. Vascular discoloration in stems is yellowish to brown. The outer layer (epidermis) | Infected plants and debris. Bacteria spread in water. Favoured by high temperatures. | Use disease-free cuttings and treat soil as above. Bacteria can spread quickly. Avoid use of cutting dips and avoid splashing water. Break cuttings from stock |

| | | | |
|---|---|---|---|
| | separates easily from stem, which is sticky to the touch. | | plants. Disinfect tools. |
| <u>Chrysanthemum</u><br><br>Bacterial Blight *Erwinia chrysanthemi* | Cuttings turn dark brown and collapse. Surviving cuttings may be infected but have no symptoms. Established plants wilt during the day when infected and recover at night. | In plant debris. Favored by high temperatures (80° to 90°F), absence of free water, and high humidity. | Use disease-free cuttings. Reduce humidity in growing areas. Dip cuttings in streptomycin. Disinfect propagation beds between crops. Destroy infected cuttings. |
| Bacterial Leaf Spot *Pseudomonas cichorii* | Small dark brown to black spots on lower leaves enlarge and become irregular in shape. When infected leaves dry, the spots become brittle and crack. The disease often spreads up plants in one side of the pot, eventually to the flowers. | | Do not plant infected cuttings. Avoid overhead irrigation. Water in a manner that keeps leaf surfaces dry at all times. Protect plants grown outdoors from splashing. |
| Chlorotic Mottle (Chrysanthemum chlorotic mottle viroid) | Leaves, at first mottled, become completely yellow. Infected plants grown under low light conditions and when temperatures average less than 20°C exhibit no symptoms. Symptoms vary with the cultivar infected. Young leaves are light green and very upright. | Spread by handling, cutting tools, and vegetative propagation. | Use disease-free cuttings. Remove infected plants. Disinfest tools used to handle them. Do not handle healthy chrysanthemums after handling infected plants. |
| Stunt (Chrysanthemum stunt viroid) | Plants are stunted to half their normal height at maturity. Infected plants flower prematurely and flower size is reduced. Some cultivars exhibit small dead spots or flecks on the leaves. | Viroid is easily transmitted by handling, cutting knives, vegetative propagation, etc. Not spread by aphids or thrips. Viroid has a wide host range. Symptomless in some plants. | Obtain disease-free plants. Destroy infected plants and disinfest tools used to handle them. Do not handle healthy chrysanthemums after handling infected plants. |
| Spotted wilt (Tomato spotted wilt virus, Impatiens necrotic | Frequently one-sided in plant. Ring patterns on leaves of some cultivars. Leaf distortion and | Thrips-transmitted; not spread by cutting knife, but can be spread by vegetative | Eliminate nearby susceptible ornamental plants and weeds. Control thrips inside and outside |

| | | | |
|---|---|---|---|
| spot virus) | necrosis. Dark necrotic streaks on stems. Flowers may be distorted and with some necrosis. | propagation. Many weeds and perennial ornamental plants act as reservoirs of virus (dahlia, calla, nasturtium, mallow, knotweed, plantain, and others). | growing areas. Use virus - free cuttings. |
| Mosaic (Chrysanthemum mosaic virus) | Mosaic mottling of foliage. Dwarfing and deforming of leaves, stems, and flowers. Symptoms variable, depending upon strain of virus and cultivar. | Transmitted by vegetative propagation and by aphids. | Obtain disease-free plants. Control aphids. |
| Aspermy (Tomato aspermy virus) | Flower distortion and reduction in flower size. Color break in florets of red, bronze, and pink cultivars. Foliar symptoms not apparent. Some cultivars are symptomless. | Transmitted by handling, cutting tools, and vegetative propagation. Also transmitted by aphids. | Use disease-free cuttings. Control aphids; remove infected plants. |
| <u>Narcissus</u><br><br>White streak virus | Narrow, dark green to purple streaks, which later become white to yellow-white, appear in leaves and flower stalk after flowering. Bulb size and yields are reduced by premature senescence. | Narcissus is the only host of the virus, which is transmitted by aphids. Symptoms do not occur until air temperatures exceed 64°F. | Eliminate infected plants. Control aphids. |
| Yellow stripe virus | Conspicuous light green to yellow streaks and mottling of leaves and flower stalk, which occurs shortly after emergence. Sometimes leaves are distorted and a colour-break of flowers occurs. Bulb yields are reduced. | Narcissus is the only host of the virus, which is spread by aphids. Symptoms appear early in growing season. | Eliminate infected plants. Control aphids. |

Chapter 5

# MEDICINAL AND AROMATIC CROPS

## MEDICINAL CROPS

In medicinal crops sector, the WHO has estimated that about 80% of the population in developing countries rely on traditional medicines mostly plant drugs for primary health-care needs. Even modern medicines contain about 25% drugs derived from plants. A large number of people are earning their livelihood from different activities of medicinal plants such as collection, cultivation, marketing, processing, etc. The area under cultivation of the above crops is low, but the cash returns are high because of intensive cropping.

In recent times, due to increasing realization of health hazards and toxicity caused by synthetic drugs (side effects), there has been a renewal of interest in the use of plant based drugs throughout the world. A campaign is going on worldwide to utilize more and more plant derived chemicals in the human health care/personal care system. As a result, the demand for plant-based raw materials has increased enormously in both national and international markets. Herbal plants are finding diverse uses as raw materials, not only for medicines but also as biopesticides.

Medicinal crops form one of the important groups due to their demand in various pharmaceutical industries and also to earn foreign exchange by way of export. The potential for earning foreign exchange by India from the exports of medicinal and aromatic plants is estimated over US$ 3,000 million per annum. The export of these plants and their products has a tremendous potential in advanced countries like Europe, USA and Japan. The international market for medicinal plants related trade is estimated at US$ 60 billion/year having a growth rate of 7% per annum. Destination of foreign tourists is now aimed at India for health treatment. Kerala and Gujarat have become their favourite destinations since Ayurvedic treatment is very popular in these two states.

Unless we start cultivating the important medicinal plants, it will not be possible to meet the increased demands. However, cultivation of medicinal plants has to be taken up in polyculture models rather than in monoculture models. There is a huge task to be carried out at every level to make the Indian system of medicine a viable, sustainable and modern. For this task, urgency is felt which can only be established through regular and systematic scientific research on collection, conservation, evaluation and development of good agricultural practices to compete in the world market.

## 5.1. SOLANUM

Solanum (*Solanum viarum*), also called as Khasi Kateri, is a steroid-bearing perennial, tall bush distributed all over Assam, Sikkim and Manipur between 1,600-2,000m elevations. Its berry pulp is rich in Solasodine alkaloid, which is a starting chemical for production of steroids. About 22 species including *Solanum viarum* contain an alkaloid Solasodine in their berries. Solasodine serves as an excellent raw material for the manufacture of Cortisone and Cortisonal hormones. It is used in the production of contraceptive pills, corticosteroids and sex hormones. It is cultivated in 3,000-5,000 ha, mainly in Maharashtra and Andhra Pradesh. It is also cultivated in West Bengal, Assam, Madhya Pradesh and Gujarat on a small scale.

### 5.1.1. Bacterial Diseases

#### 5.1.1.1. Bacterial Blight

**i) Symptoms:** Streaks of yellow alternated with green stripes appear on the leaves.

**ii) Management:** Dissolve 150 g of streptocycline and 750 g copper sulphate separately in 2.5 litres of warm water. Add both solutions to 750 litres of water to spray one ha.

## 5.2. PERIWINKLE

Periwinkle (*Catharanthus roseus*), commonly known as vinca, is a common bedding plant. It is popular in container gardens and new flower colours produced on compact cultivars, have increased its use in many public and home gardens.

It contains several alkaloids found to be effective in the treatment of leukemia and other diseases such as diabetes, hypertension, etc. The more important alkaloids obtained from Vinca are Leurosine, Vinblastin, Vincristin, etc.

### 5.2.1. Viral Diseases

#### 5.2.1.1. Little Leaf (Mycoplasma)

**i) Symptoms:** Occasionally some plants have been found to suffer from little leaf disease due to mycoplasma infection resulting in stunted growth and rosetting of leaves on the plant. The disease spreads through insect vectors.

**ii) Management:** The disease can be checked from spreading by uprooting and destroying the affected plants. The disease can be minimized by controlling the vector by spraying Metacid or Nuvacron at 15 days interval when the infection is prevalent.

## 5.3. BELLADONNA

Indian belladonna (*Atropa accuminata*) and European belladonna (*Atropa belladonna*) are the source for tropane alkaloids which include hyoscine, hyoscyamine and atropine (0.13-0.7%) which are used in the manufacture of tinctures, treatment of whooping cough, epilepsy, Parkinson's disease, renal and bilary colic. Indian belladonna is grown in the farms of Kashmir State Forest Department which is a remunerative crop for North-Western hilly region of India.

### 5.3.1. Viral Diseases

#### 5.3.1.1. Mottle Virus

**i) Management:** Collection and destruction of affected plants helps to reduce further spread of the disease.

## AROMATIC CROPS

In recent years, there has been an increased interest in the cultivation of aromatic plants to meet the requirements of cosmetic, flavouring and perfumery industries. The aromatic plants provide raw material for the production of flavours, herbal cosmetics, perfumery etc. Aromatic plants and their products, particularly the essential oils, are now becoming one of the more important export items from many developing countries of Asia. Chemically essential oils are terpenes, which act as carriers of the aromatic substances. Most essential oils also contain camphors, and the more odiferous compounds present in them consist of oxygen derivatives of terpenes, alcohols, esters, aldehydes and ketones. India has enjoyed a prominent position in the manufacture of superior perfumes and aromatics by using essential oils. The important aromatic plants are lemon grass, vetiver, patchouli, palmarosa, citronella, mints, geranium, lavender, basil, jasmine etc.

In the world market, India stands in second position. Most of the essential oils produced though marketed within the country, but a sizeable amount has also been exported. Among the essential oils exported from India are Japanese mint oil, ambrette seed essence, sandal wood oil, citronella oil, lemon grass oil, palmarosa oil, ginger oil, clove oil, eucalyptus oil, vetiver oil, pepper oil, etc. The annual exports of the derivatives of aromatic plants are to the tune of Rs. 600-700 million.

## 5.4. MINTS

Mints (*Mentha* spp.) are a group of aromatic herbs belonging to the family Lamiaceae, which are considered to be the most important cash crops in Indo-Gangetic plains. Cultivation of mints has been done in a large scale in many tropical and sub-tropical countries of the world including India, China, Brazil, Japan and USA. Indian farmer grow it as a bonus crop on an area of more than 1,50,000 ha in Central and Northern parts of Indo-Gangetic plains. The crop fit well in the traditional food based cropping system with other crops like paddy, wheat, potato, mustard, maize, okra, carrot, onion, spinach, pigeon pea, cowpea, etc. Similarly it can be grown as an intercrop with sugarcane and some legumes. In India, the highest area under commercial cultivation of mint has been in Uttar Pradesh (Terai region), but during last few years it has spread through out the country, especially in North and North-Western states including Madhya Pradesh.

Among different oil yielding species of mint, Japanese mint (*M. arvensis*) occupies the highest area under mint cultivation followed by Peppermint (*Mentha spicata*), Spear mint (*M. piperata*), Bergamot mint (*M. citrata*) and Scotch spearmint (*M. cardiaca*).

Japanese mint essential oil is used as a source of menthol, menthyl acetate, menthone and terpenes which finds wide use in medicine, perfumes, food and cosmetics. The oil of peppermint, which contains less menthol but has a sweeter aroma and taste, is used in tooth paste, chewing gums, candies, high grade liquors, medicines and other pharmaceutical preparations. Spearmint oil having carvone and limonene, is mainly used for flavouring; whereas, Bergamot mint oil is mainly used in perfumery and flavouring due to higher content of linayl acetate and linalool.

### 5.4.1. Viral Diseases

#### 5.4.1.1. Begomo Virus

**i) Symptoms:** Typical symptoms include foliar rugosity, light yellow mosaic, upward curling, crinkling and retarded growth leading to drastic reduction in the herb's yield (Fig. 5.1). In 2007, the disease incidence was recorded to be in the range of 50 to 60%. The disease also appeared in the Terai region of Uttarakhand province. From the naturally infected plants, the disease was transmitted to healthy seedlings of *M. spicata* cv. Viridis by white flies (*Bemisia tabaci*), but not by aphids or mechanical inoculation. Viruliferous white flies did not transmit the infection to other mint species (*M. arvensis; M. citrata* or *M. piperita*) in glasshouse tests.

**Fig. 5.1.** Symptoms of Begomo virus on mentha leaves.

**ii) Management:** Out of four samplings in field from March to May, the Begomo virus infection was found minimum in variety Kushal as compared to others. The maximum infestation of Begomo virus was found in cultivar Shivalik.

## 5.5. BASIL

Among several species of basils, sweet basil (*Ocimum basilicum*) is considered to be the most important crop for its high quality of essential oil. Major components of its oil are linalool (43-50%), menthyl chavicol (18-33%), eugenol and isoeugenol (5-6%) and the minor constituents are alpha and beta pinene, camphor, geraniol, etc. The oil of sweet basil owes its importance to its extensive use in condimentary products, cosmetics, toiletry, perfumery and confectionary industries, particularly in European countries.

### 5.5.1. Bacterial Diseases

#### 5.5.1.1. Bacterial Leaf Spot, *Pseudomonas cichorii, P. syringae*

**i) Symptoms:** Spots on leaves are water-soaked, dark, and may be both angular and delineated by small veins in the leaves or show as irregular leaf spots. A wet stem rot may occur. Bacterium is reported to be seed-borne. The disease is favoured by wet, humid conditions and is disseminated by splashing water or by handling and spreading infected tissue.

**ii) Management:** Decrease moisture on plants with low humidity and sufficient plant spacing for adequate air movement to reduce leaf wetness periods. Keep the foliage dry. Use disease-free seed and transplants. Remove diseased leaves and plants to reduce inoculum levels. Avoid overhead irrigation. Use clean, sterile equipment and do not move between infected and healthy plants. Spray ridomil or mancozeb for the control of the disease.

## 5.6. SCENTED GERANIUM

Scented geranium (*Pelargonium graveolens*) is an aromatic perennial herb. It is grown in cooler, sub-tropical climate of Mysore and Bangalore (Karnataka), and extended to Hyderabad (Andhra Pradesh). Its oil has refreshingly delicate long-lasting rosy odour with fruity under-note, containing 66-78% primary alcohols (rhodinol). It is used in manufacturing of perfumes, creams, talcum powder and body lotion. It is stable in alkaline medium and is therefore used for scenting soaps.

### 5.6.1. Bacterial Diseases

#### 5.6.1.1. Bacterial Wilt/Blight, *Xanthomonas campestris* pv. *pelargonii*

This disease, caused by *Xanthomonas campestris* pv. *pelargonii*, is a major problem of geraniums during warm, humid weather. It has caused losses of 10 to 100 percent, depending on the cultivar.

**i) Symptoms:** Disease symptoms usually appear in warm, sunny weather when the plants are under moisture stress. Leaves infected with bacterial blight show two types of symptoms:

Small, round, water-soaked spots form on the underside of the leaf. Several days later, the spots develop into angular, slightly sunken, well-defined lesions about 1/16 to 1/8 inch (up to 3 mm) in diameter that may be surrounded by a diffuse yellow halo. The spots eventually turn dark brown to black and become hard and dry. Leaves quickly wilt, die, and may fall off, or hang on the plant for a week or more (Fig. 5.2). The bacteria in these leaf lesions may migrate into the water-conducting tissues, move down through the petiole and into the stem, eventually causing the plant to die from the stem rot phase of the disease.

A wilting of the leaf margin occurs in some cultivars, resulting in large angular (V-shaped), yellow or dead areas bounded by the veins. These leaves soon wither and hang on the petiole or drop off. Other factors, such as potash deficiency and Botrytis blight, may cause angular, dead patches, in geranium leaves; but the wilted condition is always associated with bacterial blight infections (Fig. 5.2).

The stem rot phase of the disease is commonly called "black rot" by growers. Water-conducting tissues within the branches and stems of infected plants turn brown to black, 2 to 4 weeks after infection. At this stage, one or more leaves on the branch will usually wilt and develop angular dead areas. The bacteria soon spread from the water-conducting tissue to other stem tissues. The stem rapidly turns a dull gray to blackish brown. The roots are blackened but not decayed (Fig. 5.2).

Plants gradually become defoliated, except for a few dwarfed leaves at the tips of the upright blackened branches (Fig. 5.2). The stem and branches rapidly blacken and shrivel into a dry rot with all the tissues destroyed, except for the fibers and epidermis. Some plants may partially recover from initial infection and produce new, healthy-looking branches. Since these branches harbour the bacterium, such plants usually die within 3 to 6 months.

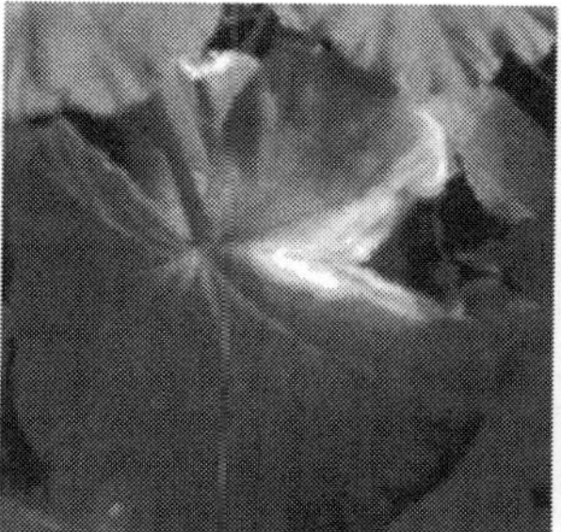

**Fig. 5.2.** Bacterial wilt symptoms on geranium.

Cuttings infected with bacterial blight fail to root, the stems slowly rot, turn a dull blackish brown from the base upwards (Fig. 5.2), and eventually die. The leaves wilt, often showing the typical symptoms described earlier. Cuttings serve as a source of infection in the propagation bed.

**ii) Disease Cycle:** Infection occurs through plant openings or through wounds that are created when cuttings are taken from stock plants. A cutting knife that is not disinfected is one of the surest means of transmitting the bacteria. The bacterium survives in water-conducting and other tissues of infected cuttings that are taken from symptomless stock plants. Infected plants are symptomless during cool weather when the temperatures are below 21°C or at very high temperatures 32-38°C. The bacteria are very infectious and are spread easily from plant to plant through plant contact and contaminated propagation media, by splashing during hose watering or misting, cultural operations, contaminated cutting knives, and insects. The bacteria are spread by handling them. The organism has been isolated from white flies (*Trialeurodes vaporariorum*), which may transmit the bacteria to healthy geraniums. The organism can survive in moist soil from 3 to 6 months, or until infected plant residues are completely decayed. Symptoms and disease spread occur most rapidly at temperatures of 21-29°C and in plants overfed with nitrogen fertilizer.

### 5.6.1.2. Southern Bacterial Wilt, *Ralstonia solanacearum*

Southern bacterial wilt caused by *Ralstonia solanacearum* was first reported in geranium in 1979. The symptoms are similar to bacterial blight or wilt. Although it is not a major disease of geraniums, it could pose a threat to geranium growers in the South and the Midwest, due to the exchange of cuttings and plants from these geographic regions. The common geranium (*Pelargonium hortorum*), is susceptible to this disease. No geranium cultivars are resistant.

**i) Symptoms:** The initial symptoms include a wilting of the lower leaves followed by chlorosis and finally necrosis of the leaves. The stem eventually turns brown and then black at the soil line, and the plant collapses when 1/2 to 1 inch of the basal stem is infected. The roots can also become necrotic. Flowers fail to open normally. Stem sections reveal vascular discolouration. The symptoms of southern bacterial wilt always progress upward from the soil line, whereas those of bacterial blight or wilt move in either direction-up or down the stem. It is unknown if infected plants can remain symptomless.

**ii) Disease Cycle:** *Ralstonia solanacearum* is a soil-borne bacterium that may persist in soil for 6 years or more in the absence of a suitable host plant. The pathogen survives and spreads within susceptible plants and cuttings. Infection normally occurs through the roots. The organism is transmitted on infected cutting knives, in soils and other planting media, in crop debris, by splashing or flowing water and by insects. *R. solanacearum* has a very wide host range.

Symptoms develop rapidly at 23-39°C, and plants die within 2 to 3 weeks under these conditions, whereas it may take 5 weeks for similar development of bacterial blight or wilt.

### 5.6.1.3. Bacterial Leaf Spot, *Pseudomonas cichorii*

A bacterial leaf spot caused by *Pseudomonas cichorii* has been known to only attack geranium in the past few years. It has been reported only in Florida, but due to the exchange of plant material, the disease is a potential threat to the florist industry in the Midwest. This bacterium also has a wide host range.

**i) Symptoms:** The appearance of leaf spot symptoms varies, depending on the weather conditions. Plants growing outside exposed to rains exhibit dark brown to black, irregularly shaped areas 5 to 10 mm or larger. The margins of the spots appear water-soaked a few hours after a rain. The lesions may extend or enlarge along the veins and may coalesce until the entire leaf may eventually die and curl up.

Under less favourable conditions for disease development, such as when plants are exposed to dews or only occasional overhead watering, the lesions that develop are sunken on both the upper and lower leaf surfaces. These spots may have tan centres and may be slightly raised. A lesion may have a dark margin and a yellow halo that may be as wide as the diameter of the lesion. The lesions commonly enlarge in wet weather but are checked by dry weather. Eventually most spots turn dark brown to black and large lesions, formed by the merging of smaller lesions, may appear to have "eyes" 1 to 2 mm in diameter.

Flower clusters are susceptible and individual flowers fail to open when infection occurs in the bud stage. Young seedlings can also become infected, causing a soft, dark, wet decay of the developing leaves.

**ii) Disease Cycle:** The bacterium is thought to be carried on clothes, hands, sprayers, and other equipment. Environmental parameters specific for this disease on geranium are presently unknown. Free moisture on the leaves and rainy periods are known to promote disease development.

**iii) Management**

The control measures outlined will control all three diseases: bacterial blight or wilt, southern bacterial wilt, and bacterial leaf spot.

- Procure disease-free cuttings or plants from a reputed nursery. Florists should obtain CVI (culture-virus-indexed) cuttings from a commercial propagator who has established disease-free planting stock and who practices strict sanitary measures or uses culture-indexed cuttings to set up a mother block system. Start with new stock plants each year; do not keep them from year to year.
- Follow strict sanitary procedures to prevent the contamination of healthy plants.
- Immediately remove and destroy all infected plants, preferably by burning. Remove all old leaf and plant debris.
- Florists should use an automatic watering system. If this is impossible, hang the host sc that the nozzle does not touch the

greenhouse floor. Avoid over watering and sprinkling or misting water on the foliage.

- When propagating, select a raised bench away from other areas of geranium production. The bench surface should be periodically wiped or sprayed with a fresh household bleach (Clorox, Purex, Sunny Sol) solution prepared by adding 1 part of bleach to 9 parts of clean tap water. Also disinfect any equipment used to transport, hold, or work with geraniums.
- Steam the propagating medium, soil mix, clay pots, and tools for 30 minutes at 82°C or 72°C for one hour.
- Wash hands thoroughly with soap and hot running water before handling plants. Take cuttings by (i) breaking off succulent tips of disease-free stock plants; or (ii) sterilizing the cutting knife or razor blade by dipping in a 70% solution of rubbing alcohol and flaming or soaking for 5 minutes in a liquid bleach solution (add 1 part of bleach to 9 parts of water).
- Knives or razor blades should be changed between stock plants. Do not dip cuttings in liquid solutions. Place cuttings in clean, sterile flats lined with new newspapers.
- After establishment, take cuttings from flats and plant them directly into steam-treated soil mix in clean pots.
- Maintain good cultural conditions and ventilation. Space the plants to avoid plant contact. The greenhouse bench surface should be covered with woven wire mesh.
- Control white flies and other insects to lessen spread of bacteria.
- Avoid forcing the plants too rapidly, especially during warm 21-29°C and humid weather. Maintain a balanced fertility based on a soil test.
- In a field, flower bed, or nursery, wait at least a year before replanting with geraniums.
- Infrequent sprays containing basic copper sulfate or other fixed copper fungicide may aid in control of bacterial leaf spot caused by *P. cichorii*. Frequent use (every 7 to 10 days), however, causes a surface glazing of the leaves and a stunting of growth.
- Sanitation: Pull weeds and remove plant debris.
- Use soil amendments to develop well-drained soil.
- Use disease-free plants.
- Avoid overhead irrigation and water early in the day and not later than 11 a.m.
- Avoid growing perennial geraniums (*Geranium* spp.) near *Pelargonium* spp.
- Remove diseased plants as soon as possible.
- Space plants to promote good air circulation.

- Avoid unnecessary handling of plant material and wash hands frequently.

#### 5.6.1.4. Bacterial Fasciation, *Rhodococcus fascians*

**i) Symptoms:** Short, thick, fleshy, aborted stems that form at base of main stem at or below soil level are pale green or green-yellow. The rest of the plant appears healthy.

**ii) Management**

- Purchase culture-indexed plants.
- Discard infected stock plants and cuttings and infested media.

### 5.6.2. Viral Diseases

#### 5.6.2.1. Tomato Spotted Wilt and Impatiens Necrotic Spot Viruses

**i) Symptoms:** Symptoms of these viruses are highly variable and include stunting, ring spots, and sunken purple-brown lesions on leaves, stems and petioles. The symptoms vary with plant age, physiological condition, level of infestation and timing.

**ii) Management**

- Use disease-free, virus-indexed cuttings.
- Destroy diseased plants.
- Avoid unnecessary handling of plant material.
- Pull and control weeds.
- Pay attention to insect control to avoid spreading of the disease. Monitoring and managing thrips is very important.

**Table 5.1. Bacterial and viral diseases of geranium and their management**

| Disease/ Pathogen | Symptoms | Management |
|---|---|---|
| Bacterial Blight, *Xanthomonas campestris* pv. *pelargonii* | Two types of symptoms may develop: (1) Small spots, less than 1/4 inch, develop on underside of the leaf and become sunken and well defined. The leaf wilts and dies as the bacterium spreads through water-conducting vessels of vein and petiole. V-shaped areas form with wide part of V on leaf margins and point of V on veins; (2) Lower | Purchase culture-indexed cuttings or grow plants from seed. Immediately discard infected plants after a positive diagnosis is made. Avoid overhead irrigation. Discard all unwanted geraniums at season's end. Do not take cuttings with knives. Wash hands frequently when breaking out cuttings. Or disinfest the cutting knife by dipping in alcohol and flaming or |

| | | |
|---|---|---|
| | leaves wilt at margins while the blade and petiole remain turgid. The vascular tissue of the main stem on the side of the affected portion of the plant is discolored gray-brown and may ooze bacteria when cut. Lower leaves die and fall. | leaving in ZeroTol* or 1 Clorox* to 9 parts water for 10 minutes. |
| Bacterial Fasciation, *Rhodococcus fascians* | Short, thick, fleshy, aborted stems that form at base of main stem at or below soil level are pale green or green-yellow. The rest of the plant appears healthy. | Purchase culture-indexed plants. Discard infected stock plants and cuttings and infested media. |
| Pseudomonas Leaf Spot, *Pseudomonas cichorii* | Elliptical, water-soaked 1/4- to 1/2-inch spots form on leaves. Spots become dark brown to black and irregularly shaped. A yellow halo may or may not surround each spot. | Purchase disease-free cuttings. Do not irrigate overhead. Keep leaf surfaces dry at all times. |
| Southern Bacterial Wilt, *Ralstonia solanacearum* | Lower leaves wilt, yellow, and fall. Vascular tissue of affected stems turns brown or black. | Purchase culture-indexed plants. Do not bring ground-planted geraniums into the production area or propagate from them. Destroy infected plants. |
| Viruses | Symptoms vary with virus, cultivar, and growing conditions. Symptoms include: mottling, ring spotting, leaf cupping, yellowing of veins, leaf malformation, leaf spotting, and loss of zonation. | Purchase virus-indexed cuttings. Maintain strict insect and mite control. |

Chapter 6

# TUBER CROPS

Tuber crops form an important part of the diet in many areas of the tropics and subtropics. They are most important after cereals and grain legumes. They are mostly grown for the carbohydrates stored in underground storage organs viz., roots or rhizomes. In addition, the leaves or leaf stalks of some of the tuber crops are rich in protein, vitamins and minerals and are used for either human consumption or cattle feed. Tuber crops are a major source of food and calories in many tropical countries. The annual tropical root crop production (sweet potato and cassava) is in the range of 8.921 million metric tonnes in 0.367 million ha with productivity of 20.1 tonnes/ha (Table 6.1). Root and tuber crops are predominantly grown in the upland or rain fed areas, mainly by resource poor small and marginal farmers. The major tuber crops include cassava (tapioca), sweet potato, colocasia, amorphophyllus, yams and coleus. Cassava and sweet potato are grown to a larger extent in southern and eastern India. Their cultivation has spread to Maharashtra, Gujarat and North-Eastern states as well. The Salem belt in Tamil Nadu and the Samalkot belt in Andhra Pradesh are known for cassava as an industrial crop. Yams and Colocasia are popular in Andhra Pradesh, Tamil Nadu, West Bengal, Uttar Pradesh and Orissa states.

**Table 6.1. Area, production and productivity of tuber crops in India (2005-06)**

| Tuber crop | Area ('000 ha) | Production ('000 tonnes) | Productivity (tonnes/ha) |
|---|---|---|---|
| Sweet potato | 122.8 | 1066.5 | 8.9 |
| Cassava (Tapioca) | 244.6 | 7854.9 | 31.4 |
| Total | 367.4 | 8921.4 | 20.1 |

(Source: Indian Horticulture Database 2006)

Tuber crops, especially cassava can be successfully substituted as alternative source of feed for livestock in both tropical and subtropical countries. Cassava has also been used as a substrate for the production of ethyl alcohol. The tropical root crops, in general, have a great potential in meeting basic food and energy needs of the developing world. Tuber crops are very efficient in solar energy transfer, having a clear superiority over cereals in biological efficiency and are comparatively free from severe pests and diseases. Cassava and sweet potatoes account for about 30% of the total production of root crops from developing countries. The potential of root crops as nutritionally rich sources of ß-carotene, anti-oxidants, dietary fibre and minerals like calcium have begun to be recognized as a result of the multifarious research programmes worldwide.

## 6.1. SWEET POTATO

Sweet potato (*Ipomea batatas*) is an economically important member of the family Convolvulaceae and probably originated in Central America, where it has long been cultivated. It is now grown extensively throughout the tropics and subtropics. It is one of the life sustaining crops rich in carbohydrates that stand between man and starvation. Sweet potato ranks fourth and sixth on the list of dry matter production per hectare and edible energy production per hectare, respectively. It is also used for making starch, syrup and alcohol. It occupies a prime place in terms of calories production/ unit area and time. It is grown as a starchy food crop throughout the tropical, subtropical and frost-free temperate climatic zones in the world. It is being cultivated in 0.129 million hectares producing 1.148 million tonnes of sweet potatoes with an average yield of 8.87 tonnes per hectare (Table 6.2). India ranks sixth in area but the productivity is very low (8.9 t/ha), lower than the world productivity level.

**Table 6.2. Area, production and productivity of leading sweet potato growing states in India (2005-06)**

| State | Area (in '000 ha) | Production (in '000 MT) | Productivity (in MT/ha) |
|---|---|---|---|
| Orissa | 47.1 | 394.3 | 8.4 |
| Uttar Pradesh | 20.0 | 240.8 | 12.1 |
| West Bengal | 25.8 | 208.0 | 8.1 |
| Maharashtra | 5.0 | 70.0 | 14.0 |
| Karnataka | 3.6 | 42.1 | 11.6 |
| Chhattisgarh | 3.0 | 38.2 | 12.6 |
| Tamil Nadu | 1.4 | 29.6 | 21.1 |
| Assam | 8.5 | 29.5 | 3.5 |
| Madhya Pradesh | 4.7 | 28.2 | 6.0 |
| Meghalaya | 5.4 | 18.9 | 3.5 |
| Others | 4.9 | 49.1 | 10.0 |
| Total | 129.4 | 1148.6 | 8.87 |

(Source: Indian Horticulture Database 2006)

Sweet potatoes are affected by several different diseases in the field, some of which also affect the storage life of the tubers. Virus diseases are probably responsible for the greatest field losses and some fungus diseases can also cause economic losses. The use of clean planting material is the most important disease control measure and crop rotation aids control of many fungal diseases.

### 6.1.1. Bacterial Diseases

#### 6.1.1.1. Soil Rot, *Streptomyces ipomoeae*

**i) Symptoms:** A heavy infection results in dwarfed plants which only produce one or two short vines with small, pale green and thin leaves. These symptoms result from damage to the underground parts of the plant. Lateral roots are attacked and many are destroyed, others are often distorted. The most obvious symptom is the development of pits on the tubers, the first signs of which are water-soaked dark spots.

As the spots on the tubers develop, the overlying skin breaks and the necrotic tissue underneath falls out leaving the characteristic pits with jagged margins. The pits range in diameter from 5-30 mm. The tubers may become girdled by lesions but continue to grow on either side of the constricted infected area.

**ii) Development and Spread:** The organism which causes soil rot can survive in the soil for several years and the disease is most severe in soils that are alkaline or only slightly acidic. The causal organism is spread through farm implements, by people and animals and by transport of diseased tubers. Dry seasons and poor soils are conducive to severe losses.

**iii) Management:** The most effective method to control this disease is to prevent the transfer of infected materials (soil and tubers) to uninfected fields. Planting material should be healthy. Fields known to be infested with the disease should be used for other crops for longest period possible. Addition of sulphur to the soil as an acidifying agent will reduce the disease incidence.

### 6.1.2. Viral Diseases

#### 6.1.2.1. Witches' Broom (Little Leaf) (Mycoplasma)

**i) Symptoms:** Witches' broom disease has been reported from Taiwan and the Pacific. The disease is caused by a mycoplasma-like organism which is transmitted by leaf hopper, *Orosius ryukyuenis* and grafting. Mycoplasma-like organisms 200-2500 nm have been observed in the phloem of infected plants. Plants infected with witches' broom have shortened internodes giving them a bushy appearance.

Axillary buds are elongated with small leaves. Rosette, stunting, yellowing mosaic and necrosis symptoms have also been observed. Tubers are fewer and of poorer quality.

**ii) Management:** Heat therapy and meristem tip culture have been used to produce pathogen-free plants. Only tetracycline and related chemicals are effective against this disease. At present, the use of healthy planting material and eradication of diseased material are the most promising practical control measures.

### 6.1.2.2. Mild Mottle Virus

**i) Symptoms:** It is caused by a poty virus (sweet potato virus B/ sweet potato virus T/ sweet potato ipomo virus) (Fig. 6.1). The virus is transmitted by white flies (*Bemisia tabaci*). The virus has a wide host range including tomato, tobacco and ornamental species. Symptoms of the disease include leaf mottling and plant stunting. The wishbone flower (*Torenia fournieri*) is a potential wild reservoir host in East Africa.

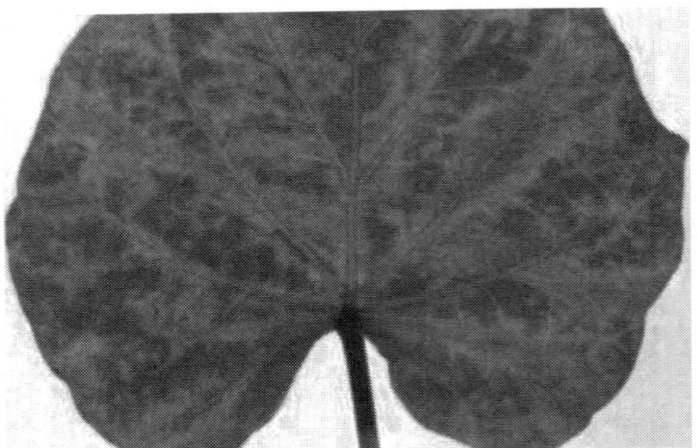

**Fig. 6.1.** Mild mottle virus on sweet potato leaf.

**ii) Management**

- Plant resistant varieties: (e.g. in Uganda the variety "Namujuna")
- Use disease-free planting material
- Isolate new plantings from old plantings
- Control the virus vector (white flies)

### 6.1.2.3. Sweet Potato Virus Complex

This disease is caused by a combination of sweet potato feathery mottle virus (SPFMV) and sweet potato chlorotic stunt virus (SPCSV).

**i) Symptoms:** Symptoms include severe stunting of plants, the production of small distorted leaves, excessive branching, yellowing of vines and dark, brown to blackish corky spots in the roots. SPFMV is a poty virus transmitted by aphids in a non-persistent manner whereas SPCSV is a crini virus transmitted in a semi-persistent manner by the white fly *Bemisia tabaci*.

**ii) Management**

- Use resistant or tolerant varieties, if available.
- Use disease-free planting material.
- Avoid diseased plants as sources of planting material.
- Practice proper field sanitation.

### 6.1.2.4. Other Virus Diseases

**i) Symptoms:** Several viruses can cause a mottling or so-called mosaic patterns to the leaves of sweet potatoes (Fig. 6.2). These include internal cork, Russet crack, mild mottle and tobacco mosaic. Diseased leaves may be distorted and have a variegated pattern of dark green and lighter yellowish-green areas. Affected plants have shortened internodes and may have a substantial yield reduction.

**Fig. 6.2.** Virus infection on sweet potato leaves.

Virus diseases are a major problem in sweet potato production, largely because these viruses are transmitted from infected roots to slips during propagation. The most important virus disease of sweet potato is the feathery mottle virus. Symptoms of this disease are quite variable and cultivar dependent. Symptoms have been listed under various disease names including russet-crack, internal cork, and chlorotic mottle. Initially, infected plants show a yellowing or vein clearing of the leaves. In some cases, small feathery yellow patterns develop along leaf veins. Affected plants commonly are stunted and appear unthrifty.

Internal cork appears as dry, corky, dark-coloured clumps of tissue scattered through the flesh of affected sweet potatoes. Small, corky lesions may be found in the roots at digging time. Usually extensive corking does not occur until the roots are stored for several weeks. The symptoms are more serious when storage temperatures are high (above 60°F).

The corky spots may occur singly or in groups and are most readily detected by cutting the roots into thin slices. Severely diseased roots may develop so many corky lesions that they become practically useless for food. Russet crack symptoms are most noticeable on the surface of affected sweet potatoes and consist of areas containing numerous fine or coarser irregular cracks and browning of the surface. The cracked areas are irregular and they may be in bands or spots. The lesions may cover much or only a little of root's surface. Lesions may also develop on the stems to which the sweet potatoes are attached or on the underground parts of sprouts in the hotbed.

**ii) Management :** The virus persists in the roots of diseased plants. If these roots are used to produce slips for new plantings, then plants derived from the infected root will also be diseased. Therefore, it is extremely important to use virus-free seed stock. Purchase seed certified to be free of viruses. If seed is saved from last year's crop, use only those roots that show no symptoms of russet-crack or internal cork. Unfortunately, some infected roots remain symptomless, and there is no guarantee that these roots will be free of the virus. The virus also may be transmitted from plant-to-plant in the field by aphids. However, efforts to control the disease by decreasing aphid populations have had only limited success and are costly.

## 6.2. CASSAVA (TAPIOCA)

Cassava (*Manihot esculenta*), popularly known as tapioca, is a member of the family Euphorbiaceae, which has long been cultivated in Central America and Brazil and was later carried to Africa (where it is now an important subsistence crop), Asia and the Pacific Islands. An economic crop can be obtained even under marginal conditions or on exhausted soils. The crop requires very little attention after planting. It has draught tolerance (mainly due to in-built mechanism to shed or drop the leaves under adverse moisture conditions) and produces more calories/unit area. The tubers contain 25-30% starch, 1.5% protein and 2.1% minerals. It is being cultivated in 0.252 million hectares producing 7.925 million tonnes of cassava with an average yield of 31.36 tonnes per hectare (Table 6.3). The Salem belt of Tamil Nadu and Samalkot belt of Andhra Pradesh are known for cassava as an industrial crop.

**Table 6.3. Area, production and productivity of leading cassava growing states in India (2005-06)**

| State | Area (in '000 ha) | Production (in '000 MT) | Productivity (in MT/ha) |
|---|---|---|---|
| Tamil Nadu | 127.1 | 4857.4 | 38.2 |
| Kerala | 95.0 | 2585.1 | 27.2 |
| Andhra Pradesh | 19.0 | 388.7 | 20.5 |
| Meghalaya | 4.2 | 23.1 | 5.5 |
| Pondicherry | 0.7 | 19.5 | 27.9 |

| | | | |
|---|---|---|---|
| Nagaland | 1.3 | 15.9 | 12.2 |
| Assam | 2.9 | 14.0 | 4.8 |
| Karnataka | 1.1 | 12.0 | 10.7 |
| Others | 1.4 | 9.2 | 6.6 |
| Total | 252.7 | 7924.9 | 31.26 |

(Source: Indian Horticulture Database 2006)

Even though cassava is a hardy crop, it has been shown that diseases can cause serious losses. As cassava is propagated vegetatively, the use of disease-free planting material is the most important single factor in disease control. Under traditional farming systems, yields have been more than doubled by using disease-free planting material and improving cultural practices. Cassava bacterial blight is the most important disease of cassava and virus diseases are also of considerable threat to the crop.

### 6.2.1. Bacterial Diseases

#### 6.2.1.1. Bacterial Blight, *Xanthomonas campestris* pv. *manihotis*

Bacterial blight, caused by the bacterium *Xanthomonus campestris* pv. *manihotis*, has perhaps caused more damage on a worldwide basis than any other disease of cassava. It has been seen nearly every year in Miami-Dade County, although severe outbreaks are sporadic and are usually confined to a limited number of plantings. Today, sufficient technology exists to control bacterial blight through the integration of several recommended measures. This disease could be reduced to one of minor importance if these steps were taken on an industry-wide scale. Yield loss due to the disease may range from 20 to 100% depending on variety, bacterial strain and environmental conditions.

**i) Distribution:** The bacterial blight is the most important disease of cassava and is widely distributed in Argentina, Brazil, Cameroon, Colombia, Ghana, India, Indonesia, Madagascar, Malawi, Malaysia, Mauritius, Mexico, Nicaragua, Nigeria, Panama, Philippines, Taiwan, Thailand, Togo, Uganda, Venezuela and Zaire.

**ii) Symptoms:** It is particularly devastating during wet seasons. The characteristic symptoms are leaf spotting and blight (Fig. 6.3), wilting, dieback, gum exudation and stem and root vascular necrosis. Primary symptoms, shown by infected planting material, are wilting of the young germinated shoots followed by dieback.

Symptoms of secondary infection are angular leaf spotting, followed by blight, defoliation, wilting and dieback. At first the leaf spots are small, angular and water-soaked but they enlarge, covering all or part of the leaf, and turn brown or dark brown. The initial spots are clearly visible on the undersides of leaves. The necrotic leaves roll and dry up and after remaining attached to the stem for a short while they fall off. Gum is often exuded from young infected stems, petioles and leaf spots, and collects as droplets. On leaves these droplets occur mainly on undersides along veins or vein lets. On petioles and stems the gum is exuded from cracks. The gum forms a yellowish glistening scab when dried.

**Fig. 6.3.** Bacterial blight on tapioca leaf.

Vascular tissue of infected petioles and stems becomes necrotic and appears as brown strings. Leaves supplied by these vascular tissues wilt and young stem tissues rot. The young green stems rot faster than the older greenish-brown ones which maintain a healthy appearance. The rotting of the young stems causes characteristic dieback symptoms.

Root rot may appear when susceptible cultivars with a good water balance are infected. Roots have dry vascular necrosis.

On young immature plants, complete destruction of the aerial portions may be caused. New shoots are usually produced but these rapidly become infected, especially during the rainy season.

All symptoms appear mainly during the rainy season and vary according to the susceptibility of the variety and the period of infection.

### iii) Development and Spread

Primary infection is due to the use of infected planting material. Secondary infection results from spread of the bacteria in the gummy exudates by rain splash. Movement of man, animals and insects through the crop and use of contaminated tools are also likely to spread the disease. Rain splash is the most important means of spread over short distances.

Development of the disease is retarded during dry periods and little gum is exuded thus halting the spread of the disease. However, the bacteria remain viable in the plant and are reactivated in later wet periods.

The bacteria enter the plant through stomata or wounds and penetrate to the vascular system from where they can spread to the rest of the plant. Symptoms generally develop within 7 to 13 days after infection.

**iv) Management:** This potentially devastating disease can be managed quite well through the adoption of a series of integrated control measures. Crop rotation is an important control. The bacteria can only survive in the soil for a month or two enabling land to be cleared of infection by a 6 month rotation or fallow period before planting disease-free material. Infected crop debris should be incorporated promptly into soil where the bacteria die rapidly: Six months is sufficient to prevent pathogen carry-over. Weeds should be controlled because *X. campestris* pv. *manihotis* can survive on the leaves and in the root zone of some weeds. Remove and burn all infected plant debris and weeds. Alternatively plough them into the soil.

Only healthy planting material should be used for planting. Clean seed material can be obtained by pruning most of the above ground portion of infected plants. This method is most effective with moderately resistant cultivars which are only lightly infected. Tools used for cutting planting material should be regularly disinfected to avoid mechanical dissemination of the pathogen. Infected stem cuttings can be rendered pathogen-free if rooted from stem tips. Intercrop cassava with maize or melon. This has been reported to reduce cassava bacterial blight significantly.

### 6.2.1.2. Bacterial Stem Rot, *Erwinia cassavae*

**i) Symptoms:** The disease causes shoots to wilt and on woody parts of the stem a characteristic pungent, soft rot or discolouration is observed. Buds on infected stems become necrosed. The bacterium causing the disease enters by wounds, usually caused by the insects, *Anastrepha* spp. The bacterium has been isolated from adults and larvae of this insect. The holes caused by *Anastrepha* can be easily located by traces of dried exudate on the stem surface.

**ii) Management:** The best method to control stem rot disease is to use healthy planting material and varieties resistant to the insect vector.

## 6.2.2. Viral Diseases

### 6.2.2.1. Cassava Mosaic Virus

**i) Distribution:** The disease occurs in East, West and Central Africa and India. Estimate of losses caused by mosaic ranges from 30-80%. It is the most serious cassava disease in Africa and is the biggest constraint in cassava production.

**ii) Symptoms:** The virus disease causes characteristic leaf mosaic symptoms that affect discrete areas and are determined at an early stage of leaf development. Young leaves have chlorotic areas and are often deformed. Leaves may also be reduced in size. When the infection is severe, leaves are small, misshapen, distorted and have bright yellow areas with normal green tissue in between (Fig. 6.4). In milder cases of infection, leaves may show only a mild chlorotic mosaic. The disease is transmitted by white fly. Insect transmission alone does not cause much spread of the disease-man appears to be the main factor.

**Fig. 6.4.** Cassava mosaic virus infection on cassava leaves.

**iii) Transmission:** It is spread through infected cuttings and by white flies (*Bemisia tabaci*).

**iv) Management**

- Use disease-free cuttings. If it is not possible to find cassava plants that are completely free from the disease, select cuttings from stem branches instead of from the main stem. Stem cuttings from the branches are more likely to sprout into disease-free plants than stem cuttings from the main stems (James *et al.*, 2000).
- Use resistant/tolerant varieties (e.g. SS 4, TMS 60142, TMS 30337 and TMS 30572).

### 6.2.2.2. Cassava Common Mosaic Virus

**i) Distribution:** The disease is distributed in Brazil, Colombia, Indonesia, Ivory Coast and Sumatra. The disease has been found to cause losses of 10-20%.

**ii) Symptoms:** The common mosaic causes characteristic mosaic symptoms of yellowish areas on the leaf lamina, distortion and stunting of the plant. The yellowish areas are not as clearly defined as those of African mosaic but otherwise the symptoms of these two diseases are similar.

**iii) Transmission:** The common mosaic is transmitted mechanically and by grafting but no natural insect vector has been identified.

**iv) Host Range:** The virus is able infect other *Manihot* spp., *Euphorbia prunofolia, Chenopodium amaranticolor, C. guinea, Malva parviflora* and *Gossypium hirsutum*.

**v) Pathogen:** The causal agent is a virus which has elongated flexuous rods (15 nm in diameter and about 500 nm in length).

**vi) Management:** As there is no insect vector, the disease can be easily controlled by the use of clean planting material and rouging of infected plants.

### 6.2.2.3. Cassava Brown Streak Virus (Potyvirus - Potyviridae)

**i) Distribution:** The cassava brown streak virus occurs only on the east coast of Africa at altitudes under 1,066 m and is uncommon. It is particularly serious in coastal areas of Kenya, Zanzibar, Mozambique and Tanzania and lake shore region of Malawi and in Uganda and is a threat to the whole of sub-Saharan Africa.

**ii) Symptoms:** Infected plants develop leaf chlorosis and roots become necrotic in storage. Leaf scars persist longer than usual after leaf drop and brown lesions sometimes occur on young stems.

Symptoms include yellowing (leaf chlorosis) and brown streaks in the stem bark (cortex) (Fig. 6.5). Infected tubers have brown streaks (root necrosis). It is a stealth virus, which destroys everything in the field. The leaves may appear healthy even when the roots have rotted away.

Fig. 6.5. Cassava brown streak virus disease (Potyvirus - Potyviridae)

**iii) Transmission:** The virus can be transmitted mechanically and by grafting. The virus is vectored by white flies (*Bemisia* spp.) and also transmitted through infected cuttings.

**iv) Management**

- Use disease-free cuttings.
- Use tolerant/resistant varieties (e.g. 5543/156, TMS 30572)
- Remove diseased plants from the field.

### 6.2.2.4. Cassava Leaf Vein Mosaic Virus

**i) Distribution:** The disease is distributed in Brazil and Venezuela.

**ii) Symptoms:** The leaf vein mosaic is of limited economic importance. The characteristic symptoms of the disease are vein clearing and leaf curling.

**iii) Transmission:** The disease can be transmitted mechanically and by grafting.

**iv) Pathogen:** Spheroid particles of 50-60 nm in diameter have been found associated with the disease.

**v) Management:** Infected plants should be rouged and only disease-free planting material should be used for planting.

#### 6.2.2.5. Witches' Broom

**i) Distribution:** The disease is distributed in Brazil, Mexico, Peru and Venezuela. The disease can cause up to 80% reduction in yield. The incidence of infection is low but in affected plantings there are a high percentage of diseased plants.

**ii) Symptoms:** Diseased plants are stunted with short internodes and excessive proliferation of branches. The disease may cause symptoms of proliferation of shoots from the cutting but in other cases only a few stunted shoots develop.

**iii) Transmission:** The disease is transmitted mechanically and by the use of infected cuttings.

**iv) Management:** Infected plants should be rouged and only disease-free planting material should be used for planting.

## 6.3. COLOCASIA

Colocasia (*Colocasia esculenta*), also known as cocoyam, taro, dasheen and eddoe, is a member of the family Araceae and occurs wild in S.E. Asia. It is cultivated throughout the humid tropics and is of great importance in the Pacific Islands. It is mostly a staple food or subsistence crop grown commercially in some countries.

### 6.3.1. Bacterial Diseases

#### 6.3.1.1. Bacterial Leaf Spot, *Xanthomonas campestris pv. dieffenbachiae*

Bacterial leaf spot of colocasia, caused by *Xanthomonas campestris* pv. *dieffenbachiae*, was described and characterized in the United States in the mid-1980s. It is the most commonly encountered disease problem in Miami-Dade County fields, with nearly 100% of the plants during the rainy season showing at least a few lesions.

**i) Symptoms:** Disease symptoms begin as small, water-soaked spots on the lower surface of leaves. Spots enlarge and substantial areas of brown, dead tissue surrounded by yellow halos develop on upper leaf surfaces (Fig. 6.6). Pronounced water-soaking continues on the lower leaf surface, sometimes with a cream to light yellow bacterial exudate (sticky ooze) in the centre of the water-soaking. Lesions along the margins of leaves are often seen (Fig. 6.6). Apparently, infection occurs readily through hydathodes (water pores) on leaf margins.

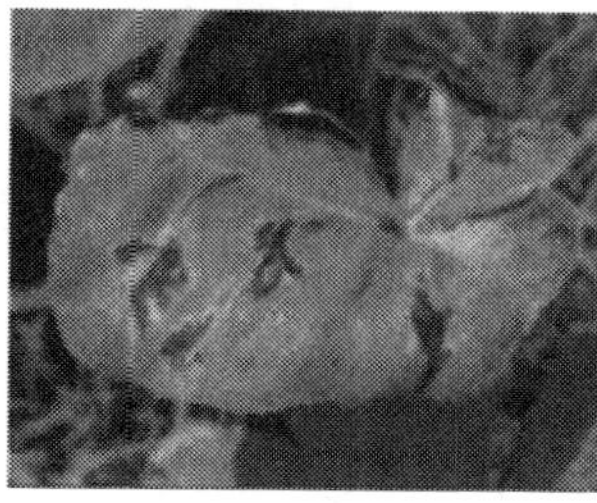
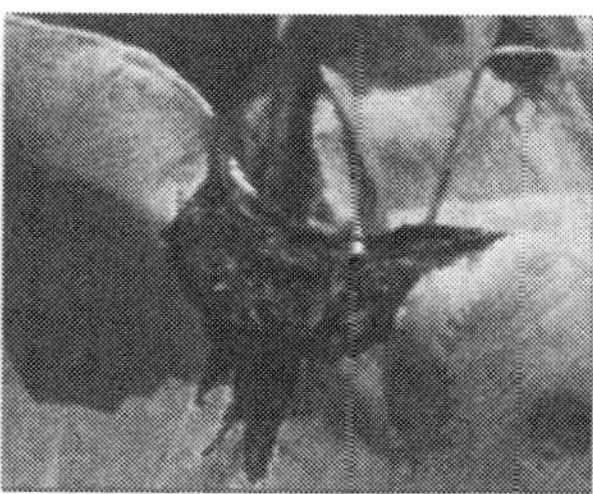
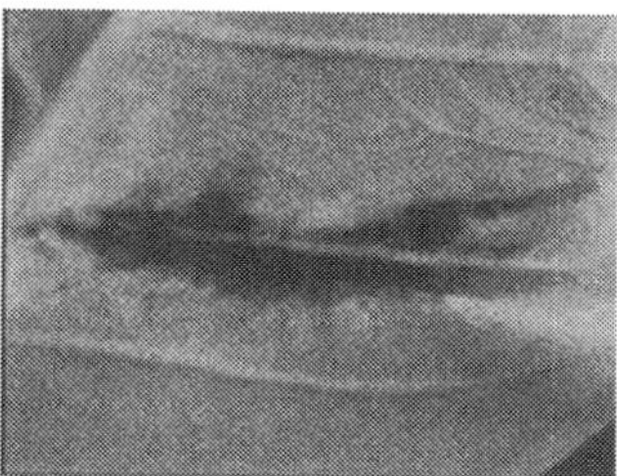

**Fig. 6.6.** Bacterial leaf spot on colocasia.

The pathogen can invade vascular tissue of leaves, causing water-soaked streaks along major veins (Fig. 6.6). Once inside the vascular tissue, bacteria may be able to reach the corms and cormels, thus contaminating potential planting material.

The main mode of bacterial movement is probably by splashing rain. Disease development is favoured by high temperatures in the range of 29-32°C. Thus, the problem is most evident in the summer months. Entry into the plant is facilitated by wounding. The relationship between bacterial leaf spot severity and yield is unknown.

**ii) Management :** Growers should stay out of fields when plants are wet, as *X. campestris* pv. *dieffenbachiae* is readily transmitted whenever workers, tools, or farm machinery contact infected, sodden plant material. Since planting stock likely is contaminated with the pathogen, the same tissue culture methods that someday may be used for dasheen mosaic virus control, may also help reduce bacterial spot.

## 6.3.2. Viral Diseases

### 6.3.2.1. Dasheen Mosaic Virus

Dasheen mosaic is a widespread disease of colocasia caused by dasheen mosaic virus (DMV). DMV is common in Florida, also affecting taro (Colocasia) and many ornamental aroid species. It occurs in virtually all colocasia plantings, affecting as many as 90% of the plants in some fields. The disease is thought to reduce yield, but no systematic studies have ever been done to actually quantify losses from dasheen mosaic.

**i) Distribution:** The disease occurs in Egypt, Hawaii, India, Japan, Papua New Guinea, Puerto Rico, Trinidad, USA and Venezuela.

**ii) Symptoms:** As with many virus diseases, symptoms of DMV infection are quite subtle and can be overlooked by the casual observer. Often, the only evidence of infection is a mild, light and dark mottled appearance of the foliage (Fig. 6.7). In some cases, more striking symptoms will appear. Feather-like patterns of light yellow tissue may develop along veins (Fig. 6.7).

Severely infected plants may be noticeably slower in growth than surrounding plants in the field.

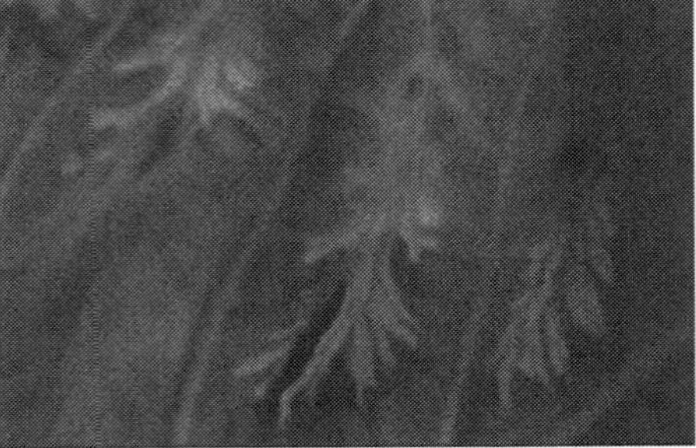

**Fig. 6.7.** Dasheen mosaic virus on colocasia leaves.

The virus has long flexuous rod-shaped particles measuring about 75 nm in length. The infection results in a feathery mosaic pattern adjacent to the veins. Yellow-green streaking of the petioles may also be observed in *Xanthosoma* spp. In severely affected plants, petioles are short and leaves are dwarfed and fail to unfurl properly.

DMV is transmitted by several species of aphids. The virus is transmitted non-persistently by the aphids, *Myzus persicae, Aphis gossypii* and *Pentalonia nigronervosa*. This virus is also sap transmissible. Aphid-mediated field spread can be very rapid. The virus particles become attached to the stylet (feeding probe) of the aphid when it feeds on an infected plant. Within a short time (probably less than one hour), the virus must then be transferred to another host if infection is to take place.

The virus is carried in infected planting stock. Since colocasia is propagated exclusively by means of vegetative cuttings of corms and cormels, the virus is continually introduced into new plantings.

**iii) Management:** Techniques to provide high levels of DMV control are known but are at present cost-prohibitive for the edible aroids. The virus can be eliminated from planting stock through the use of meristem tip tissue culture. Tissue culture propagation is now widely used in the ornamental aroid industry. Currently costs are sufficiently high to prohibit its use by colocasia growers. Even more sophisticated techniques may one day lower costs enough to make possible virus-free propagation of colocasia.

Insecticidal control of aphids is of little value in DMV control. Aphids can transmit the virus before they are killed by insecticide.

### 6.3.2.2. Alomae and Bobone Virus

**i) Distribution:** These virus diseases are distributed in Cook Islands, Papua New Guinea and Solomon Islands.

**ii) Symptoms:** Alomae is a systemic necrosis which leads to death of the plant. The first symptom is a feathery mosaic. Young leaves fail to open properly and have a crinkled appearance. The leaves become abnormally thickened and the veins prominent. Later on the leaves do not open and begin to die. The necrosis spreads from the tips of the leaves and down the petioles and finally the whole plant dies.

Symptoms of Bobone are similar to those of Alomae but milder. Affected plants tend to be more stunted and the leaves are twisted and curled. The foliage does not become necrotic but remains green and the plants eventually recover.

**iii) Pathogen:** One of these has large bacilliform particles (280 x 55 nm) and the other has small bacilliform particles (140 x 30 nm). Both types of particles are found in plants with Almoae whereas in plants with Bobone only large particles are present.

**iv) Transmission:** The larger bacilliform particle is transmitted persistently by the plant hopper, *Tarophagus proserpina*. The smaller bacilliform particle is transmitted by the mealy bugs, *Planococcus citri* and *Pseudococcus longispinus*. Almaoe may be apparently caused by *T. proserpina* transmitting the larger bacilliform particle to male plants already infected with the smaller particle.

**v) Management:** The infected plants should be rouged and burnt and the planting material should be carefully selected. Although the female cultivars can survive infection, corm yield losses of about 25% still occur. So it is still advisable to rogue infected plants as they provide a reservoir of virus particles.

## 6.4. ELEPHANT FOOT YAM/AMORPHOPHALLUS

Elephant-foot yam (*Amorphophallus paeoniifolius*) is a popular tuber crop of humid tropics. Among root and tuber crops, *Amorphophallus* is one of the most important tuber crops that occupy special significance as high value cash crop due to its high production potential and popularity as a vegetable. The genus *Amorphophallus* is indigenous to tropical Asia and Africa and A. paeoniifolius is found widely distributed in thickets and secondary growth forests at low and medium altitude in the Philippines, Malaysia, Indonesia, Sri Lanka and the South East Asia subcontinent. In India, it is widely grown in Andhra Pradesh, Bihar, West Bengal, Assam, Jharkhand, Gujarat, Tamil Nadu, Kerala, Uttar Pradesh, Maharashtra, Chhattisgarh, Meghalaya and Andaman and Nicobar Island. Chemical composition of Elephant-foot yam corms: moisture 60-80%, starch 25-28%, sugars 0.5-0.6%, crude protein 1.0-2.8%, fat 0.1-0.3%, ash 0.5-2.6%, Vitamin A-0.8 mg, Vitamin B1 - 0.08-0.09 mg, Vitamin B2 - 0.02-0.03 mg, Vitamin C - 5-12 mg.

### 6.4.1. Bacterial Diseases

#### 6.4.1.1. Bacterial Leaf Blight/Leaf Rot, *Xanthomonas amorphophalli*

**i) Symptoms:** The leaf blight/leaf rot is a major and widespread disease of elephant foot yam causing considerable loss in yield (Fig. 6.8).

**ii) Management:** Three sprays of 100 ppm Agrimycin at 15 days interval after the first appearance of the disease significantly reduced the disease incidence (24.65% as compared to 50.88% in control) and increased the yield (36.25 t/ha as compared to 17.81 t/ha in control).

Seed tuber dip treatment with 0.05% Agrimycin for 12 hrs decreased the disease incidence (19.73% as compared to 65.91% in control) and increased the tuber yield (23.95 MT/ha as compared to 7.81 MT/ha in control).

**Fig. 6.8.** Leaf blight damage on leaves of elephant foot yam.

## 6.5. YAM

Yam (*Dioscorea* spp.) belongs to the family Dioscoreaceae, which is widespread throughout the tropics and subtropics. Normally yams are consumed as boiled, baked or fried vegetables. They contain 18-20% starch with mucilaginous substance. Yam also contains alkaloids, tannins and steroids which have pharmaceutical value. *Dioscorea alata* (greater yam), *D. rotundata* (white or white Guinea yam), *D. cayensis* (yellow Guinea yam) and *D. esculenta* (lesser yam) produce edible tubers. *D. alata* was first cultivated in S.E. Asia and is widely grown in Asia and the Pacific, is less favoured in Africa, but much grown in the Caribbean. *D. rotundata* is the most important species in W. Africa, where it is originated. *D. cayenensis* is a native of W. Africa, where it is extensively cultivated. *D. esculenta* originated in Indo-China and is extensively grown in S.E. Asia and elsewhere in the tropics.

In India, greater yam and lesser yam are under cultivation. Madhya Pradesh, North-Eastern states, West Bengal, Bihar, Orissa, Uttar Pradesh, Kerala, Tamil Nadu, Gujarat and Maharashtra are major yam-growing states.

### 6.5.1. Viral Diseases

#### 6.5.1.1. Yam Mosaic Virus

**i) Distribution:** The mosaic disease of yam has been reported from West Africa and the Caribbean but it probably occur worldwide.

**ii) Symptoms:** Leaves become mottled with dark green and chlorotic patches. Vein-clearing frequently occurs. Leaves may be distorted at the edges. When the infection is severe the whole plant appears stunted and may be less than a metre height. Many shoots are produced both at the base and from the lateral buds, giving the plant bushy appearance.

**iii) Transmission:** The virus is transmitted mechanically and by aphid vector, *Aphis gossypii* in a non-persistent manner.

**iv) Host Range:** The virus infects *Dioscorea rotundata, D. cayenensis, D. alata, D. trifida, D. esculenta, D. composita, D. floribunda* and wild *D. minutiflora.*

**v) Management:** The best way of preventing serious outbreaks is to select planting material from apparently disease-free plants. Severely infected plants should be rouged.

Chapter 7

# PLANTATION CROPS

Plantation crops serve a variety of human needs such as food, oil, industrial raw materials, beverages and confectionary items. They generate huge employment opportunities directly or indirectly to several million people in their production, processing, marketing and international trade sectors. Plantation crops are a large group of crops. These are grown over an area of 3.282 million ha with production of 11.262 million tonnes in India (Bijay Kumar, 2007) (Tables 7.1 and 7.2). The plantation crops contributed 6.09% of total horticulture production, in 2006-2007. India is the largest producer and consumer of areca nut, cashew nut, tea, third largest producer of coconut, fourth largest producer and consumer of rubber and sixth largest producer of coffee in the world. While the major plantation crops include coconut, areca nut, oil palm, cashew, coffee, tea and rubber, minor plantation crops are cocoa and betel vine. They play an important role in export as well as domestic requirement, employment generation and poverty alleviation particularly in rural sector. The annual income generated is more than Rs. 5,52,590 million and export earning is approximately Rs. 93,350 millions.

**Table 7.1. Area, production and productivity of plantation crops in India (2005-06)**

| Plantation crops | Area ('000 ha) | Production ('000 MT) | Productivity (MT/ha) | % Share of India production |
|---|---|---|---|---|
| Coconut | 2028.9 | 10190.1 | 5.0 | 90.5 |
| Areca nut | 381.1 | 483.1 | 1.3 | 4.3 |
| Cocoa | 29.5 | 10.2 | 0.3 | 0.1 |
| Cashew | 843.4 | 579.4 | 0.7 | 5.1 |
| Total | 3282.9 | 11262.8 | 3.4 | |

**Table 7.2. State-wise area, production and productivity of plantation crops in India (2005-06)**

| State | Area in '000 ha | Production in '000 MT | Productivity in MT/ha |
|---|---|---|---|
| Kerala | 1096.6 | 4545.1 | 4.1 |
| Tamil Nadu | 497.7 | 3415.0 | 6.8 |
| Karnataka | 735.1 | 1095.9 | 1.5 |
| Andhra Pradesh | 286.1 | 706.6 | 2.4 |
| Maharashtra | 180.0 | 376.1 | 2.0 |
| Orissa | 170.8 | 267.0 | 1.5 |
| West Bengal | 44.3 | 261.9 | 5.9 |
| Assam | 104.3 | 217.8 | 2.0 |
| Goa | 81.9 | 115.8 | 1.4 |
| Gujarat | 20.4 | 99.1 | 4.8 |
| Andaman & Nicobar | 30.0 | 64.7 | 2.1 |
| Lakshadweep | 2.7 | 36.5 | 13.5 |
| Meghalaya | 17.9 | 21.9 | 1.2 |
| Pondicherry | 5.2 | 20.3 | 3.9 |
| Tripura | 6.7 | 11.7 | 1.7 |
| Mizoram | 2.0 | 5.3 | 2.6 |
| Nagaland | 1.1 | 2.0 | 1.8 |
| Total | 3282.9 | 11262.8 | 3.4 |

There is a good scope for switching over the production of plantation crops viz., coffee, tea, coconut and areca nut to organic farming, as it is possible to realize higher returns from the unit quantity exported when grown organically. The plantation crops are generally grown in ecologically fragile hilly tracts, adopting organic farming methods would protect environment and also prevent contamination thereby producing products of high nutritional quality. Further, the plantation crops are amenable for organic farming as they produce huge amount of waste biomass for recycling, which can meet major portion of nutrients required.

## Export of Plantation Produce

India is the largest producer in the world of cashew nuts, coconuts and tea. India is largest producer, importer, processor, exporter and second largest consumer of cashew. Indian exports during 2006-2007 were US$ 866 million in tea and coffee (Table 7.3).

**Table 7.3. Export of plantation crops**

| Commodity | Apr-Mar 2006 | Apr-Mar 2007 | % Growth | % Share |
|---|---|---|---|---|
| Tea | 390.92 | 432.00 | 10.51 | 0.34 |
| Coffee | 358.83 | 434.77 | 21.16 | 0.34 |
| Cashew | 585.76 | 553.41 | -5.52 | 0.44 |

## 7.1. COFFEE

Coffee (*Coffea arabica, C. canephora*) is an important commercial plantation crop mostly grown on the hilly slopes of the traditional coffee tracts in South India (Karnataka, Kerala, Tamil Nadu and Andhra Pradesh) and non-traditional Orissa, Maharashtra, Madhya Pradesh and North-Eastern Council States. The coffee industry is well established and flouris-hing enterprise. The major coffee plantations in India are in the Southern States with Karnataka being the major coffee producing state. Recently some coffee plantations have come up in the North Eastern States of India as well. Besides being one of the major foreign exchange earner to a tune of around Rs. 2000 million annually, coffee industry supports more than four lakh people to earn their livelihood. The two main species of coffee in cultivation are *Coffea Arabica* (Arabica) and *Coffea canephora* (Robusta). *C. arabica* produces finer seeds which are more valued and grows best at higher elevations (1000 and 1500 m MSL). *C. canephora* is sturdier and better adapted to the warm climate prevailing in the lower elevations (500 and 1000 m MSL). Robusta, in general, is more tolerant to many diseases and pests including nematodes, as compared to Arabica. Coffee in India is grown under a canopy of shade trees of various species along with other plantation crops like black pepper, cardamom, banana, cocoa, etc. The present area under coffee in India is above 100,000 hectares, nearly three-fourth of which is under the fine quality Arabica species with most of the remaining area under Robusta.

India produced 290,000 tonnes of coffee accounting for approximately 4% of the global production. Over 80% of the coffee produced in India is exported. Although over 50 varieties are produced, the popular ones are the Indian Robusta which is used for blending and Arabica. There is a strong demand in the world market for production of high value, high quality, especially organic coffee. Today, the area under certified organic coffee is 2500 ha with an estimated production of 1500 MT of which 1000 MT is being exported currently.

### 7.1. Bacterial Diseases

#### 7.1.1. Bacterial Blight, *Pseudomonas syringae* pv. *garcae*

Bacterial blight of coffee (BBC) is confined to a few coffee growing areas in Kenya (Solai in Nakuru District and around Mt. Elgon). Recently a few cases have been reported in the East of Rift Valley. A more severe dieback occurs at high altitudes (over 1800 m) in Kenya. The pathogen has also been reported from Brazil.

**i) Symptoms:** This bacterium is favoured by cool wet weather. Lesions appear on leaves with water soaked margins at start of infection. Leaves eventually dry up and roll inwards turning brown (Fig. 7.1). Dead and dried leaves do not shed but remain attached to the plant. Symptoms occur on secondary or tertiary branches and in severe cases the heads may be affected.

On twigs the terminal bud is attacked. Infection then extends downwards, resulting in dieback. Flowers and pin-heads shrivel and turn black.

**Fig. 7.1.** Bacterial blight on coffee.

The bacterium is an epiphyte on coffee shoots. Under wet and cool conditions, the pathogen gains entry through young tissues to produce water-soaked, dark, and necrotic lesions on the leaf, stem and berry. The infection spreads backwards from the shoot, tip of twigs causing a dieback. Affected bushes appear as if they have been scorched by a fire. The initial symptom is the blackening of a node often accompanied by blackening of the petiole and basal part of the leaf. This leads to death of the branch above the node and leaves on the branch turn brown and eventually necrotic (burnt). Severely affected trees appear scorched as if by fire. The bacterium persists on leaves, berries and on green and matured bark. During cool, wet weather, the bacterium multiplies and initiates an epidemic. The entire crop may be lost.

### ii) Management

- Desucker or cut off infected twigs (at least 5 cm from the point of infection) and collecting them in buckets for burning.
- Observe field hygiene by disinfecting tools used for pruning infected trees with a suitable disinfectant such as spirits with 70% alcohol. Implements used must be sterilized after making each cut to avoid the risk of infecting the cut surfaces of the twig with a carry-over inoculum.
- Use of copper sprays especially during wet season. Spray at 2 week interval during the rains and 3 week interval after the rains.
- Plant shade trees and wind breaks on exposed sides of BBC areas.

- Avoid planting material from known BBC areas. In emergencies spray with copper just before, during and after flowering especially when this coincides with cool wet weather.

The current control measures involve frequent spraying with copper fungicides and/or tank mixture of copper and organic fungicides during the rainy season. More recently, it has been observed that pruning of the infected branches before application of bactericides enhances the control (Kairu, 1991).

## 7.2. TEA

India is the largest producer and consumer of tea (*Camellia sinensis*) in the world. During 2006-2007, India produced approximately 950,000 tonnes of tea of which around 218,000 tonnes was exported. The total area under tea is 5,07,200 ha. It accounts for 20% of the total area under tea in the world, 28% of world production, 22% of global tea consumption and 15% of the total global tea exports. Tea accounts for approximately £1 billion in turnover in India. Tea is grown mainly in three major regions in India - Assam, Darjeeling in North East and the Nilgiris region in Southern India. The largest producers of tea are Assam (50.7%), West Bengal (22.1%), Tamil Nadu (15.9%) and Kerala (8.3%).

Other areas where tea is grown to a small extent are Karnataka, Tripura, Himachal Pradesh, Uttarakhand, Arunachal Pradesh, Manipur, Sikkim, Nagaland, Meghalaya, Mizoram, Bihar and Orissa. The Northern States of India account for around 75% of the tea grown in India and the Southern States accounting for around 25% of the production. The tea industry provides direct employment for one million workers of which sizeable number are women.

More than half of the one million hectares of tea cultivation is in India and Sri Lanka, on areas ranging from 225,000 to 350,000 hectares. Between them, these two major producers account for more than three-quarters of the world's export trade of 600,000 metric tonnes of this commodity. This apparently disproportionate share in the export trade is because in these two countries, domestic consumption is far outstripped by total production.

With the current low price trends which are highly unremunerative to the growers, that most of the growers have shifted the cultivation practices to organic farming system. Today, more than 10,000 ha have been certified under organic tea cultivation.

### 7.2.1. Viral Diseases

#### 7.2.1.1. Camellia Yellow Mottle Virus

**i) Symptoms:** Camellia yellow mottle virus is transmitted by root grafts and propagation of diseased stock. This virus causes irregular, yellow, mottled or splotchy patterns of various sizes and shapes on the leaves. Some leaves may turn entirely yellow. Irregular white blotches will appear on infected flowers.

**ii) Management:** There is no chemical that will cure the virus. Plant only virus-free plants obtained from reputed nurseries.

## 7.3. COCONUT

The coconut palm (*Cocos nucifera*), originated in South East Asia, is now cultivated widely throughout the tropics of the world. The largest producers of coconut are the Philippines, Indonesia, India and Sri Lanka. In all the coconut producing countries, it is an important source of export earning despite being a regular constituent of the food. Coconut is an important small holders' plantation crop. The total area under coconut cultivation in India is 2.029 million ha with an annual production of 10.190 million tonnes and the average productivity of 5.0 tonnes/ha is the highest in the world (Bijay Kumar, 2007) (Table 7.4). It provides proteins, fats and some vitamins. Coconut contributes 700 billion rupees to the GDP of the country. Coir and coir products exported from India earned Rs. 2260 million during the above period. The contribution of the crop to the total edible oil pool in India is around 6 per cent. It is supporting the livelihood of 10 million people across the country. The four Southern States – Kerala, Tamil Nadu, Karnataka and Andhra Pradesh together accounts for 90% of the total area and production. Coconut is predominantly grown under rainfed conditions in Kerala and parts of coastal Karnataka, Maharashtra and Tamil Nadu. In rest of the country, it is mainly grown under irrigated conditions.

**Table 7.4. State-wise area, production and productivity of coconut in India (2005-06)**

| State | Area in '000 ha | Production in '000 MT | Productivity in MT/ha |
|---|---|---|---|
| Kerala | 897.8 | 4352.3 | 4.847 |
| Tamil Nadu | 370.5 | 3348.5 | 9.037 |
| Karnataka | 467.5 | 832.5 | 1.780 |
| Andhra Pradesh | 104.0 | 613.7 | 5.900 |
| Maharashtra | 18.0 | 188.1 | 10.450 |
| Orissa | 50.8 | 189.0 | 3.720 |
| West Bengal | 24.9 | 222.6 | 8.939 |
| Assam | 19.1 | 141.0 | 7.382 |
| Goa | 25.3 | 86.2 | 3.407 |
| Gujarat | 16.4 | 95.1 | 5.798 |
| Andaman & Nicobar | 25.6 | 59.9 | 2.339 |
| Lakshadweep | 2.7 | 36.5 | 13.518 |
| Pondicherry | 2.1 | 19.2 | 9.142 |
| Tripura | 3.3 | 4.8 | 1.454 |
| Nagaland | 0.9 | 0.8 | 0.888 |
| Total | 2028.9 | 10190.1 | 5.0 |

### 7.3.1. Viral Diseases

#### 7.3.1.1. Tatipaka Disease (Pytoplasma)

**i) Symptoms:** The disease first appeared in Tatipaka village, East Godavari district, Andhra Pradesh, following a cyclone in 1949. Palms in the age group of 25 to 60 years are more susceptible. Development of an abnormally large crown with dark green inner leaves and higher yield is the precursor of disease incidence. Subsequently the crown becomes smaller in size producing progressively shorter leaves (Fig. 7.2). The stem begins to taper. The leaves give a fascinated appearance due to improper unfolding of leaflets. The affected tree produces smaller bunches with atrophied barren nuts. The causal agent is suspected to be phytoplasma.

**Fig. 7.2.** Tatipaka disease on coconut.

**ii) Management**

- Destroy the affected plants to prevent the spread of the disease.
- Deterioration in condition of the palms was arrested with benomyl, antiviral compounds and Kasumin (Ramapandu and Rajamannar, 1981).
- Root feeding with tetracycline was found to have ameliorative effect on the diseased palms (Rajamannar *et al.*, 1989, 1991).

### 7.3.1.2. Root Wilt (Phytoplasma)

The Disease is non-lethal but debilitating and palms of all age groups are affected. A reduction in yield of nuts up to 80% has been reported in palms in advanced stage of the disease. The oil content is very much reduced and the oil loses its flavour as well.

**i) Distribution:** it is prevalent in a contiguous manner in 410,000 ha in the eight southern districts of Kerala with an intensity ranging between 75.6%, the highest being in Kottayam district, to 2.6 and 1.5% in Trichur and Trivandrum districts, respectively.

**ii) Symptoms:** The symptoms of Root (wilt) disease are flaccidity (characteristic bending or ribbing of leaves), foliar yellowing and marginal necrosis of leaflets (Fig. 7.3). Delayed flowering and reduction in yield are observed in palms contracting the disease in the pre-bearing age. Extensive rotting of the root system was considered as the major symptom. The leaves of diseased palms showed slight downward curling of leaf lets.

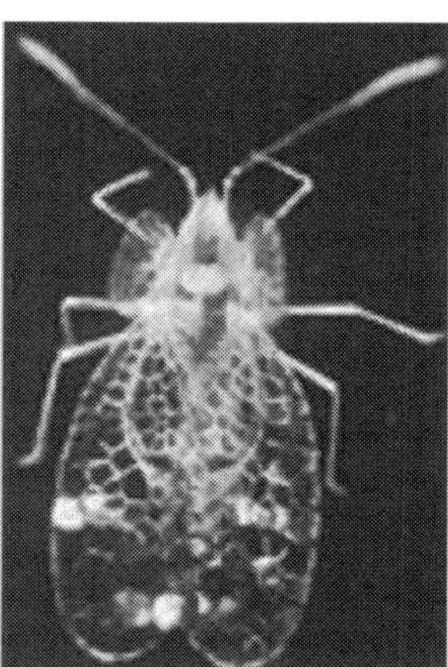

**Fig. 7.3.** Root wilt of coconut and its lace wing vector.

**iii) Etiology:** Research conducted at CPCRI Regional Station, Kayangulam using electron microscope revealed the causal organism as phytoplasma.

**iv) Epidemiology:** The disease is spread through insects like lace bug, *Stephanitis typicus* and/or plant hopper, *Proutista moesta*. The spread is faster in sandy, sandy-loam, alluvial and in heavy textured clayey soils than in laterites. The disease incidence is relatively higher in water-logged low lying areas adjacent to rivers and canals.

**v) Management:** The strategy comprises eradication of disease advanced palms, gap filling with elite disease tolerant seedlings, adoption of mixed farming/inter cropping, organic recycling, providing summer irrigation, application of Mg $SO_4$ along with fertilizer, control of leaf rot etc. CPCRI identified disease tolerant palms, among Chowghat Green Dwarf and local talls. Progenies of crosses involving CGD x WCT have been found to be promising.

The yield of palms can be sustained through adoption of integrated management practices like removal of disease advanced and juvenile palms, addition of organic matter, raising green manure crops in the basin and their incorporation into the soil, irrigation during summer months and adopting inter and mixed cropping (Muralidharan *et al.*, 1986; Rajagopal *et al.*, 1987). Mixed farming in diseased gardens involving the raising of fodder crops in the interspaces, maintaining milch cows and recycling of organic waste has helped in increasing the yield of palms by 26% (Sahasranaman *et al.*, 1983). Mixed cropping with cacao also increased the yield by 30% and slowed down the decline of palms. Cultivation of tapioca, elephant foot yam and yam in the interspaces of palms in disease affected gardens for a period of 3 years increased the nut yield by 4.96, 15.57 and 8.07%, respectively (Menon and Nair, 1978).

### 7.3.1.3. Coconut Lethal Yellowing (Phytoplasma)

**i) Symptoms:** Symptoms include necrosis of inflorescences, yellowing of leaves and premature shedding of all fruits (nut fall) regardless of their developmental stage. Aborted nuts often develop a brown-black calyx-end rot reducing seed viability. Premature nut fall is accompanied or followed by inflorescence necrosis. This symptom is most readily observed as newly mature inflorescences emerge from the ensheathing spathe. Normally light yellow to creamy white in colour; affected inflorescences are instead partially blackened (necrotic) usually at the tips of flower spikelets. As disease progresses, additional emergent or unemerged inflorescences show more extensive necrosis and may be totally discoloured. Such symptom intensification results in the death of most male flowers and an associated lack of fruit set. Yellowing of the leaves usually starts once necrosis has developed on two or more inflorescences and discolouration is more rapid than that associated with normal leaf senescence. Yellowing begins with the older (lower most) leaves and progresses upward to involve the entire crown. Yellowed leaves turn brown, desiccate and die. In some cases, the advent of this symptom is seen as a single yellow leaf (flag leaf) in the mid-crown. Affected leaves often hang down forming a skirt around the trunk for several days before falling.

A putrid basal soft rot of the newly emerged spear (youngest leaf) occurs once foliar yellowing is advanced. Spear leaf collapse and rot of the apical meristem invariably precedes death of the palm at which point the crown topples away leaving a bare trunk. Infected palms usually die within 3 to 6 months after the appearance of the first symptoms.

Lethal yellowing symptomatology may be complicated by other factors. For example, non-bearing palms lack fruit and flower symptoms. Foliar discolouration also varies markedly among coconut ecotypes and hybrids. For most tall-type coconut palms, leaves turn a golden yellow before dying whereas on dwarf ecotypes leaves generally turn reddish to grayish-brown.

Phytoplasmas are transmitted in a persistent (circulative-propagative) manner primarily by insect vectors belonging to the families *Cicadelloidea* (leaf hoppers) and *Fulgoroidea* (plant hoppers).

**ii) Management :** Use genetically resistant ecotypes (Malayan Yellow Dwarf) and hybrids (Malayan Yellow Dwarf x Panama Tall). This is the only practical long-term solution to Lethal Yellowing.

## 7.4. ARECANUT

Areca nut (*Areca catechu*), otherwise known as the betel nut, has its origin in the humid regions of Asia and Malay Islands. India continues to dominate the world in area, production and productivity of arecanut and has achieved self sufficiency in arecanut production. It is an important cash crop of India. India is the largest producer of areca nut in the world covering an area of 3.811 lakh ha with annual production of 4.831 lakh tonnes (Bijay Kumar, 2007) (Table 7.5). Most of the production is domestically consumed. It is masticatory and the ripe fruits are sometimes used as an antihelminthic and astringent in Europe. It has a toning effect and stimulates the nerves. The extract after boiling the nuts contains tannins which are used for tanning leather. Areca nut industry forms the economic back bone of nearly 6 million people in India and for many of them it is the sole means of livelihood.

**Table 7.5. State-wise area, production and productivity of arecanut in India (2005-06)**

| State | Area in '000 ha | Production in '000 MT | Productivity in MT/ha |
|---|---|---|---|
| Kerala | 108.6 | 119.3 | 1.098 |
| Tamil Nadu | 4.8 | 10.3 | 2.140 |
| Karnataka | 161.6 | 215.7 | 1.334 |
| Andhra Pradesh | 0.3 | 0.2 | 0.666 |
| Maharashtra | 2.0 | 5.0 | 2.500 |
| West Bengal | 9.4 | 29.3 | 3.128 |
| Assam | 71.2 | 66.8 | 0.938 |
| Goa | 1.6 | 2.6 | 1.625 |
| Andaman & Nicobar | 4.4 | 4.8 | 1.090 |
| Meghalaya | 11.5 | 15.5 | 1.356 |
| Pondicherry | 0.1 | 0.1 | 1.000 |
| Tripura | 3.4 | 6.9 | 2.029 |
| Mizoram | 2.0 | 5.3 | 2.650 |
| Nagaland | 0.2 | 1.3 | 6.500 |
| Total | 381.1 | 483.1 | 1.3 |

### 7.4.1. Bacterial Diseases

#### 7.4.1.1. Bacterial Leaf Stripe, *Xanthomonas campestris* pv. *arecae*

**i) Symptoms:** The disease is caused by *Xanthomonas campestris* pv. *arecae*. The symptoms are purely parenchymatous in nature causing water-soaked linear lesions parallel to the midrib of the leaflets. The lesions are covered with abundant creamy-white bacterial exudates on the under surface which is a striking feature of the disease. The entire leaflet in a frond may be affected resulting in complete or partial blighting. In severe cases, entire crown may be affected. When growing buds are affected, death of palm takes place. The disease is aggressive during monsoon. Younger palms (3-5 years old) are highly susceptible.

**ii) Management :** Spraying or stem injection with tetracycline group of antibiotics at 500 ppm concentration is effective.

### 7.4.2. Viral Diseases

#### 7.4.2.1. Yellow Leaf Disease (Phytoplasma)

**i) Symptoms:** The disease is rampant in Kerala and Karnataka. Yellowing of leaves begin in the inner whorl, gradually spreading to the outer parts of the crown (Fig. 7.4). Chlorosis is finally observed on almost all leaves in the whorl from the edges of the individual leaflets to the midrib region. Withering of the tips (Fig. 7.4) starts and gradually spreads to the older portions of the leaf. The freshly formed leaves grow shorter and their laminae show unequal growth and flaccidity. In a few cases, wilting and shedding of leaves are also observed. The nuts are reduced in size, shriveled with their kernels often turning black and there is severe reduction in yield. Stem of the affected palms becomes spongy and friable, the conducting strands getting destroyed. In advanced stages, the stem breaks off at the top. Rotting of the roots is also observed. There is a gradual decline in yield. Association of mycoplasma-like organisms (phytoplasma) with the disease has been confirmed. It is spread through plant hoppers. Water logging is a predisposing factor in the incidence of the disease. Lack of balance nutrition and unscientific cultivation practices make the palm susceptible to the disease.

**Fig. 7.4.** Yellow leaf disease of areca nut.

**ii) Management:** Destroy affected plants to prevent the spread of disease. Apply FYM at 12 kg/palm/year with provision of summer irrigation and drainage to improve the condition of the palm.

Maintain the garden properly to keep affected palms in a healthy condition by adopting recommended manurial, cultural, plant protection and other management practices. Improve drainage conditions in the garden.

Since the disease is not amenable to management by conventional plant protection measures, other means of containing the disease should be looked into. They are:

- Physical condition of the soil should be improved by deep digging and adding sand/organic matter. In heavy sticky soils, gypsum at 500 kg/ha may be incorporated, once in two years prior to normal fertilizer application.
- Adequate drainage should be provided, especially during monsoon season.
- Phyto-sanitary and plant protection measures should be adopted to control Anabe, bud rot, spindle bud and mite infestation.
- Sun-scorching of the stem should be avoided by covering stem with areca nut leaves or painting with lime slurry.
- Application of NPK fertilizers as per schedule along with lime and zinc at 8.5 g each per palm.

## 7.5. BETEL VINE

Betel vine (*Piper betel*) is a perennial, dioecious, evergreen creeper grown in India. It is a chewing stimulant of brain, lungs and heart. The leaves contain Vitamins B and C. It is an important cash crop in Andhra Pradesh, Assam, Bihar, Karnataka, Kerala, Madhya Pradesh, Maharashtra, Orissa, Tamil Nadu, Tripura, Uttar Pradesh and West Bengal with an annual turnover of about Rs.700 crores. Its leaves are exported to Pakistan, Bangladesh, Indonesia, Malaysia, Burma and Thailand.

### 7.5.1. Bacterial Diseases

#### 7.5.1.1. Bacterial Leaf Spot, *Xanthomonas betlicola*

**i) Symptoms:** The bacterial leaf spot caused by *Xanthomonas betlicola* is most serious in southern Kerala. The initial symptoms appear as small water-soaked regions on the under surface of leaves, which enlarge and turn dark brown in the centre with yellow halo. Later these appear on upper surface also. The centre of spots are mottled brown and later turn black and in severe conditions leaves turn yellow and fall. Profuse bacterial ooze may be seen on the lower side of the lesion. In severe cases, defoliation and stem injury occur, leading to wilting of the plant.

**ii) Management:** Drenching with wettable ceresan 0.1% or Agallol 0.1% or spraying 1% Bordeaux mixture at every month, staring from August to January gives effective control of the disease.

## 7.6. OIL PALM

Oil palm (*Elaeis guineensis*) is the highest oil-yielding plant (4-6 tonnes oil/ha) among perennial oil-yielding crops, producing palm oil and palm kernel oil. These are used for culinary as well as industrial purposes. It occupies an area of 44,413 ha in Andhra Pradesh (80% area), Karnataka,

Assam, Gujarat, Goa, Kerala, Maharashtra, Orissa, Tamil Nadu, Tripura and West Bengal. Average productivity in oil palm plantations is now 4-5 tonnes/ha which compares favourably with other countries.

### 7.6.1. Bacterial Diseases

#### 7.6.1.1. Spear Rot, *Erwinia* spp.

**i) Damage:** This bacterial disease is noticed to affect oil palms of all ages. The incidence is less than one per cent. Yellowing starts from tip of the innermost whorl of leaves. Small lesions occur at the distal portions of spear and rotting extends downwards. As the disease advances, new leaves become rudimentary and show rotting. General decline in vigour and production is then noticed. Occurrence of spear rot without yellowing has also been noticed. Distinguishable marginal yellowing of leaflets and sudden drying of leaves showing yellowing are other symptoms.

**ii) Management:** Rouging of all the affected palms may be adopted to prevent further spread of the disease. In early stages of the disease, the affected portions of leaves may be removed and burnt.

Chapter 8

# SPICE CROPS

From time immemorial, India is regarded by the rest of the world as the "magic land of spices". Spices are natural plant products used to improve flavour, aroma, taste and colour of food products, beverages, liquors, pharmaceutical, cosmetic and perfumery products. No other country in the world has such a diverse variety of spice crops as in India. Indian spices are also noted for their excellent aroma, flavour and pungency not so easily matched by any other country. Even in minute quantities, spices are a real delight to all our sensory organs, making the food dishes more palatable. Usage of spices in food processing industries is increasing quite fast in all the countries. Hence, the world demand for spices in recent years is continuously on the rise.

India is the largest producer, consumer and exporter of spices in the world. India is the largest producer in the world of ginger, turmeric and black pepper. The annual production of spices is more than 5 million tonnes valued at US$ 1,500 million. India's share of the world spices trade is estimated at 45 to 50% by volume and 25% by value. The present annual production of spices in the country is 5.107 million tonnes from over 2.857 million hectares (Bijay Kumar, 2007) (Table 8.1). The spices contributed 2.76% of total horticulture production in 2005-2006. About 8.5% of India's export earnings from agricultural and allied products come from spices. Indian spices flavour foods in over 134 countries. The important spice crops include black pepper, cardamom, ginger, garlic, turmeric and chilli.

**Table 8.1. State-wise area, production and productivity of major spices in India (2005-06)**

| State | Area in '000 ha | Production in '000 MT | Productivity in MT/ha |
|---|---|---|---|
| Andhra Pradesh | 326.9 | 920.5 | 2.8 |
| Tamil Nadu | 130.0 | 784.0 | 6.0 |
| Karnataka | 258.8 | 619.8 | 2.4 |
| Gujarat | 426.1 | 505.6 | 1.2 |

| | | | |
|---|---|---|---|
| Madhya Pradesh | 292.3 | 345.7 | 1.2 |
| Rajasthan | 347.9 | 297.1 | 0.9 |
| Uttar Pradesh | 91.0 | 283.8 | 3.1 |
| West Bengal | 108.0 | 226.7 | 2.1 |
| Orissa | 145.6 | 195.3 | 1.3 |
| Kerala | 428.1 | 191.8 | 0.4 |
| Chhattisgarh | 28.9 | 168.2 | 5.8 |
| Maharashtra | 112.5 | 106.0 | 0.9 |
| Others | 162.1 | 463.2 | 2.9 |
| Total | 2857.7 | 5107.7 | 1.8 |

Since spices form a part of traditional/ethnic medicines, the demand for organically produced spices is growing. Export of organic spices from India has started in right earnest. The country at present exports around 50 tonnes of different varieties of organic spices. Export will get a significant boost in the coming years as more and more farmers switch to organic methods.

**Export of Spice Produce**

India is the largest producer, consumer and exporter of spices and spice products in the world and produces more than 50 spices. Spice export figures for the year 1996-97 and 2006-07 shows the phenomenal growth of the Indian spice industry. The Indian spice export for the year 1996-97 was 2,25,000 metric tonnes which was worth $350 million; in the year 2006-07 the spice exports touched the figure of 3,75,000 MT which was worth US$ 800 million. Thus, these figures show the exponential growth of the Indian spice industry in a very short time. Spice exports from India have touched US $ one billion mark during 2007-08. The major destination of Indian spice exports is USA followed by European Union, Sri Lanka, Japan and Middle East Countries.

## 8.1. BLACK PEPPER

Black pepper (*Piper nigrum*), known as the "King of spices", has its origin in the South-Western hills of India. It is a branching and climbing perennial shrub belonging to the family Piperaceae and is grown in the hot and humid parts of the world. Pepper is loved for its spicy factor, piperine. It is used in various culinary seasonings, as a preservative for meat and fish, in medicine and perfumery (pepper oil). In India, it is a common practice to use live trees as standards for climbing pepper. Pepper is the most important spice produced in the country and accounts for 50-60% of the total export earnings. It is being cultivated in 0.223 million hectares producing 0.070 million tonnes of black pepper with an average yield of 1.3 tonnes per hectare.

### 8.1.1. Viral Diseases

#### 8.1.1.1. Stunt Disease (Little leaf)

**i) Symptoms:** Leaves become small and narrow with varying degree of deformation. Leaves appear leathery, puckered and crinkled. Chlorotic spots and streaks on leaves give appearance of mosaic. Vines show shortening of internodes to varying degree.

### ii) Management

- Affected plants should be destroyed.
- Planting material should be collected from the disease-free plants.

### 8.1.1.2. Phyllody Disease (Phytoplasma)

**i) Symptoms:** Floral buds are transformed into narrow leaf like structures. In advanced stages, leaves become small and chlorotic. Internodes are also shortened. The affected fruiting laterals give a witches' broom appearance. Severely affected plants become unproductive in 2-3 years

**ii) Management:** The infected vines should be collected and destroyed to prevent further spread of the disease.

## 8.2. CARDAMOM

Cardamom (*Elettaria cardamomum*), also known as the "Queen of Spices" enjoys a unique position in the world market has its origin in the evergreen rain forests of Western Ghats of South India. India enjoyed near monopoly in the field till recently. Guatemala has emerged as a competitor to India. India and Guatemala are the main producers and exporters of cardamom. Other cardamom growing countries are Tanzania, Sri Lanka, El Salvador, Vietnam, Laos, Kampuchea and Papua New Guinea. The area under cultivation in India during 2004-05 was 73,237 ha, with an annual production of 11,415 metric tonnes. The export earning during 2006-07 was Rs. 274.7 million by exporting 650 million tonnes of small cardamom. Kerala, Karnataka, and Tamil Nadu produce cardamom in India. While Kerala accounts for 60 per cent of the production, Karnataka contributes 30 per cent, and Tamil Nadu 10 per cent. In Karnataka, cardamom is cultivated mostly in Kodagu, Hassan, and Chikmagalur districts. The plants last for 15 years. They grow in moist, elevated, and shady forest loams. They require a fairly distributed annual rainfall ranging from 2,500 mm to 5,000 mm and are found within an altitude ranging between 600 m and 1,200 m above the mean sea level. It is used for flavouring various food preparations, confectionery, beverages, liquors and medicines.

### 8.2.1. Viral Diseases

### 8.2.1.1. Katte or Mosaic

**i) Symptoms:** The disease generally appears as chlorotic specks on tender leaves. The symptoms consist of discontinuous stripes of light green colour running almost parallel to each other from the mid-rib to the margin of the leaves. Later typical mosaic symptoms appear on mature leaves. On young shoots, such stripes are seen on the leaf sheath also. The affected plants gradually show decline symptoms producing clumps of smaller size with fewer slender tillers.

**ii) Transmission:** This is a virus disease, which is transmitted by the banana aphid, *Pentalonia nigronervosa* in a non-persistent manner. *P. nigronervosa* f.sp. *caladii* is the major vector in India (Siddappaji and Reddy, 1972). The aphid acquired the virus within a few minutes and transmitted immediately without any incubation period. A positive correlation was obtained between vector population and disease incidence (Naidu and Venugopal, 1984). The virus spreads through infected suckers when vegetative propagation is adopted. The virus has been identified as 'poty virus' with flexuous rods measuring 650 nm long and 10-12 nm in diameter (Naidu *et al.*, 1985).

**iii) Host Range:** Natural infection was recorded on *Ammomum cannecarpum, A. microstephanum, A. nutans, Maranta arundanacea* and *Curcuma nilgherrensis*. These might serve as alternate hosts (Naidu *et al.*, 1985; Siddaramaiah *et al.*, 1986).

**iv) Management:** Eradication of the source of inoculum by destroying infected plants and destruction of the vector by insecticide application is effective. Regular application of insecticide against cardamom thrips controls the aphids also. Avoid using katte-infected rhizome for planting.

The virus being not seed transmissible, the importance of using disease-free seedlings raised through seed is advocated.

Destruction of plants showing symptoms of the disease should be done promptly once in two months. Removal of all alternate hosts of virus is also recommended.

## 8.3. GINGER

Ginger (*Zingiber officinale*), an herbaceous perennial, belonging to the family Zingiberaceae, is valued for its rhizome. The country of its origin is not definitely known but is presumed to be either from India or China. India is the largest producer (3,05,900 metric tonnes) and exporter of dry ginger in the world (export earning of Rs. 191 million). The other ginger producing countries are Jamaica, Sierra Leone, Nigeria, Southern China, Japan, Taiwan and Australia. In India, It is being cultivated in 85,100 hectares producing 0.306 million tonnes of ginger with an average yield of 3.6 tonnes per hectare.

### 8.3.1. Bacterial Diseases

#### 8.3.1.1. Bacterial Wilt, *Ralstonia solanacearum*

A bacterial wilt of ginger caused by *Ralstonia solanacearum* was recorded from Australia, Hawaii, India, Mauritius and Malaysia. Bacterial wilt is a very serious disease that can spread very rapidly through an area, causing complete destruction.

**i) Symptoms:** At first, symptoms occur on a single plant or on plants in a small patch. There is a downward curling of the leaves, without any marked yellowing, and quickly the leaves turn orange-brown or golden-brown. They then collapse and hang down around the stem. This gives the appearance of wilt. If these stems are placed in water, a white substance streams from the cut ends. Stems become infected at the base, near the rhizome. Dark, water-soaked patches occur, and soon after the stems can be pulled easily from the rhizomes. Usually, the rotten stems have a foul smell. The young, tender, sprouts of the rhizomes are also infected. The vascular tissues become dark brown or black. Water-soaked areas occur and a milky liquid oozes out when rhizomes are cut. At harvest, there is nothing left of the plant or only the skin remains; everything else has been destroyed by *Ralstonia* and other organisms. Biotype 4 was found to produce the most severe symptoms.

**ii) Management :** It is possible to grow ginger in land normally used for paddy in order to escape infection from bacterial wilt (the bacterium is not likely to survive in the irrigation water of the previous rice crop). Seed must be taken from a bacterial wilt-free source; this means the crop from which the seed is taken must have been monitored for the disease. Good drainage is essential, and water from one bay should not drain directly into the one where ginger is growing. Hygiene is important: people should not walk in the field where ginger is growing, in case they bring soil on their shoes from bacterial wilt-contaminated fields. As long as the crop is free from disease, seed can be saved for planting the following season.

Solarization of ginger rhizomes for 2 to 4 hr from 9.00 a.m. to 1.00 p.m. (which raised the rhizome temperature to 55°C during summer) just before planting, eliminated the seed-borne bacterial infection from ginger rhizomes. The seed rhizomes may be treated with 200 ppm streptocycline for 30 min and shade dried before planting. Once the disease is noticed in the field, all beds should be drenched with 1% Bordeaux mixture or 0.2% copper oxychloride.

Infested areas should not be replanted with ginger, but planted with a perennial grass and fenced off to stop animals and people walking through and spreading the disease. All crop residues where the disease is known to occur should be destroyed, by burning if possible. Implements and boots should be washed down with a commercial agricultural disinfectant.

### 8.3.1.2. Bacterial Soft Rot, *Erwinia* sp.

**i) Symptoms:** Normally bacterial soft rot is only a storage rot. The bacteria are present in most soils but field infection usually only occurs in waterlogged areas. *Erwinia* sp. has caused severe losses of stored rhizomes on some farms, but is not considered a serious storage problem where precautions are taken. Softening of the tissue is accompanied by production of a strong odour and the rhizome eventually collapses completely. Bacterial soft rot differs from other rhizome rots in that putrid odour is produced in this soft rot.

**ii) Management :** Make sure the area is well drained. When harvesting ginger for seed purposes, do not commence harvesting until tops have dried down naturally, as the stalk scar is an ideal point for infection to occur. Only harvest when it is dry and store in new bags or clean bins in a dry and well-ventilated shed.

### 8.3.2. Viral Diseases

#### 8.3.2.1. Big Bud (Bunchy Top)

**i) Symptoms:** The mycoplasma-like organism which causes this disease also causes big bud on a wide range of plants and is similar in characteristic to a virus. The disease is spread by leaf hoppers.

**ii) Management :** All diseased plants should be detected and rouged out. Control leaf hopper vectors.

## 8.4. CORIANDER

Coriander (*Coriandrum sativum*) is used as a common flavouring substance. The stems, leaves and fruits have a pleasant aroma. The whole young plant is used in preparing chutney. Its leaves are used for flavouring curries, sauces and soups. Dry seeds are extensively used in preparation of curry powder, pickling spices, sauces and seasonings. The seeds contain 0.5-1% essential oil which has an active principle, an isomer of geraniol. This plant contains Vitamin A. In medicines, its seeds are used as a carminative, refrigerant, diuretic, flatulence, vomiting and intestinal disorders. The decoction is used in indigestion, cold and other diseases. The oil from seeds is used in sweet meats, coco eatables, liquor and for agreable smell and taste. Coriander seeds contain thalides which increase the levels of anticancer protective enzymes. The oil is also useful as an insecticide.

**Fig. 8.1.** Bacterial leaf spot on coriander.

It is being cultivated in 0.282 million hectares producing 0.172 million tonnes of coriander seeds with an average yield of 0.6 tonnes per hectare. In India, coriander is cultivated in Andhra Pradesh, Madhya Pradesh, Karnataka, Tamil Nadu and Uttar Pradesh.

### 8.4.1. Bacterial Diseases

#### 8.4.1.1. Bacterial Leaf Spot, *Pseudomonas syringae* pv. *coriandricola*

**i) Symptoms:** This is a seed-borne, bacterial disease against which there is no commercially available preventive treatment. This disease remains dormant in the plant until 'stress' is applied to growing plants (e.g. frost, hail damage). The pathogen causes leaf spots (Fig. 8.1) and can cause total crop loss.

**ii) Management:** Growers should use only seed that has been tested for freedom from the disease or seed with infection levels below 0.2%. Once the disease is evident in irrigated crops, it is suggested to stop watering for 2–4 weeks in an effort to reduce the spread caused by water droplets.

## Chapter 9

# ALLIED TOPICS IN PEST MANAGEMENT

### 9.1. Use of Power Sprayers

Power sprayers require less spray fluid per unit area as compared to high volume sprayers like gator, back pack, etc. While high volume spray require about 800-1000 litres of spray fluid per hectare, low volume spray with power sprayers require only 200-300 litres. Hence, while using power sprayers, spray solutions should be at least three times more concentrated than high volume sprays and spray solutions also should be reduced three times. For example, if 450 ml of monocrotophos is to be mixed with 300 litres of water for Gator sprayer, then mix 450 ml of monocrotophos in 100 litres of water for power sprayer.

### 9.2. Use of Stickers/Spreaders

The addition of stickers/spreaders to the spray solution improves the retention, even spreading of the spray solution over sprayed surface.

Commonly used stickers and spreaders are as follows:

| | | |
|---|---|---|
| Sandovit | : | 0.5 ml/lit |
| Teepol | : | 0.5 ml/lit |
| Triton | : | 0.5 ml/lit |
| Citowet | : | 0.5 ml/lit |

### 9.3. Preparation of Bordeaux Mixture and Bordeaux Paste

#### 9.3.1. Preparation of 1% Bordeaux Mixture

Dissolve one kg of copper sulphate (98% purity) in 50 litres of water in wooden barrel or plastic drum. Dissolve one kg of fresh hydrated lime in 50 litres of water separately. Both the solutions should be sieved. Pour slowly copper sulphate solution and lime water simultaneously by continuous stirring into the tank.

It should be ensured that there is no free copper in the mixture. To check this, a clean iron knife/blade should be held/dipped in the mixture for a few minutes, removed and observed for any reddish brown deposit of copper on it. If the copper deposit is noticed, add a little lime solution to the mixture to neutralize excess copper. The best way is to use litmus paper and check pH. The Bordeaux mixture should be neutral or slightly alkaline, otherwise it will injure the leaves/plants. Bordeaux mixture should always be prepared fresh and should not be stored for more than one day. However, mixture can be stabilized by adding 200 gm of jaggery per 100 litres of mixture.

### 9.3.2. Preparation of 10% Bordeaux Paste

- Dissolve 1 kg of copper sulphate in 5 litres of water.
- Dissolve 1 kg of quick lime in another 5 litres of water separately.
- Mix the above 2 solutions to get the Bordeaux paste.

### 9.3.3. Caution While Using Bordeaux Mixture

- Always spray freshly prepared mixture for effective management. Therefore, prepare only required quantity for the day.
- Use only wooden, earthen or plastic vassals, to avoid the corrosion of the metal vassals by the mixture.
- Do not use the Bordeaux mixture in combination with any other chemical or pesticide.
- To avoid choking of nozzle, it is advisable to strain the Bordeaux mixture through a cloth or a sieve before putting into the spray tank.
- The Bordeaux mixture tends to sediment easily. Therefore, its stirring while using is desirable.
- In exceptionally hot days, when the plants are showing signs of temporary wilting or when it is raining, the Bordeaux mixture should not be sprayed, particularly on newly emerged tender foliage. Exercise this caution particularly on nursery plants.
- After carrying out spray operations, the appliances should be thoroughly washed with plenty of water to remove any copper deposits.
- The left over Bordeaux mixture should not be dumped in the field as this may prove toxic to the subsequent sowings.

## 9.4. Seed Treatment for Storage

Diseases affect seeds in storage. Following seed treatments may be adopted for protecting seeds during storage (Table 9.1):

**Table 9.1. Effect of different pesticides on storage of seeds**

| Causal organism | Pesticide | Dose |
|---|---|---|
| Diseases | Subeej/Derosal | 2 g / kg seed |
| | Chlorothalonil/Captan/ Iprodione | 2 g kg seed |
| | Thiride/Vitavax | 2 g kg seed |

## 9.5. Ready Reckoner for Making Pesticide Spray Solutions

**Table 9.2. Ready reckoner for making pesticide spray solutions**

| % a.i. of pesticides in commercial products | Quantity of pesticide formulation to be added per litre of water in gram or ml to get the desired strength of solution (%) | | | | | | | | | | | |
|---|---|---|---|---|---|---|---|---|---|---|---|---|
| | 0.01 | 0.02 | 0.03 | 0.04 | 0.05 | 0.06 | 0.07 | 0.08 | 0.09 | 0.1 | 0.2 | 0.3 |
| 2.5 | 4.00 | 8.00 | 12.00 | 16.00 | 20.00 | 24.00 | 28.00 | 32.00 | 36.00 | 40.00 | 80.00 | 120.00 |
| 5 | 2.00 | 4.00 | 6.00 | 8.00 | 10.00 | 12.00 | 14.00 | 16.00 | 18.00 | 20.00 | 40.00 | 60.00 |
| 10 | 1.00 | 2.00 | 3.00 | 4.00 | 5.00 | 6.00 | 7.00 | 8.00 | 9.00 | 10.00 | 20.00 | 30.00 |
| 15 | 0.67 | 1.30 | 2.00 | 2.60 | 3.30 | 4.00 | 4.66 | 5.20 | 6.00 | 6.66 | 13.30 | 20.00 |
| 20 | 0.50 | 1.00 | 1.50 | 2.00 | 2.50 | 3.00 | 3.50 | 4.00 | 4.50 | 5.00 | 10.00 | 15.00 |
| 25 | 0.40 | 0.80 | 1.20 | 1.60 | 2.00 | 2.40 | 2.80 | 3.20 | 3.60 | 4.00 | 8.00 | 12.00 |
| 30 | 0.33 | 0.66 | 1.00 | 1.30 | 1.65 | 1.98 | 2.31 | 2.64 | 2.97 | 3.33 | 6.60 | 9.90 |
| 35 | 0.29 | 0.58 | 0.87 | 1.16 | 1.45 | 1.74 | 2.03 | 2.32 | 2.61 | 2.85 | 5.70 | 8.55 |
| 40 | 0.25 | 0.50 | 0.75 | 1.00 | 1.25 | 1.50 | 1.75 | 2.00 | 2.25 | 2.50 | 5.00 | 7.50 |
| 50 | 0.20 | 0.40 | 0.60 | 0.80 | 1.00 | 1.20 | 1.40 | 1.60 | 1.80 | 2.00 | 4.00 | 6.00 |
| 75 | 0.13 | 0.27 | 0.40 | 0.53 | 0.66 | 0.79 | 0.92 | 1.07 | 1.20 | 1.33 | 2.66 | 4.00 |
| 80 | 0.12 | 0.25 | 0.37 | 0.50 | 0.63 | 0.75 | 0.88 | 1.00 | 1.13 | 1.25 | 2.50 | 3.75 |
| 85 | 0.12 | 0.25 | 0.36 | 0.48 | 0.59 | 0.70 | 0.82 | 0.96 | 1.06 | 1.18 | 2.35 | 3.54 |
| 100 | 0.10 | 0.20 | 0.30 | 0.40 | 0.50 | 0.60 | 0.70 | 0.80 | 0.90 | 1.00 | 2.00 | 3.00 |

**Note:** The pesticides should be measured in g or ml. e.g. For preparation of malathion 0.05% spray solution from 50% malathion, the value at the point of intersection of 10th row and 5th column should be taken i.e. 1.0 ml should be dissolved in 1 litre of water.

### Preparation of Spray Solution

The spray solution must be prepared to the exact concentrations as recommended. The quantity of the commercial pesticide formulation required for preparation of the spray solution is calculated by using the following formula:

$$\text{Quantity of the pesticide formulation} = \frac{\text{Qty. of spray soln. Reqd.} \times \text{Conc. of soln. desired}}{\text{\% of a.i. of pesticide in the formulation}}$$

Prepare spray solution first in small quantity and then increase the volume to desired level by adding water. In case of wettable powder, take required quantity of pesticide, add a little quantity of water, mix it thoroughly to prepare the paste and then add remaining quantity of water to this paste with constant stirring.

## 9.6. Compatibility Chart

**Table 9.3. Compatibility Chart: Precautions to be taken while using insecticides, fungicides and other agrochemicals**

| No. | Pesticides/Agrochemicals | 1 | 2 | 3 | 4 | 5 | 6 | 7 | 8 | 9 | 10 | 11 | 12 | 13 | 14 | 15 | 16 | 17 | 18 | 19 | 20 | 21 | 22 |
|---|---|---|---|---|---|---|---|---|---|---|---|---|---|---|---|---|---|---|---|---|---|---|---|
| 1 | Plant origin insecticides@ | a | a | a | a | a | a | a | a | a | a | b | a | a | b | a | a | a | a | b | b | c | a |
| 2 | Carbaryl | a | a | a | b | a | c | a | d | a | a | a | d | a | d | a | d | a | a | a | d | a | a |
| 3 | DDT Group | a | a | a | a | a | a | a | a | a | a | a | d | a | a | a | b | a | a | a | b | a | a |
| 4 | Monocrotophos | a | b | a | a | a | a | a | a | a | a | a | b | a | a | a | a | a | a | a | b | a | a |
| 5 | Dimethoate | a | a | a | a | a | a | a | a | a | c | a | c | a | a | a | b | a | a | a | b | a | a |
| 6 | Endosulfan | c | a | a | a | a | a | a | d | a | a | a | b | a | c | a | b | a | a | a | c | a | c |
| 7 | Fenitrothion | a | a | a | a | a | a | a | a | a | a | a | b | d | a | a | b | a | a | a | b | a | a |
| 8 | Kerosene | a | d | a | a | a | d | a | a | a | a | a | b | b | a | b | b | b | b | b | b | a | c |
| 9 | Malathion, Parathion, Demeton | a | a | a | a | a | a | a | a | a | a | a | b | d | a | a | a | d | a | a | a | a | a |
| 10 | Oxy-demeton methyl | a | a | a | a | c | a | a | a | a | a | a | a | c | a | a | a | c | c | a | a | c | c |
| 11 | Phoslane | d | a | a | a | a | a | a | a | a | a | a | a | d | a | a | a | a | a | b | b | b | b |
| 12 | Bordeaux mixture | b | b | d | b | b | b | b | b | b | b | d | a | b | d | a | b | b | a | a | a | b | a |
| 13 | Captan, Difolatan | a | a | a | a | a | a | d | b | d | a | a | b | a | d | a | b | a | d | a | b | a | a |
| 14 | Copper oxychloride | a | b | a | a | a | c | a | a | a | c | a | a | d | d | d | b | b | a | b | a | a | a |
| 15 | Dithiocarbamate group# | a | a | a | a | a | a | b | a | a | a | a | a | d | a | b | d | a | a | a | a | a | - |
| 16 | Lime sulphur | b | b | b | b | b | b | b | b | b | b | d | b | b | b | b | a | b | a | a | a | a | b |
| 17 | Karathane | a | a | a | a | a | a | a | b | a | a | a | b | e | b | a | b | a | b | c | c | b | c |
| 18 | Wettable sulphur | a | a | a | a | a | a | a | b | a | a | a | a | d | a | a | a | a | a | a | a | a | a |
| 19 | 2,4-D | b | - | a | a | a | c | a | b | a | c | c | a | a | b | a | a | c | a | a | a | c | a |
| 20 | Lime | b | b | b | b | b | c | b | b | b | c | a | b | a | a | a | a | a | a | a | a | a | a |
| 21 | Streptomycin | c | a | a | a | a | a | a | c | a | c | c | a | a | a | a | a | a | a | c | c | a | c |
| 22 | Urea | b | a | a | a | a | c | a | a | a | c | c | a | a | a | b | a | a | a | a | b | c | a |
| 23 | Carbendazim | a | a | a | a | - | a | - | - | a | - | - | a | a | - | a | - | - | a | a | - | - | a |

@ = Plant origin insecticides: Rotenene, Pyrethrum. # = Dithiocarbamates: Maneb, Zineb, Nabam, Thiram, Ziram, Mancozeb

a = Can be mixed.; b = Not to be mixed.; c = Information not available; d = Use immediately. e = Usually not mixed.

## 9.7. Soil Solarization

Soil solarization is a method of killing soil-borne bacterial pathogens (fungi, bacteria, nematodes), soil insects and weed seeds by heat. This technique consists of covering moist soil with a plastic film during periods of intense sunshine and heat, thereby capturing radiant heat energy from the sun, causing physical, chemical and biological changes in the soil. The soil temperature is increased (by 5-15°C) to levels lethal to many soil-borne fungi, bacteria, nematodes and weeds. The short wave incident solar radiation penetrates the polyethylene sheet but the long wave radiation is prevented from soil, thus trapping the heat resulting in thermal inactivation and production of heat shock proteins and irreversible heat injury. Solarization also improves plant nutrition by increasing the availability of N and other essential elements. This practice can be profitably used on a small scale and is especially useful for nursery beds.

Soil solarization for 40-45 days using 400 gauge transparent LDPE was proved ideal pre-sowing treatment for the management of nematodes and other soil-borne diseases in cardamom nurseries. Mean soil temperature during the solarization period increased by 8.7°C in solarized beds and the germination was enhanced by 25.5%, while weed growth was suppressed by 82%.

## 9.8. Mass production of biocontrol agents

### 9.8.1. *Pseudomonas fluorescens*

A loopful of *P. fluorescens* should be transferred to 100 ml of King's B broth in a 250 ml conical flask under aseptic conditions. The flask can be incubated at room temperature for 48 hours. The population of *P. fluorescens* in the broth can be assessed (a minimum of 9 x $10^9$ cfu/ml is necessary).

The pH of the substrate (talc powder) can be adjusted to 7.0 by addition of 150g calcium carbonate/kg of talc powder and sterilized at 1.1 kg/$cm^2$ for 30 min. for 2 successive days. One kg of the sterilized substrate (talc powder) and 5g of carboxy methyl cellulose should be placed in a polythene bag or any sterile container under aseptic conditions, 400 ml of *P. fluorescens* suspension to be added and mixed thoroughly. The formulation can be stored in a polythene bags for one month. The product should contain a minimum of 2.5 x $10^8$ cfu/g.

The product should have the following quality parameters:

- Fresh product should contain 2.5 x $10^8$ cfu/g.
- Three months after storage at room temperature, the population should be 8-9 x $10^7$ cfu/g.
- Storage period is 3-4 months.
- Minimum population load for use is 1x$10^8$ cfu/g.
- The product should be packed in white polythene bags.
- Moisture content of the final product should not be more than 20%.
- Population in broth should be 9 ± 2 x $10^8$ cfu/ml.

The composition of King's B medium is as follows:

| | | |
|---|---|---|
| Peptone | - | 20.00 g |
| Dipotassium hydrogen phosphate | - | 1.50 g |
| Magnesium sulphate | - | 1.50 g |
| Glycerol | - | 10.00 ml |
| Agar | - | 15.00 g |
| Distilled water | - | 1000 ml |
| Cycloheximide | - | 100 ppm |

Ampicillin - 50 ppm

Chloramphenicol - 12.50 ppm

(King's B broth is prepared without the agar and antibiotics)

**Liquid Formulation Technology to extend the Shelf Life of *P. fluorescens***

Poly Vinyl Pyrrolidone (PVP) (2%) was tested as an effective preservation agent in protecting the cells of *P. fluorescens* for more than a year without affecting the natural properties of bacteria. The antagonistic efficacy of preserved cells was also confirmed *in vitro* and *in vivo*.

**9.8.2. *Trichoderma* spp.**

Among the various fungal biological control agents which are effective against nematode and fungal diseases of vegetable crops, *Trichoderma viride* and *T. harzianum* have a good market. They can be easily mass produced on several organic substrates.

**9.8.2.1. Small Scale**

The mycelial discs grown in potato dextrose agar medium were transferred to 250 ml conical flasks containing sterilized molasses yeast medium (Yeast extract – 5 g, Molasses – 30 g, Distilled water – 1 litre sterilized at 1.42 kg/cm$^2$ for 20 min.) at 100 ml /flask and incubated for 10 days. The mycelial mat along with the broth was homogenized and mixed with talc powder (500 mesh) at 1: 2 ratio. Carboxy methyl cellulose (CMC) was added at 5 g/kg of the product as sticker. The product was shade dried until it attained 12% moisture level. The clumps were broken, homogenized and packed in polythene bags of 100g capacity. The required population level at the time of packing should be 28 x $10^7$ cfu/g of the product. The shelf life of the product was 4 months. At the end of expiry, the cfu should not be less than 20 x $10^6$ cfu/g of the product.

**9.8.2.2. Large Scale**

Add 70 ml of molasses yeast medium in each 250 ml conical flasks and plug with non-absorbent cotton wool. Autoclave at 1.42 kg/cm$^2$ for 1 hr. Transfer 8 ml mycelial disc of *T. viride/T. harzianum* to each flask and incubate for 10 days. This mother culture is used for inoculating the medium in fermentor.

Add 50 litres of molasses yeast medium in 50 litre fermentor, sterilize at 1.42 kg/cm$^2$ for 20 min. and cool to room temperature. Add 2.5 litres of inoculum (from mother culture) into the sterilized medium and incubate for 10 days. Provide aeration in the fermentor for 4-8 hr per day by agitation. Presence of greenish fungal growth indicates *T. viride/T. harzianum*. This broth culture contains fungal mat, spores and antibiotics. Draw samples 10 days after incubation and assess the spore load in the broth which should be at least $10^8$ spores/ml. Mix one litre of broth culture with 2 kg talc powder, add CMC at 5g/kg of the product, air dry (12% moisture content) and sieve. The population should not be less than 20 x $10^6$ cfu/g of the product. It is packed in opaque polythene bags and stored in a cool dry place before use. The product can be safely stored for 4 months without losing its viability.

### 9.8.3. Arbuscular Mycorrhizal Fungi (*Glomus mosseae, G. fasciculatum*)

#### 9.8.3.1. In Glasshouse

Cement pots or tanks are filled with sterilized sand: soil (1: 1 by volume) or soilrite: peat (1: 1 by volume) mixture. A standard starter culture of AMF obtained from any authentic source is placed as a band 3 cm below the soil surface. The seeds of Rhoades grass/finger millet/maize are sown on them and covered with soil. Pots are irrigated normally. When the plants are about 10-12 weeks old, the top shoots are cut off and the root system colonized by the fungus along with the soil containing chlamydospores is used as the inoculum.

#### 9.8.3.2. On-farm Production

A suitable land (even waste land) is selected near the field where the fungus inoculum has to be used. The land is brought to the fine tilth by ploughing, harrowing and cultivator operations. A raised bed (20 cm height) of soil: sand (1: 1 by volume) mixture is prepared on an even land surface. The raised bed is sterilized by soil solarization for 2 to 3 weeks using polythene sheet. The starter culture of AMF is applied at a depth of 3-4 cm from the top by removing the top soil. The seeds of Rhoades grass/ finger millet/maize are sown on them and covered with soil. Irrigation is done normally. When the crop attains a height of 10-15 cm, 12 weeks after sowing, the shoot portion is cut off at ground level. The soil is dug up to a depth of 20 cm along with root system and is used as the inoculum.

#### 9.8.3.3. Method of Inoculation

Vegetable crops can be inoculated with AMF in the nursery beds. The AMF inoculum consisting of chlamydospores, soil and infected root bits is either spread uniformly in the nursery beds or placed in rows over which seeds of vegetable crops (tomato, chilli, capsicum, brinjal) are sown. The seedlings get colonized by the fungus in about 4-5 weeks after which they are transplanted in the main field. Inoculated plants show better growth and improved nutrient content.

### 9.9. Enrichment of FYM with Bioagents (*Trichoderma harzianum, Paecilomyces lilacinus*)

Commercial formulation of the bioagent (*T. harzianum/P. lilacinus*) (1 kg/ton) was mixed with moist FYM along with 50 kg of neem/pongamia cake. Mixture was spread 15 cm high heap under shade, covered with polyethylene sheet and incubated for 3 weeks at 25 to 30°C and sprayed with water regularly to maintain moisture.

Colonized FYM can be used at 6 g/kg soil in pot culture experiments (12 g/2 kg soil in 15 cm diameter pots). It can also be used for main field treatment at 5 tonnes/ha before sowing of seeds or transplanting of seedlings. It can be applied at 2 kg/plant of perennial crops along with 400g of neem/pongamia cake once in every 6 months after planting for the management of nematodes and other soil-borne pathogens.

*T. harzianum* colonized compost significantly enhanced germination and growth of both tomato and okra as compared to non-colonized FYM. Population of *T. harzianum* in colonized compost was several folds higher ($48.6 \times 10^{12}$) than non-colonized FYM ($3 \times 10^{4}$).

Colonization of FYM by *T. harzianum* helps to improve the quality of compost. It significantly enhanced total and water soluble humic matter. A number of macro and micro-nutrients like P, K, S, Zn, Cu and Fe (both total and water soluble content) were significantly higher in *T. harzianum* colonized compost. Six times increase in water soluble humic content due to colonization of *T. harzianum* seems quite important as it is reported to enhance the plant growth.

## 9.10. Seed Biopriming/Solid Matrix Priming

Seed biopriming (treatment of seeds with a slurry of biocontrol agents mixed with an organic carrier and incubated for 4 days under warm and moist conditions with relative humidity of 80-100% until just prior to radical emergence) has potential advantages over simple coating of seeds as it results in rapid and uniform seedling emergence. *Trichoderma* required less moisture for growth. *Trichoderma* conidia germinate on the seed surface and form a layer around bioprimed seeds. Such seeds tolerate adverse soil conditions better. Biopriming could also reduce the amount of biocontrol agents that is applied to the seed. Population of *T. harzianum* and *T. virens* on the surface of treated seeds of tomato and brinjal increases from $10^{4}$ cfu per seed up to $10^{6}$-$10^{7}$ cfu per seed after biopriming.

## 9.11. Formulation of Alginated Pellets of Biocontrol Agents

Alginate is a water soluble polysaccharide gum that has excellent gel-forming properties. Living organisms that are capable of biologically controlling a variety of pathogens can easily be entrapped in a calcium or sodium alginate matrix in aqueous form at room temperature. A pellet formulation of biocontrol agents was prepared by mixing sodium or calcium alginate, clay and streptomycin sulfate. The mixture was homogenized in a blender and then mixed with bioagent propagules. The gel beads, granules or seed coatings that are produced by the process are biodegradable. The encapsulation in Ca or Na alginate maintained the viability of the bioagents for 12 months at 7°C and 5 months at room temperature. Antagonistic fungi (species of *Paecilomyces, Trichoderma, Gliocladium, Talaromyces* and *Pencil-*

*lium*) and bacteria (species of *Pseudomonas* and *Bacillus*) that can control soil-borne pathogens can be formulated in Ca or Na alginate (Connick, 1988). Alginated pellets can be used for soil application at 400 to 800 pellets/sq.m.

## 9.12. Production of Bioagent Colonized Seedlings

The horticultural crops are invariably attacked by soil borne pathogens (nematodes, fungi, bacteria) at the seedling stage itself. Securing healthy seedlings of horticultural crops is essential to ensure freedom from soil-borne pathogens, optimum plant stand, good growth and higher yields. The following methods may be adopted to produce bioagent colonized seedlings of horticultural crops:

### 9.12.1. Seed Treatment

Seeds have to be treated with talc based formulations of *Pseudomonas fluorescens, Trichoderma harzianum* and *T. viride* at 10g/kg seed before sowing.

### 9.12.2. Nursery Bed Treatment

The nursery beds may be treated with *P. fluorescens/T. harzianum/T. viride* at 50g/$m^2$ along with neem/ pongamia cake at 200g/$m^2$ before sowing.

### 9.12.3. Treatment of Coco Peat or any other Substrate

The coco peat or any other substrate used for producing the nursery seedlings in portrays is treated with *P. fluorescens/T. harzianum/T. viride* at 1 kg/tonne before filling with the substrate.

### 9.12.4. Treatment of Soil Mixture

The soil mixture used for producing the nursery seedlings in polythene bags in a conventional manner is treated with *P. fluorescens/T. harzianum/ T. viride* at 2 kg/tonne along with neem/pongamia cake at 50kg/tonne before sowing.

Chapter 10

# SOURCES OF CRITICAL INPUTS FOR THE MANAGEMENT OF DISEASES

## 10.1. BACTERICIDES

**Table 10.1. Common and trade names and source of bactericides**

| Group | Common Name | Trade Name/s (Source) |
|---|---|---|
| Copper bactericides | Copper hydroxide | Kocide 101 (4) |
| | Copper oxychloride | Blitox (12), Blue Copper (47), Captox (1), Cupramar (10), Cuprastar (13), Devicopper (2), Dhanucop (3), Fytolan (6) Hexamar Copper (10), Hil Copper (5), Blue Vit (9), Trucop (21). |
| | Cuprous oxide | Cuprous oxide |
| Mercury bactericides | Methoxy ethyl mercuric chloride | Bagalol (16). |
| | Phenyl mercury acetate | |
| | Vapam | |
| | Zineb | Discon-Z (1), Devizeb (2), Hexathane (10), Indofil Z-78 (8). |
| | Ziram | Cumin L (14), Hexazir (10), Tagziron (15), Ziride (7), egfru Zitox (11), Dhanuka Z-27 (3). |

**Source of Bactericides**

1. AIMCO; 2. Devi Dayal; 3. Dhanuka; 4. Du Pont; 5. Hindusthan Insecticides; 6. ICI Zeneca; 7. IIL; 8. Indofil; 9. Kilpest; 10. Parry Chemicals; 11. Pesticides India; 12. Rallis India; 13. Shaw Wallace; 14. Syngenta; 15. Tropical Agro; 16. United Phosphorus Ltd.

## 10.2. PATHOGENS

### 10.2.1. Arbuscular Mycorrhizal Fungi

**Table 10.2. Trade names and source of availability of Arbuscular Mycorrhizal Fungi**

| Arbuscular Mycorrhizal Fungi | Trade Name | State | Source of Availability |
|---|---|---|---|
| *Glomus mosseae* | Symbion VAM | Karnataka | Agri Technol Infmn Centre, IIHR, B'lore |
| | | Maharashtra | Bharatiya Agro Industries Foundation, Pune |
| | | Tamil Nadu | T Stanes & Co, Coimbatore |
| *G. fasciculatum* | Symbion VAM | Karnataka | Agri Technol Infmn Centre, IIHR, B'lore |
| | | Maharashtra | Bharatiya Agro Industries Foundation, Pune |
| | | Tamil Nadu | T Stanes & Co, Coimbatore |
| *G. aggregatum* | | Maharashtra | Bharatiya Agro Industries Foundation, Pune |
| *Gigaspora margarita* | | Maharashtra | Bharatiya Agro Industries Foundation, Pune |
| Arbuscular mycorrhizal fungi | Multiplex Trishul | Karnataka | Karnataka Agro Chemicals, B'lore |

### 10.2.2. Antagonistic Fungi Against Plant Diseases and Nematodes

**Table 10.3. Trade names and sources of availability of antagonistic fungi against Plant Diseases and nematodes**

| Antagonistic Fungi / Trade Name | State | Source of Availability |
|---|---|---|
| *Ampelomyces quisqualis* (Bio-Dewcon) | Tamil Nadu | T. Stanes & Co, Coimbatore |
| *Arthrobotrys cladodes* var. *macroides* | Karnataka | Project Directorate of Biol Control, Bangalore |
| *Arthrobotrys oligospora* | Karnataka | Project Directorate of Biol Control, Bangalore |
| *Aspergillus nizer* (Kalisena SD, Kalisena SL, Sanjeevni) | Gujarat | Cadila Pharmaceuticals Ltd., Ahmedabad |
| | New Delhi | Plant Path Divn, Indian Agri Res Inst, New Delhi |
| *Dactylella brochophaga* | Karnataka | Project Directorate of Biol Control, Bangalore |

| | | |
|---|---|---|
| *Fusarium oxysporum* | Karnataka | Project Directorate of Biol Control, Bangalore |
| *Fusarium sporotrichoides* | Karnataka | Project Directorate of Biol Control, Bangalore |
| *Gliocladium catenulatum* | Karnataka | Project Directorate of Biol Control, Bangalore |
| *G. deliquescens* | Karnataka | Project Directorate of Biol Control, Bangalore |
| *G. roseum* | Karnataka | Project Directorate of Biol Control, Bangalore |
| *Phoma glomerata* | Karnataka | Project Directorate of Biol Control, Bangalore |
| *Pochonia chlamydosporia* (Bionema) | Karnataka | Project Directorate of Biol Control, Bangalore |
| | | Agri Technol Infmn Centre, IIHR, Bangalore |
| | | Central Sericultural Res &Trng Inst, Mysore |
| *Paecilomyces lilacinus* Bio-Nemator, Biocon, Niyantran, Multiplex Sparsha, Abtec Paecilomyces) | Karnataka | Project Directorate of Biol Control, Bangalore |
| | | Agri Technol Infmn Centre, IIHR, Bangalore |
| *T. aureoviride* | Karnataka | Project Directorate of Biol Control, Bangalore |
| *T. hamatum* | Karnataka | Project Directorate of Biol Control, Bangalore |
| *T. harzianum* (Binab-T, Ecoderma, F-Stop, Supravit, Tricodex, Trichoderma) | Andhra Pradesh | ADA (BCL), P.M. Palem |
| | | ADA (BCL), Kakinada |
| | | ADA (BCL), Nidadavole |
| | | ADA (BCL), Ibrahimpatnam |
| | | ADA (BCL), Ongole |
| | | ADA (BCL), Nellore |
| | | DDA (FTC), Nandyal |
| | | ADA (BCL), Anantapur |
| | | DDA (FTC), Rajendranagar |
| | | ADA (Oilseeds) (BCL), Mahabubanagar |
| | | ADA (BCL), Nalagonda |
| | | ADA (BCL), Mulugu Road |
| | | ADA (BCL), Karimnagar |
| | | ADA (BCL), Adilabad |
| | | Pest Control (India) Pvt Ltd, Hyderabad |
| | Assam | State Biocontrol Laboratory, Guwahati |
| | | Pest Control (India) Pvt Ltd, Guwahati |

| | |
|---|---|
| Bihar & Jharkhand | Pest Control (India) Pvt Ltd, Jamshedpur |
| Gujarat | Pest Control (India) Pvt Ltd, Vadodara |
| Haryana | State Biocontrol Laboratory, Sirsa |
| | Pest Control (India) Pvt Ltd, Chandigarh |
| Karnataka | Project Directorate of Biol Control, Bangalore |
| | Agri Technol Infmn Centre, IIHR, Bangalore |
| | Central IPM Centre, Bangalore |
| | Central Sericultural Res & Trng Inst, Mysore |
| | Coffee Research Sub Station, Chethalli |
| | Central Coffee Res Inst, Entomol Divn, Chikmagalur |
| | Univ of Agri Sci, Entomol Dept, Dharwad |
| | Inst of Pulses & Oilseeds Res, Gulbarga |
| | Regional Res Stn, UAS, Raichur |
| | Agri Res Stn, UAS, Bijapur |
| | Kadur Agro Pvt. Ltd., Bangalore |
| | Pest Control (India) Pvt Ltd, Bangalore |
| | P.J.Margo Pvt. Ltd., Bangalore |
| Kerala | Travancore Organic Fertilizers, Kottayam |
| | VRM Biotech Products, Changanacherry |
| Madhya Pradesh | Pest Control (India) Pvt Ltd, Bhopal |
| Maharashtra | Bio Era Technologies, Nagpur |
| | Hoechst Schering Agr. Envt, Mumbai |
| | KNS Biotech, Nanded |
| | Nathkrupa Biocontrol Lab, Nagpur |
| | Om Agro Organics, Yavatmal |
| | Pest Control (India) Pvt Ltd, Mumbai-400 063 |
| | Pest Control (India) Pvt Ltd, Mumbai-400 059 |
| | Pest Control (India) Pvt Ltd, Thane |
| | Pest Control (India) Pvt Ltd, New Mumbai |
| | Pest Control (India) Pvt Ltd, Nashik |
| | Pest Control (India) Pvt Ltd, Pune-411 033 |
| | Pest Control (India) Pvt Ltd, Pune-411 030 |
| | Sanvardhini Agro Pvt. Ltd., Satara |
| | Soman Biofertilizers, Pune |
| | Vidyas Biotech Lab, Nagpur |
| | West Coast Herbochem P Ltd, Mumbai |
| New Delhi | Biotech Intern Ltd., New Delhi |

| | |
|---|---|
| | Pest Control (India) Pvt Ltd, New Delhi |
| Orissa | Pest Control (India) Pvt Ltd, Bhubaneswar |
| Punjab | Pest Control (India) Pvt Ltd, Chandigarh |
| Rajasthan | Biocontrol Lab, ARS, RAU, Sri Ganganagar |
| | JDA (Pl Path), Biocontrol Lab, Durgapura |
| | DDA (Agro), IPM Lab, Ajmeer |
| | DDA (Agro), IPM Lab, Malikpur |
| | DDA (Agro), IPM Lab, Rampura |
| | DDA (Agro), IPM Lab, Cjjatrapura |
| | DDA (Agro), l IPM Lab, Chittorgarh |
| | DDA (Agro), IPM Lab, Hanumangarh |
| | DDA (Pl Path), IPM Lab, Banaswara |
| | Pest Control (India) Pvt Ltd, Jaipur |
| Tamil Nadu | Centre for Pl Prot Studies, TNAU, Coimbatore |
| | Bio-control Lab, Papparapatty |
| | Bio-control Lab, Coimbatore |
| | Bio-control Lab, Salem |
| | Bio-control Lab, Villupuram |
| | Bio-control Lab, Trichy |
| | Bio-control Lab, Panjupettai |
| | Bio-control Lab, Kuruppanaichenpalyam |
| | Bio-control Lab, Vinayagapuram |
| | Bio-control Lab, Tirunelveli |
| | Bio-control Lab, Tanjavur |
| | Bio-control Lab, Namakkal |
| | Basarass Biocontrol Res, Vallavar |
| | Greentech Agro Prod P Ltd, Coimbatore |
| | Jeypee Bio Techs, Virudhunagar |
| | Pest Control (India) Pvt Ltd, Chennai |
| | Rajendra Foundn for Agri Res, Chittar |
| | Shri Durga Agro Services, Coimbatore |
| | Shrishti Bioprod P Ltd, Coimbatore |
| | Sun Agro Ind India, Chennai |
| | TN Co-op Sugar Fedn Ltd, Chengalpattu |
| | TAP Industries, Chennai |
| Uttarakhand | Pl Path Dept, GB Pant Uni Agri & Technol, Pantnagar |
| | Pest Control (India) Pvt Ltd, Dehradun |
| Uttar Pradesh | Crop Health Bioprod Res Centre, Ghaziabad |
| | Pest Control (India) Pvt Ltd, Bareilly |
| | Pest Control (India) Pvt Ltd, Lucknow |
| | Pest Control (India) Pvt Ltd, Kanpur |

| | | |
|---|---|---|
| | West Bengal | Pest Control (India) Pvt Ltd, Kolkata |
| *T. koningii* | Karnataka | Project Directorate of Biol Control, Bangalore |
| *T. longibactriatum* | Karnataka | Project Directorate of Biol Control, Bangalore |
| *T. polysporum* | Karnataka | Project Directorate of Biol Control, Bangalore |
| *T. pseudokoningii* (Nursery Guard) | Karnataka | Project Directorate of Biol Control, Bangalore |
| | | Central Sericultural Res &Trng Inst, Mysore |
| *T. virens* (Gligard, WRC-GV) | Karnataka | Project Directorate of Biol Control, Bangalore |
| *T. viride* (Bio-Cure, Bio-Cure F, Agroderma, Antogon TV, Bioderma, Dermapak, Ecofit, Monitor, Multiplex Nisarga, Rakshak, Trichosan, Trichoft, Trichoderm, Trichonik, Tricho-X, Trichodermin) | Andhra Pradesh | ADA (BCL), P.M. Palem |
| | | ADA (BCL), Kakinada |
| | | ADA (BCL), Nidadavole |
| | | ADA (BCL), Ibrahimpatnam |
| | | ADA (BCL), Ongole |
| | | ADA (BCL), Nellore |
| | | DDA (FTC), Nandyal |
| | | ADA (BCL), Anantapur |
| | | DDA (FTC), Rajendranagar |
| | | ADA (Oilseeds) (BCL), Mahabubanagar |
| | | ADA (BCL), Nalagonda |
| | | ADA (BCL), Mulugu Road |
| | | ADA (BCL), Karimnagar |
| | | ADA (BCL), Adilabad |
| | | Pest Control (India) Pvt Ltd, Hyderabad |
| | Assam | State Biocontrol Laboratory, Guwahati |
| | | Pest Control (India) Pvt Ltd, Guwahati |
| | Bihar & Jarkhand | Pest Control (India) Pvt Ltd, Jamshedpur |
| | Gujarat | Pest Control (India) Pvt Ltd, Vadodara |
| | Haryana | State Biocontrol Laboratory, Sirsa |
| | | Pest Control (India) Pvt Ltd, Chandigarh |
| | Karnataka | Project Directorate of Biol Control, Bangalore |
| | | Coffee Research Sub Station, Chethalli |
| | | Central Coffee Res Inst, Entomol Divn, Chikmagalur |
| | | Kadur Agro Pvt. Ltd., Bangalore |
| | | Pest Control (India) Pvt Ltd, Bangalore |
| | | P.J.Margo Pvt. Ltd., Bangalore |

| | |
|---|---|
| | Travancore Org Fert Co P Ltd., Bangalore |
| Kerala | Travancore Organic Fertilizers, Kottayam |
| | VRM Biotech Products, Changanacherry |
| Madhya Pradesh | Pest Control (India) Pvt Ltd, Bhopal |
| Maharashtra | Bio Era Technologies, Nagpur |
| | Hoechst Schering Agr. Envt, Mumbai |
| | KNS Biotech, Nanded |
| | Nathkrupa Biocontrol Lab, Nagpur |
| | Om Agro Organics, Yavatmal |
| | Pest Control (India) P. Ltd, Mumbai-400 063 |
| | Pest Control (India) P. Ltd, Mumbai-400 059 |
| | Pest Control (India) Pvt Ltd, Thane |
| | Pest Control (India) Pvt Ltd, New Mumbai |
| | Pest Control (India) Pvt Ltd, Nashik |
| | Pest Control (India) Pvt Ltd, Pune-411 033 |
| | Pest Control (India) Pvt Ltd, Pune-411 030 |
| | Sanvardhini Agro Pvt. Ltd., Satara |
| | Soman Biofertilizers, Pune |
| | Vidyas Biotech Lab, Nagpur |
| | West Coast Herbochem P Ltd, Mumbai |
| New Delhi | Biotech Intern Ltd., New Delhi |
| | Pest Control (India) Pvt Ltd, New Delhi |
| Orissa | Pest Control (India) Pvt Ltd, Bhubaneswar |
| Punjab | Pest Control (India) Pvt Ltd, Chandigarh |
| Rajasthan | Biocontrol Lab, ARS, RAU, Sri Ganganagar |
| | JDA (Pl Path), Biocontrol Lab, Durgapura |
| | DDA (Agro), IPM Lab, Ajmeer |
| | DDA (Agro), IPM Lab, Malikpur |
| | DDA (Agro), IPM Lab, Rampura |
| | DDA (Agro), IPM Lab, Cjjatrapura |
| | DDA (Agro), l IPM Lab, Chittorgarh |
| | DDA (Agro), IPM Lab, Hanumangarh |
| | DDA (Pl Path), IPM Lab, Banaswara |
| | Pest Control (India) Pvt Ltd, Jaipur |
| Tamil Nadu | Centre for Pl Prot Studies, TNAU, Coimbatore |
| | Bio-control Lab, Papparapatty |
| | Bio-control Lab, Coimbatore |
| | Bio-control Lab, Salem |
| | Bio-control Lab, Villupuram |
| | Bio-control Lab, Trichy |
| | Bio-control Lab, Panjupettai |
| | Bio-control Lab, Kuruppanaichenpalyam |

| | | |
|---|---|---|
| | | Bio-control Lab, Vinayagapuram<br>Bio-control Lab, Tirunelveli<br>Bio-control Lab, Tanjavur<br>Bio-control Lab, Namakkal<br>Basarass Biocontrol Res, Vallavar<br>Greentech Agro Prod P Ltd, Coimbatore<br>Jeypee Bio Techs, Virudhunagar<br>Pest Control (India) Pvt Ltd, Chennai<br>Rajendra Foundn for Agri Res, Chittar<br>Shri Durga Agro Services, Coimbatore<br>Shrishti Bioprod P Ltd, Coimbatore<br>Sun Agro Ind India, Chennai<br>TN Co-op Sugar Fedn Ltd, Chengalpattu<br>TAP Industries, Chennai<br>T. Stanes & Co, Coimbatore |
| | Uttarakhand | Pl Path Dept, GB Pant Univ Agri & Technol, Pantnagar<br>Pest Control (India) Pvt Ltd, Dehradun |
| | Uttar Pradesh | Crop Health Bioprod Res Centre, Ghaziabad<br>Pest Control (India) Pvt Ltd, Bareilly<br>Pest Control (India) Pvt Ltd, Lucknow<br>Pest Control (India) Pvt Ltd, Kanpur |
| | West Bengal | Pest Control (India) Pvt Ltd, Kolkata |
| *Verticillium lecanii* | Karnataka | Project Directorate of Biol Control, Bangalore<br>Pest Control (India) Pvt Ltd, Bangalore |
| | Maharashtra | Sio Agro Research Lab, Mumbai |
| | Tamil Nadu | Sakthi Biocontrol & Biofert Centre, Kallipatti Gobi |
| *V. suchlosporium* | Karnataka | Project Directorate of Biol Control, Bangalore |
| *Arthrobotrys* sp. + *P. lilacinus* + *Verticillium* sp. (Nemastin) | Maharashtra | Kumar Krishimitra Bioproducts, Pune |
| *T. harzianum* + *T. viride* (Ecoderma, NIPROT) | Karnataka | Bio Control Res Lab, PCI P. Ltd., Bangalore<br>P.J. Margo Pvt. Ltd., Bangalore |
| *T. harzianum* + *T. viride* + *T. virens* (Combat) | Maharashtra | Kumar Krishimitra Bioproducts, Pune |

### 10.2.3. Bacterial Biopesticides

**Table 10.4. Trade names and sources of availability of bacterial biopesticides**

| Bacterial Biopesticides/ Trade Name's | State | Source of Availability |
|---|---|---|
| *B. sphaericus* | New Delhi | Biotech Intern Ltd, New Delhi |
| | Tamil Nadu | Sinagro Industries India, Chennai |
| *B. subtilis* (Kill Dew DP Epic, Kodiac) | Maharashtra | Kumar Krishimitra Bioproducts, Pune |
| *B. thuringiensis* var. *kurstaki* (Biolep, Biosap, Delfin, Dipel, Bibot, Spicturin, Halt) | Andhra Pradesh | Indian Inst of Chem Technol, Hyderabad |
| | Madhya Pradesh | Indore Biotech Inputs & Res (Pvt.) Ltd., Indore |
| | Maharashtra | Lupin Laboratories (I) P Ltd., Mumbai |
| | | Sandoz (I) Ltd., Mumbai |
| | | Wockhardt Ltd., Mumbai |
| | New Delhi | Biotech Intern Ltd., New Delhi |
| | | Hindustan Insecticides, New Delhi |
| | Tamil Nadu | Tuticorn Alkali Chem & Fert Ltd, Chennai |
| *Pseudomonas fluorescens* (Bio-Cure B, Bioshield, Tracomonas, Sudocel, Dagger-G) | Assam | State Biocontrol Laboratory, Guwahati |
| | Haryana | State Biocontrol Laboratory, Sirsa |
| | Karnataka | Agri Technol Infmn Centre, IIHR, Bangalore |
| | | Travancore Org Fert Co P Ltd., Bangalore |
| | Kerala | Plantrich Chem & Fert Ltd, Kottayam |
| | New Delhi | Biotech Intern Ltd., New Delhi |
| | Tamil Nadu | Bio-control Lab, Papparapatty |
| | | Bio-control Lab, Coimbatore |
| | | Bio-control Lab, Salem |
| | | Bio-control Lab, Villupuram |
| | | Bio-control Lab, Trichy |
| | | Bio-control Lab, Panjupettai |
| | | Bio-control Lab, Kuruppanaichenpalyam |
| | | Bio-control Lab, Vinayagapuram |

| | | |
|---|---|---|
| | | Bio-control Lab, Tirunelveli<br>Bio-control Lab, Tanjavur<br>Bio-control Lab, Namakkal<br>Basarass Biocontrol Res, Vallavar<br>Greentech Agro Prod P Ltd, Coimbatore<br>Jeypee Bio Techs, Virudhunagar<br>Rajendra Foundn for Agri Res, Chittar<br>Sun Agro Ind India, Chennai<br>T. Stanes & Co, Coimbatore |

### 10.2.4. Combination Products of Biocontrol Agents

**Table 10.5. Source of availability of combination products of biocontrol agents**

| Combination Products of Biocontrol Agents / Trade Name's | State | Source of Availability |
|---|---|---|
| *B. pumilis* + *P. chlamydosporia* (Bacillus Plus-P) | Karnataka | Agri Technol Infmn Centre, IIHR, Bangalore |
| *B. subtilis* + *P. chlamydosporia* (Bacillus Plus-S) | Karnataka | Agri Technol Infmn Centre, IIHR, Bangalore |
| *P. fluorescens* + *P. chlamydosporia* (Pseudomonas Plus) | Karnataka | Agri Technol Infmn Centre, IIHR, Bangalore |
| *Arthrobotrys* sp. + *P. lilacinus* + *Verticillium* sp. (Nemastin) | Maharashtra | Kumar Krishimitra Bioproducts, Pune |
| *Aspergillus niger* + *T. viride* + *P. lilacinus* (Multiplex Shock) | Karnataka | Karnataka Agro Chemicals, B'lore |
| *P. chlamydosporia* + *B. pumilis* (Bacillus Plus-P) | Karnataka | Agri Technol Infmn Centre, IIHR, Bangalore |
| *P. chlamydosporia* + *B. subtilis* (Bacillus Plus-S) | Karnataka | Agri Technol Infmn Centre, IIHR, Bangalore |
| *P. chlamydosporia* + *P. lilacinus* (Biovert Plus) | Karnataka | Agri Technol Infmn Centre, IIHR, Bangalore |
| *P. chlamydosporia* + *P. fluorescens* (Pseudomonas Plus) | Karnataka | Agri Technol Infmn Centre, IIHR, Bangalore |
| *P. chlamydosporia* + *T. harzianum* (Trichorich) | Karnataka | Agri Technol Infmn Centre, IIHR, Bangalore |
| *P. chlamydosporia* + *T. viride* (Trichovert) | Karnataka | Agri Technol Infmn Centre, IIHR, Bangalore |
| *T. harzianum* + *T. viride* (Ecoderma, NIPROT) | Karnataka | Bio Control Res Lab, PCI Pvt Ltd., Bangalore<br>Margo Pvt Ltd., Bangalore |
| *T. harzianum* + *T. viride* + *T. virens* (Combat) | Maharashtra | Kumar Krishimitra Bioproducts, Pune |

# REFERENCES


Ahlawat, Y.S. and Chenulu, V.V. 1982. Studies on transmission of radish mosaic virus by the aphid. *Indian Phytopath.* **35**: 633-638.

Ahlawat, Y.S. and Chenulu, V.V. 1984. Radish mosaic – A new disease caused by turnip mosaic virus in India. *Trop. Agric.* (Trinidad) **61**: 188-192.

Anjaneyulu, A. and Ramakrishnan, K. 1972a. Chemotherapy of egg plant little leaf disease. *Mysore J. Agric. Sci.* **6**: 135.

Anjaneyulu, A. and Ramakrishnan, K. 1972b. Symptomatology of little leaf disease on egg plant. *Mysore J. Agric. Sci.* **6**: 147-159.

Anjaneyulu, A. and Ramakrishnan, K. 1973. Host range of egg plant little leaf disease. *Mysore J. Agric. Sci.* **7**: 568-579.

Balaraman, K. and Ramakrishnan, K. 1977. *Studies on Strains and Strain Interaction in Citrus Tristeza Virus.* UAS Tech. Series Bull. 19, Univ. of Agri. Sci., Bangalore

Balaraman, K. and Ramakrishnan, K. 1979. Transmission studies with strains of citrus tristeza virus on acid lime. *Zeishrifit Pflanzenkrankheiten und Pflanzenschutz.* **86**: 653-661.

Bandopadhyay, S. and Chattopadhyay, S.B. 1986a. Survival of *Xanthomonas campestris pv. campestris* in the host tissue. *Indian J. Mycol. Res.* **24**: 57-63.

Bandopadhyay, S. and Chattopadhyay, S.B.1986b. Study on incidence of black rot disease of cabbage and cauliflower under different months of planting following natural conditions of infection. *Indian J. Mycol. Res.* **24**: 125-133.

Bera, S.C. 1986. Control of black rot disease of cauliflower by some chemicals. *Pesticides* **20**: 51-52.

Bhargava, K.S. and Joshi, R.D. 1960. A mosaic disease of vegetable marrow (*Cucurbita pepo* L.) in Uttar Pradesh. *Proc. 47th Indian Sci. Cong. Part III*, p. 314.

Bhattiprolu, S.L. 1991. Seed-borne cucumber mosaic virus infecting French bean (*Phaseolus vulgaris*) in India. *Indian J. Virol.* **7**: 67-76.

Bijay Kumar. 2007. *Indian Horticulture Database 2006.* National Hort. Board, Gurgaon, Haryana, India, 324 pp.

Black, L.L., Bond, W.P., Story, R.N. and Batti, J.M. Jr. 1986. Tomato spotted wilt virus in Louisiana: Epidemiological aspects. *Proc. Workshop on Epidemiology of Plant Virus Diseases,* Orlando, Florida.

Bose, P.C. and Singh, C. 1980. *Pesticides* **14**(8): 30-31.

Capoor, S.P. 1967. Important Virus Diseases of Field and Garden Crops in India and their Control. ICAR Tech. Bull. (Agric.) 12, pp. 16-17.

Capoor, S.P. and Sawant, D.M. 1987. Studies on bell pepper yellow mosaic virus. *Indian Phytopath.* **40**: 443-450.

Capoor, S.P. and Varma, P.M. 1948a. A mosaic disease of *Carica papaya* in Bombay Province. *Curr. Sci.* **17**: 57-58.

Capoor, S.P. and Varma, P.M. 1948b. Yellow mosaic of *Phaseolus lunatus* L. *Curr. Sci.* **17**: 152-153.

Capoor, S.P. and Varma, P.M. 1950. Yellow vein mosaic of *Hibiscus esculentus* L. *Indian J. Agric. Sci.* **20**: 217-230.

Capoor, S.P. and Varma, P.M. 1958. A mosaic disease of papaya in Bombay. *Indian J. Agric. Sci.* **38**: 225-233.

Chakraborty, S., Singh, M., Kalloo, G., Banerjee, M.K. and Pandey, P.K. 1999. Screening of cowpea cultivars against cowpea golden mosaic Gemini virus. *Ann. App. Biol. (Supplement Test of Agrochemicals and Cultivars)* **134**(20): 40-41.

Chakravarti, B.P., Porwal, S. and Rangarajan, M. 1966. Studies on citrus canker in Rajasthan. I. Disease incidence and survival of the pathogen. *Labdev. J. Sci. Tech.* **4**: 262-265.

Chakravarti, B.P., Shekhawat, P.S. and Choudhary, S.L. 1978. Compatibility of bavistin with agrimycin in controlling bacterial blight of cowpea. *Indian J. Mycol. Pl. Path.* **8**: 28.

Chand, R. and Kishun, R. 1988. Outbreak of grapevine bacterial canker in India. *Vitis* (In Press).

Cheema, S.S., Kapur, S.P. and Bansal, R.D. 1986. Some morphological and biological changes brought about in kagzi lime (*Citrus aurantifolia*) by citrus tristeza virus and/or citrus greening pathogen. *Int. J. Trop. Pl. Dis.* **4**: 167-171.

Cheema, S.S., Kapoor, S.P. and Chohan, J.S. 1976. *J. Res. PAU* **13**: 169-171.

Cheema, S.S., Kapur, S.P. and Sharma, O.P. 1986. Chemotherapeutic control of greening disease of citrus through bud-dip treatment. *Indian J. Virol.* **2**: 104-107.

Cheema, S.S. and Reddy, G.S. 1985. Studies on the transmission of papaya mosaic virus by *Rhophalosiphum maidis* (Fitch). *Indian J. Virol.* **1**: 49-53.

Das, C.R. and Raychaudhuri, S.P. 1953. A probable spread of tomato acuba mosaic virus. *Indian Phytopath.* **6**: 116-122.

French, E.R. 1986. ACIAR Bacterial Wilt Newsletter No. **1**: 5-6.

Gadewar, A.V., Trivedi, T.P. and Shekhawat, G.S. 1991. *Potato in Karnataka*. Tech. Bull. No. 17, Central Potato Res. Inst., Shimla, India, 72 pp.

Gandhi, S.K. and Chand, J.N. 1985a. Effect of nitrogen and phosphorus on the bacterial blight of cluster bean. *Indian J. Agric. Sci.* **55**: 376-377.

Gandhi, S.K. and Chand, J.N. 1985b. Yield losses in guar due to bacterial blight caused by *Xanthomonas campestris* pv. *cyamopsidis*. *Indian Phytopath.* **38**: 516-517.

Gandhi, S.K. and Chand, J.N. 1987. Horizontal resistance of cluster bean to *Xanthomonas campestris* pv. *cyamopsidis*. *Indian J. Agric. Sci.* **57**: 755-757.

Gandhi, S.K. and Chand, J.N. 1988. Plant age, temperature and free moisture in relation to susceptibility of cluster bean to *Xanthomonas campestris* pv. *cyamopsidis*. *Proc. Indian Acad. Sci., Plant Sci.* **98**: 49-53.

Garg, S.C. and Kasera, H.L. 1984. *Indian Perfumer* **28**(2): 95-97.

Govindaswamy, C.V., Mariappan, V., Murugesan, S., Padmanabhan, C., Thangamani, G. and Janaki, I.P. 1970. Studies on a cowpea mosaic virus disease from Coimbatore. *Madras Agric. J.* **57**: 405-415.

Gour, H.N., Gupta, V.P. and Singh, R.P. 1981. Preliminary alterations in resistant and susceptible cowpea varieties inoculated with *Xanthomonas campestris* pv. *vignicola* (Abstr.). *Proc. 3rd Intn. Symp. Pl. Path.*, New Delhi, p. 120.

Gupta, D.K. 1978. Yield losses in guar (*Cyamopsis tetragonoloba*) caused by bacterial blight (*Xanthomonas cyamopsidis*). *Indian J. Mycol. Pl. Path.* **8**: 27.

Gurfan, S.M. and Chakravarti, B.P. 1960. *Proc. Bihar Acad. Agri. Sci.* **8 & 9:** 126-127.

Howard, C.M., Overman, A.J., Price, J.F. and Albregts, E.E. 1985. *Diseases, Nematodes, Mites, and Insects affecting Strawberries in Florida.* University of Florida, Agricultural Experimental Station, Institute of Food and Agricultural Sciences, 51 pp.

Jain, A.C. 1951. A bacterial disease (canker) of tomatoes, new to India. *Sci. Cult.* **17**: 46-47.

Jayarajan, R. and Ramakrishnan, K. 1969. Potato virus Y on chilli (*Capsicum annuum* L.) in Tamil Nadu. *Madras Agric. J.* **56**: 761-766.

Jha, A. 1953. *Virus Diseases of Chilli.* Associate thesis, Indian Agri. Res. Inst., New Delhi.

Jha, A. and Raychaudhuri, S.P. 1956. Mosaic diseases of chilli. *Indian J. Agric. Sci.* **26**: 217-222.

Kairu, G.M. 1991. The Coffee Research Foundation Ann. Rept. 1991-92.

Kapur, S.P., Cheema, S.S. and Dhillon, R.S. 1988. Improvement of citrus industry in Punjab. Indexing and status of virus/virus-like diseases. *J. Res Punjab Agri. Univ.* **25**: 216-220.

Karwasra, S.S., Chand, J.N. and Shaharan, G.S. 1981. Incidence, perpetuation and control of bacterial blight of guar. *Proc. 5th Intn. Conf. Pl. Path. Bact.*, California, pp. 143-147.

Karwasra, S.S., Chand, J.N. and Shaharan, G.S. 1984. Incidence, perpetuation and control of bacterial blight of guar. *Indian J. Mycol. Pl. Path.* **14**: 123-125.

Kennedy, J.S., Day, M.F. and Estop, V.F. 1982. *A Conspectus of Aphids as Vectors of Plant Viruses.* Commonwealth Inst. Entomol., 114 pp.

Khurana, S.M.P. 1968. *A Study of Virus Diseases of Papaya* (Carica papaya) *in Gorakhpur*. Ph. D. thesis, Univ. of Gorakhpur.

Khatri, H.L. and Sekhon, I.S. 1974. Studies on a virus causing mosaic disease of chilli. *Indian J. Mycol. Pl. Path.* **4**: 121-125.

Kibata, G.N. 1979. Preliminary attempts to control insect transmitted watermelon mosaic virus (WMV-K) on baby marrow. *Kenya Entomol. Newsl.* **9:** 8.

Kishun, R. 1981. Effect of different chemicals on black rot of cabbage (Abstr.). *Proc. 3rd Intn. Symp. Pl. Path.*, New Delhi, p. 250.

Kishun, R. 1981. Bacterial canker of mango. *Ann. Rept., Indian Inst. Hort. Res., Bangalore*, 70 pp.

Kishun, R. 1983. Soft rot of radish caused by *Erwinia* sp. and its control (Abstr.). *Indian J. Mycol. Pl. Path.* **13**: 3.

Kishun, R. 1984. Seed treatment for control of cabbage black rot. *J. Turkish Phytopath.* **13**: 81-88.

Kishun, R. 1985. Stem injection of chemicals for control of bacterial canker of mango (Abstr.). *Proc. Second Intn. Symp. on Mango*, Bangalore, p. 60.

Kishun, R. 1988. *Acta Hort.* **231**: 518-521.

Kishun, R. 1989. Appraisal of loss in yield of cowpea due to *Xanthomonas campestris pv. vignicola*. *Indian Phytopath.* **42**: 241-246.

Kishun, R. 1991. Annual Report, Central Inst. of Hort. for Northern Plains, Lucknow, India, p. 35.

Kishun, R. 1995. In *Detection of Plant Pathogens and their Management* (J.P. Verma, A. Verma and Dinesh Kumar, eds.), pp. 173-182. Angkar Publishers (P) Ltd., New Delhi, India.

Kishun, R. 1997. *National Symp. on Emerging Issues in Pl. Path.*, Jawaharlal Nehru Agri. Univ., Jabalpur, Madhya Pradesh, India, pp. 11-12.

Kishun, R. 1998. The plant pathogenic bacteria. In *Proc. IX Intn. Conf.* (A. Mahadevan, ed.), pp. 470-475. Centre for Advanced Studies in Botany, Madras Univ., Chennai, India.

Kishun, R. and Chand, R. 1987. Studies on germplasm resistance and chemical control of citrus canker. *Indian J. Hort.* **44**: 126-132.

Kishun, R. and Chand, R. 1988. Survival of *Xanthomonas campestris* pv. *mangifeareindicae* on weed hosts. *Indian Phytopath.* **41**: 269.

Kishun, R. and Chand, R.1988. Epiphytotic survival of *Xanthomonas campestris pv. campestris* on *Centenella asiaticum* (L.) Orban. *Intn. J. Trop. Pl. Dis.* **6**: 189-193.

Kishun, R. and Sohi, H.S. 1983. Bacterial canker on mangoes. *Indian Farmer's Digest* **14**: 21-23.

Kishun, R. and Sohi, H.S. 1984. Control of bacterial canker of mango by chemicals. *Pesticides* **18**: 32-33.

Kishun, R., Sohi, H.S. and Rawal, R.D. 1980. Screening of cowpea germplasm against bacterial blight. *Indian Phytopath.* **33**: 506-507.

Kishun, R. and Swarup, J. 1979. Further studies on bacterial stalk rot of onion. *Indian Phytopath.* **32**: 389-392.

Kishun, R. and Swarup, J. 1982. Studies on varietal resistance and chemical control of bacterial stalk rot of onion. *Veg. Sci.* **9**: 55-59.

Kore, S.S. and Dhaliwal, A.L. 1981. Epidemiology of leaf spot disease of chilli caused by *Xanthomonas campestris pv. vesicatoria*. *Indian Phytopath.* **34**: 113.

Lakshmi, B.S. 1990. Studies on seed-borne mosaic of French bean (*Phaseolus vulgaris* L.) in India (Abstr.). *VIII^th Int. Cong. Virol.*, Berlin, p. 485.

Lamberti, F. 1981. Combating nematode vectors of viruses. *Plant Dis.* **65**: 113-117.

Legard, D.E., Ellis, M., Chandler, C.K. and Price, J.F. 2003. Integrated management of strawberry diseases in winter fruit production areas. *In The Strawberry: A Book for Growers* (N. Childers, ed.), pp 111-124. Institute of Food and Agricultural Sciences, Horticultural Sciences Department, University of Florida, Gainesville. Norm Childers Publications, 246 pp.

Lim, T.K., Sijam, K. and Ong, E.T. 1991. *Acta Hort.* **291**: 317-323.

Lodha, S. and Gupta, G.K. 1981. The relationship between bacterial blight and loss in yield components of guar (Abstr.). *Proc. 3^rd Intn. Symp. Pl. Path.*, New Delhi, p. 94.

Loebenstein, G., Alper, M., Levy, S., Pdevitch, D. and Managen, E. 1975. Protecting peppers from aphid-borne viruses with aluminium mulch. *Phytoparasitica* **3**: 43-53.

Loebenstein, G., Deutsch, M., Frankel, H. and Sobar, Z. 1966. Field tests with oil sprays for the prevention of cucumber mosaic virus in cucumbers. *Phytopathology* **56**: 512.

Machmud, 1986. In *Bacterial Wilt Disease in Asia and the South Pacific* (G.J. Persley, ed.), pp. 30-34. ACIAR Proc. No. 13.

Maheswari, S.K., Gupta, P.C. and Suhag, L S. 1988. Studies on chemical control of little leaf disease of brinjal. *Indian J. Mycol. Pl. Path.* **18**: 182.

Mali, V.R. and Kulthe, K.S. 1980. A seed-borne potyvirus causing mosaic of cowpea in India. *Plant Dis.* **64**: 926-928.

Mass, J.L. 1998. *Compendium of Strawberry Diseases.* 2nd edition. APS press, St Paul, Minn.

Menon, K.S. and Nair, T.V.R. 1978. *Proc. PLACROSYM*-**2**: 416-424.

Mishra, A. and Jha, A. 1973. Transmission of chilli mosaic virus through pollen and root contact. *Phyton* (Austria) **15**: 30-32.

Mishra, A., Chaudhari, D.P. and Jha, A. 1977. Characterization of the naturally occurring chilli mosaic virus in North Bihar, India. *Meded. Fac. Landouw. Rijks. Univ. Gent.* **42/21**: 1217-1225.

Mondel, R. and Mukherjee, N. 1978. Sensitivity of plant bacteria to some antibiotics, antibactericides and other drugs. *Z. Pfenkh. Pflaschutz.* **85**: 607-616.

Muralidharan, A., Nair, M.G. and Jayasankar, N.P. 1986. *Indian Coconut J.* **17**: 3-6.

Nagarajan, K. and Ramakrishnan, K. 1971. Studies on cucurbit viruses in Madras State-I. A new virus disease in bitter gourd. *Proc. Indian Acad. Sci. Section B* **73**: 30-35.

Naidu, R. and Govindu, H.C. 1981. Association of bacterium-like organism with citrus greening in India. *Curr. Sci.* **50**: 242-244.

Naidu, R. and Venugopal, N.M. 1984. *Proc. PLACROSYM*-6: 121-127. Indian Soc. for Plantation Crops, Central Plantation Crops Res. Inst., Kasargod, Kerala, India.

Naidu, R., Venugopal, N.M. and Rajan, P. 1985. Final Report of Research Project "Invetigations on StrainalVvariations, Epidemiology and Characterization of 'Katte' Virus Agent of Small Caedamom. Central Plantation Crops Res. Inst., Kasargod, Kerala, India.

Naik, D. 2004. Biotechnological Approaches for the Management of Wilt Disease Complex in Capsicum (*Capsicum annum* L.) and Egg plant (Solanum melongena) with Special Emphasis on Biological Control. Ph. D. thesis, Kuvempu Univ., Shimoga.

Naqvi, Q.A. and Mahmood, K. 1975. Mosaic disease of spinach – A new record for India. *Indian Phytopath.* **28**: 268-271.

Nariani, T.K. 1956. Leaf curl of papaya. *Indian Phytopath.* **9**: 151-157.

Nariani, T.K. and Sastry, K.M.S. 1964. Studies on the relationship of chilli mosaic virus and its vector, *Aphis gossypii* Glover. *Indian Phytopath.* **15**: 173183.

Nayar, K., Mathew, J., Kuriyan, J., Gnanamanickam, S.S. and Mahadevan, A. (eds.). 1988. Role and association of root-knot nematode (*Meloidogyne incognita*) in the bacterial wilt of brinjal incited by *Pseudomonas solanacearum*. In *Advances in Research on Pant Pathogenic Bacteria*. Based on the proceedings of the *National Symposium on Phytobacteriology*, University of Madras, Madras, India, pp. 45-48.

Nedumaran, S. and Vidyasekaran, P. 1982. Seed-borne infection of *Corynebacterium michiganense* in tomato. *Indian Phytopath.* **35**: 510-511.

Nirvan, R.S. 1961. Citrus canker and its control. *Hort. Adv.* **5**: 124-125.

Parvatha Reddy, P., Singh, D.B. and Ram Kishun. 1979. Effect of root-knot nematodes on the susceptibility of Pusa Purple Cluster brinjal to bacterial wilt. *Curr. Sci.* **48**: 915-916.

Patel, G.T., Patel, C.B. and Shah, A.H. 1984. Biology of Sapodilla bud borer, *Anarsia achrasella* (Lepidoptera: Gelichiidae). *Indian J. Agric. Sci.* **54**: 695-697.

Patel, K.B. and Mali, V.R. 1983. Comparative studies on three isolates of cucumber mosaic virus on banana. *Indian Phytopath.* **36**: 443-447.

Patel, M.K., Abhyankar, S.G. and Kulkarni, Y.S. 1949. Black rot of cabbage. *Indian Phytopath.* **2**: 58-61.

Patel, M.K., Kulkarni, Y.S. and Moniz, L. 1948a. *Indian Phytopath.* **1**: 147-152.

Patel, M.K., Moniz, L. and Kulkarni, Y.S. 1948b. *Curr. Sci.* **17**: 180-190.

Patel, P.N. *et al.* (eds.). 1972. *Plant Bacteriology Vol. I. Bacterial Diseases of Plants in India*. Indian Council of Agri. Res., New Delhi, 261 pp.

Patil, F.S. and Mali, V.R. 1982. Natural occurrence of tobacco ring spot virus (TRSV) on cowpea in India. *Indian Phytopath.* **35**: 399-403.

Prasad, H.H. 1932. A bacterial white soft rot of turnip. *Indian J. Agric. Sci.* **1**: 534-537.

Prasad Rao, R.D.V.J., Raghunathan, V. and Joshi, N.C. 1982. Occurrence of yellow mosaic disease of cluster bean (*Cymopsis tetragogoloba*). *Indian J. Pl. Prot.* **10**: 100.

Pruvost, O., Couteau, A. and Luisetti, J. 1989. *Fruits* **44**: 343-350.

Pruvost, O., Couteau, A. and Luisetti, J. 1990. *Fruits* **45**: 125-140.

Pruvost, O. and Luisetti, J. 1991. *Acta Hort.* **291**: 314-327.

Rajagopal, V., Patil, K.D. and Amma, B.S.K. 1987. *Indian Coconut J.* **18**: 3-6.

Rajamannar, M., Narasimhachari, C.L. and Govinda Rao, N. 1989. *Proc. of Indo-US Workshop on Viroids and Diseases of Uncertain Etiology*, Indian Agri. Res. Inst., New Delhi.

Rajamannar, M., Prasadji, J.K. and Rethinam, P. 1991. *Proc. ISOCRAD-* **2**: 62, Central Plantation Crops Res. Inst., Kasaargod.

Raju, K.S. and Rao, G.S. 1984. Studies to fix up optimum spray intervals against bacterial leaf spot and fruit rot disease of chilli. *South Indian Hort.* **32**: 319-320.

Ramakrishnan, G., Ranganathan, K., Murugesan, S., Damodaran, S.A.P. and Kandaswamy, T.K. 1971. A mosaic disease of *Amaranthus gangeticus. Madras Agric. J.* **58**: 679-683.

Ramakrishnan, T.S. 1954. *Common Diseases of Citrus in Madras*. Govt. of Madras Publn.

Ramapandu, S. and Rajamannar, M. 1981. Further studies on the etiology and control of Tatapika disease of coconut (Abstr.). *Third Int. Symp. Plant Path.*, Indian Agri. Res. Inst., New Delhi, p. 20.

Rangaswami, G. and Gowda, S.S. 1963. On some bacterial diseases of ornamentals and vegetables in Madras State. *Indian Phytopath.* **16**: 74-85.

Rankin, D.J., Hewitt, W.B. and Schmitt, R.V. 1971. Controlling fan leaf virus – dagger nematode disease complex in vineyards by soil fumigation. *Calif. Agri.* **25**: 11-14.

Rao, A.N.S. and Kishun, R. 1986. A note on the control of bacterial black rot [*Xanthomonas campestris* (Pammel) Dowson] of cabbage (*Brassica oleracea* var. *capitata* Linn.) by seed treatment. *Pesticides* **20**: 47.

Rao, M.S., Naik, D., Shylaja, M. and Parvatha Reddy, P. 2002. Prospects for the management of the root-knot nematode and bacterial wilt disease complex in capsicum by integration of fungal and bacterial biocontrol agents. *Proc. Int. Conf. on Vegetables,* Bangalore, pp. 347-351.

Rao, Y.P. and Hingorani, M.K. 1963. Survival of *Xanthomonas citri* (Hasse) Dowson in leaves and soil. *Indian Phytopath.* **16**: 362-364.

Ravi Kumar, M.R., Somasekhara, Y.M., Jagirdar, S. and Ryagi, Y.H. 2001. Field evaluation of antibiotics against citrus canker caused by *Xanthomonas axonopodis* pv. *citri*. *Agri. Sci. Digest* **21**: 253-255.

Raychaudhuri, S.P., Nariani, T.K. and Lele, V.C. 1969. Citrus die-back problem in India. *Proc. Ist Intn. Citrus Symp.* **3**: 1433-1437.

Reddy, G.S. and Murti, V.D. 1985. *Citrus Diseases and their Control.* ICAR Bull., pp. 31-57.

Reddy, G.S. and Rao, A.P. 1980. Control of citrus canker in nurseries. *Andhra Agric. J.* **7**: 110-113.

Roberts, P.D., R.S. Berger, J.B. Jones, C.K. Chandler, and R.E. Stall. 1997. Disease progress, yield loss, and control of Xanthonoma fragariae on strawberry plants. *Plant Dis.* **81**: 917-921.

Sahasranaman, K.N., Pillai, N.G., Jayasankar, N.P., Potty, V.P., Varkey, T., Amma, P.G.K. and Radha, K. 1983. In *Coconut Research and Development* (N.M. Nayar, ed.), pp. 160-165. Wiley Eastern, New Delhi, 518 pp.

Saini, L.C. and Parashar, R.D. 1981. Comparative efficacy of stable bleaching powder with other antibacterial formulations in controlling black and soft rot of cauliflower. *Indian Phytopath.* **34**: 465-469.

Sastry, K.S., Satyanarayana, A., Singh, S.J. and Rajendran, R. 1981. Screening of different bean cultivars against common bean mosaic virus. *Pulse Crops Newsl.* **1**; 30.

Sastry, K.S.M. and Singh, S.J. 1971. Effect of different insecticides on the control of white fly (*Bemisia tabaci* Gen.) population in tomato crop and the incidence of tomato leaf curl virus. *Indian J. Hort.* **28**: 304-309.

Sastry, K.S.M. and Singh, S.J. 1973a. Field evaluation of insecticides for the control of white fly (*Bemisia tabaci*) in relation to the incidence of yellow vein mosaic of okra (*Abelmoschus esculentus*). *Indian Phytopath.* **26**: 129-138.

Sastry, K.S.M. and Singh, S.J. 1973b. Assessment of losses in tomato by tomato leaf curl virus. *Indian J. Mycol. Pl. Path.* **3**: 50-54.

Sastry, K.S.M. and Singh, S.J. 1974. Effect of yellow vein mosaic virus infection on growth and yield of okra crop. *Indian Phytopath.* **27**: 316-317.

Sastry, K.S.M., Singh, S.J. and Sastry, K.S. 1977. Effect of border cropping and the use of insecticides in relation to the incidence of tomato leaf curl virus (TLCV). *Indian J. Hort.* **34**: 319-322.

Sastry, K.S.M., Singh, S.J. and Sastry, K.S. 1978. Studies on epidemiology of tomato leaf curl virus. *Indian J. Hort.* **35**: 269-277.

Shah, R., Bhatnagar, M.K. and Mali, B.L. 1989. Epidemiology of bacterial blight of cowpea (Abstr.). *Proc. 7th Intn. Conf. Pl. Path.*, Budapest, p. 85.

Sharma, B.R., Swarup, V. and Chatterjee, S.S. 1977. Resistance to black rot disease [*Xanthomonas campestris* (Pam.) Dowson] in cauliflower. *Scientia Hort.* **7**: 1-7.

Sheikh, K.M., Bajwa, M.N., Rashid, A. and Mahmood, T. 1995. *Pak. J. Phytopath.* **7**: 29-32.

Shekhawat, P.S. and Chakravarti, B.P. 1976. Factors affecting development of bacterial leaf spot of chillies caused by *Xanthomonas vesicatoria. Indian Phytopath.* **29**: 392-397.

Shekhawat, P.S. and Chakravarti, B.P. 1977. Assessment of loss, symptomatology and occurrence of bacterial leaf spot of chillies caused by *Xanthomonas vesicatoria* in Rajasthan. *Indian J. Mycol. Pl. Path.* **7**: 1-14.

Shekhawat, P.S. and Chakravarti, B.P. 1978. Survival of *Xanthomonas vesicatoria* the incitant of chilli leaf spot and a medium for its isolation. *Indian J. Pl. Path.***14**: 11-19.

Shekhawat, P.S., Jain, M.L. and Chakravarti, B.P. 1982. Detection and seed transmission of *Xanthomonas campestris* pv. *campestris* causing black rot of cabbage and cauliflower and its control by seed treatment. *Indian Phytopath.* **35**: 442-447.

Shekhawat, P.S. and Patel, P.N. 1977. Seed transmission and spread of bacterial blight of cowpea and leaf spot of green gram in summer and monsoon seasons. *Plant Dis. Reptr.* **61**: 390-392.

Siddaramaiah, A.L., Chandrasekhar, S.C., Balachandra, C.K. and Pattanashetti, H.V. 1986. *Curr. Sci.* **55**: 18.

Siddappaji, C. and Reddy, D.N. 1972. *Mysore J. Agric. Sci.* **6**: 194-195.

Singh, D.V., Masoodi, S., Banerjee, A.K., Srivastava, S.S.L. and Rai, M. 1982. An outbreak of bacterial leaf spot of tomato in Uttar Pradesh. *Indian J. Mycol. Pl. Path.* **12**: 252-354.

Singh, J. and Swarup, J. 1986. Control of bacterial blight of cluster bean by seed treatment and foliar spray. *Pesticides* **20**(6): 23-24.

Singh, J. and Swarup, J. 1987. Epidemiology of bacterial blight of cluster bean caused by *Xanthomonas campestris* pv. *cyamopsidis. Farm Sci. J.* **2**: 91-92.

Singh, R.N. 1982. Natural occurrence of leaf crinkle disease of common bean. *Indian Phytopath.* **35**: 320-321.

Singh, S.J. 1979. *Studies on Pumpkin Mosaic Virus.* Ph.D. thesis, Banaras Hindu Univ., Varanasi, 379 pp.

Singh, S.J. 1980. Studies on the epidemiology of yellow vein mosaic virus of okra (Abstr.). *Indian J. Mycol. Pl. Path.* **10**: 35.

Singh, S.J. 1981a. Occurrence and epidemic of mosaic disease of chow-chow in India (Abstr.). *Proc. Third Int. Symp. Pl. Path.,* Indian Agri. Res. Inst., New Delhi, p. 160.

Singh, S.J. 1981b. Studies on a virus causing mosaic disese of pumpkin (*Cucurbita maxima* Duch.). *Phytopath. Medit.* **20**: 184-188.

Singh, S.J. 1981. Relationship of pumpkin mosaic virus with its aphid vector, *Aphis gossypii* Glov. *J. Turkish Phytopath.* **10**: 93-109.

Singh, S.J. 1982a. Studies on the Relationship of pumpkin mosaic virus with its aphid vector, *Myzus persicae* Sulz. *Acta Entomologica Jugaslavica* **18**: 81-89.

Singh, S.J. 1982b. Studies on the transmission of pumpkin mosaic virus by aphid *Aphis craccivora* Koch. *Z. Ptlkrankh. Ptlschutz.* **89**: 79-87.

Singh, S.J. 1983. Sword bean mosaic – a new virus disease in India. *Proc. Virol. Cong.*, Haryana Agri. Univ., Hisar, p. 42.

Singh, S.J. 1985. Bittergourd witches' broom – A disease in India. *Indian J. Virol.* **1**: 320-324.

Singh, S.J. 1987. Electron microscopic evidence of the association of mycoplasma-like organisms with marginal flavescence disease of tomato (Abstr.). *3rd Annual Convention of the Indian Virological Society*, Chitteranjan National Cancer Centre, Calcutta.

Singh, S.J. 1989. Winged bean witches' broom – A new mycoplasma disease in India. *Ann. Rept., Indian Institute of Hort. Res.,* Bangalore.

Singh, S.J. 1990. Etiology and epidemiology of white fly-transmitted virus diseases of okra in India. *Plant Dis. Res.* **5**: 64-70.

Singh, S.J. 1991. Association of mycoplasma-like organisms with French bean phyllody. *Ann. Rept., Indian Inst. of Hort. Res.*, Bangalore.

Singh, S.J. and Dutta, O.P. 1986. Enation leaf curl of okra, a new virus disease. *Indian J. Virol.* **2**: 114-117.

Singh, S.J. and Sastry, K.S. 1976. Tomato leaf curl virus and its control. *Indian Hort.* **21**: 9-10.

Singh, S.J. and Sastry, K.S. 1980a. *Ann. Rept., Indian Institute of Hort. Res.,* Bangalore.

Singh, S.J. and Sastry, K.S. 1980b. Marginal flaviscence of tomato, a yellows type disease present in India. *Phytopathol. Medit.* **19**: 136-138.

Singh, S.J., Sastry, K.S. and Sastry, K.S.M. 1977. Yellow vein mosaic virus of bhindi and its control. *Indian Hort.* **22**: 14-15.

Singh, S.J., Sastry, K.S. and Sastry, K.S.M. 1979. Combating leaf curl virus in chilli. *Indian Hort.* **24**: 9-10.

Singh, S.J., Sastry, K.S. and Sastry, K.S.M. 1981. Field tests with insecticides and mineral oil for the protection of French beans from yellow mosaic virus disease. *Gartenbauwissenchaft* **46**: 669-672.

Singh, S.J., Sastry, K.S. M. and Sastry, K.S. 1973. Effect of oil spray on the control of tomato leaf curl virus in field. *Indian J. Agric. Sci.* **43**: 669-672.

Sinha, P.P. 1989. Preliminary studies on bacterial leaf spot of cucumber. *Indian Phytopath.* **42**: 146-149.

Sreekantaradhya, R., Ravishankar, G. Chandrasekaraiah, S.C. and Seshadri, V.S. 1977. S-288, a promising cowpea variety resistant to bacterial blight. *Curr. Res.* **6**: 58-59.

Srinivasan, T.N. and Nariani, T.K. 1966. Studies on a mosaic disease of pea (*Pisum sativum* L.). *Indian Phytopath.* **19**: 189-193.

Srivastava, D.N. and Rao, Y.P. 1963. Bacterial blight of guar. *Indian Phytopath.* **16**: 69-73.

Srivastava, D.N. and Rao, Y.P. 1970. Epidemiology and control of bacterial blight of guar (*Cyamopsis tetragonoloba* (L.) Taub.). *Indian Phytopath. Soc. Bull.* **6**: 1-4.

Strand, L.L. 1993. *Integrated Pest Management for Strawberries.* University of California, Statewide Integrated Pest Management Project, Division of Agriculture and Natural Resources, Publication 3351, 142 pp.

Thakur, M.S. and Dube, G.S. 1981. Studies on mosaic disease of radish (*Raphanus sativus*) (Abstr.). *Proc. 3rd Int. Symp. Pl. Path.*, New Delhi.

Thind, S.K., Kapur, S.P. and Sohi, H.S. 1987. Chemotherapeutic control of citrus greening through trunk injection. In *Mycoplasma Diseases of Woody Plants* (S.P. Raychaudhuri and N. Rishi, eds.), pp. 112-116. Malhotra Publishing House, New Delhi.

Thomas, K.M. and Krishnaswami, C.S. 1939. "Little leaf" a transmissible disease of brinjal. *Proc. Indian Acad. Sci.* **310**: 201-202.

Todd, J.M., Panniah, S. and Subramanyan, C.P. 1975. First record of tomato spotted wilt virus from the Nilgiris in India. *Madras Agric. J.* **62**: 162-163.

Trivedi, B.M. and Patel, P.N. 1972. In *Plant Bacteriology Vol. I. Bacterial Diseases of Plants in India* (P.N. Patel, ed.), Indian Council of Agri. Res., New Delhi, 260 pp.

Tzeng, K.C., Lin, Y.C. and Hsu, S.T. 1994. *Pl. Path. Bull.* **3**: 24-33.

Uppal, B.N., Varma, P.M. and Capoor, S.P. Yellow mosaic of *bhindi. Curr. Sci.* **9**: 27-28.

Vasudeva, R.S. and Lal, T.B. 1943. A mosaic disease of bottlegourd. *Indian J. Agric. Sci.* **13**: 182-191.

Vasudeva, R.S. and Nariani, T.K. 1952. Host range of bottlegourd mosaic virus and its inactivation by plant extracts. *Phytopathology* **42**: 149-152.

Varma, A., Sharma, S.R. and Mahavir, A.V. 1978. Witches' broom of cowpea – A mycoplasmal disease. *Curr. Sci.* **47**: 56-57.

Varma, P.M. 1955. Ability of the white fly to carry more than one virus simultaneously. *Curr. Sci.* **24**: 317-318.

Witcher, W. and Smith, F.H. 1980. Control of watermelon mosaic virus on summer squash with oil spray (Abstr.). *American Phytopath. Soc. Meeting.*

Zitter, T.A. 1978. Virus control with oil sprays. *Veg. Grower* **26**: 12-13.

Zitter, T.A. 1979. Methods for controlling the most common vegetable viruses in South Florida. *Belle Glade AREC Reg. Rep.*, EV – 1978-79.

# SUBJECT INDEX